SUSTAINABLE MANAGEMENT OF MINING OPERATIONS

Edited by J.A. Botin

Published by

Society for Mining,
Metallurgy, and Exploration, Inc.

Society for Mining, Metallurgy, and Exploration, Inc. (SME)
8307 Shaffer Parkway
Littleton, Colorado, USA 80127
(303) 973-9550 / (800) 763-3132
www.smenet.org

SME advances the worldwide mining and minerals community through information exchange and professional development. With members in more than 50 countries, SME is the world's largest association of mining and minerals professionals.

ISBN-13: 978-0-87335-267-3

Library of Congress Cataloging-in-Publication Data
Sustainable management of mining operations / edited by J.A. Botin.
      p. cm.
Includes bibliographical references and index.
ISBN 978-0-87335-267-3
1. Mineral industries--Management. 2. Mineral industries--Environmental aspects. 3. Sustainable development. I. Botin, J. A.

HD9506.A2S964 2009
622.068'4--dc22
                           2008054638

Contents

Contributors

SENIOR EDITOR
J. A. Botin
Professor and Chair
Division of Management, Environmental Safety and Health
Universidad Politécnica de Madrid
Madrid School of Mines
Madrid, Spain

CHAPTER EDITORS

W. Eckley
Professor Emeritus
Liberal Arts and International Studies
Colorado School of Mines
Golden, Colorado, USA

R. G. Eggert
Professor and Director
Division of Economics and Business
Colorado School of Mines
Golden, Colorado, USA

J. A. Espí
Professor
Department of Geological Engineering
Universidad Politécnica de Madrid
Madrid School of Mines
Madrid, Spain

L. W. Freeman
Principal
Downing Teal, Inc.
Denver, Colorado, USA

H. B. Miller
Associate Professor
Mining Engineering Department
Colorado School of Mines
Golden, Colorado, USA

N. Mojtabai
Associate Professor and Chair
Mineral Engineering Department
New Mexico Institute of Technology and Mining
Socorro, New Mexico, USA

J. L. Rebollo
Visiting Professor
Division of Economics and Business
Colorado School of Mines
Golden, Colorado, USA

ADVISORY BOARD

J. A. Botin
Professor and Chair
Division of Management, Environmental Safety
    and Health
Universidad Politécnica de Madrid
Madrid School of Mines
Madrid, Spain

B. Calvo
Director
Universidad Politécnica de Madrid
Madrid School of Mines
Madrid, Spain

T. Davis
Professor
Department of Geophysics
Colorado School of Mines
Golden, Colorado, USA

R. G. Eggert
Professor and Director
Division of Economics and Business
Colorado School of Mines
Golden, Colorado, USA

T. G. Rozgonyi
Professor and Head
Mining Engineering Department
Colorado School of Mines
Golden, Colorado, USA

CONTRIBUTORS

M. N. Anderson
President
Norman Anderson & Associates
Vancouver, British Columbia, Canada

A. Aubynn
Head of Corporate Affairs and Social
    Development
Gold Fields Ghana Ltd.
Accra, Ghana

T. Buchanan
Director, Energy and Extractives Practice
Business for Social Responsibility
San Francisco, California, USA

C. Castañon
Independent Mining Consultant
Oviedo, Spain

B. Cebrian
General Manager
Blast Consult, S.L.
Madrid, Spain

P. Cosmen
Manager, Environmental Systems
Cobre Las Cruces, S.A.
Gerena, Sevilla, Spain

E. Crespo
R&D Engineer
Atlantic Copper, S.A.
Huelva, Spain

G. A. Davis
Professor
Division of Economics and Business
Colorado School of Mines
Golden, Colorado, USA

M. G. Doyle
Technical Director
Cobre Las Cruces, S.A.
Gerena, Sevilla, Spain

M. G. Hudon
Senior Associate
Colby, Monet, Demers, Delage & Crevier
Montreal, British Columbia, Canada

B. Johnson
Environmental Management Services
Council for Scientific and Industrial Research
Stellenbosch, South Africa

D. Limpitlaw
Consulting Engineer
Johannesburg, South Africa

L. E. Ortega
Manager, Environmental Planning and
    Development
Servicios Industriales Peñoles
Torreón, México

G. Ovejero Zappino
External Affairs Manager
Cobre Las Cruces, S.A.
Gerena, Sevilla, Spain

M. Palacios
General Manager, Technology
Atlantic Copper, S.A.
Huelva, Spain

J. M. Quintana
Senior R&D Engineer
Atlantic Copper, S.A.
Huelva, Spain

A. S. Rodríguez-Avello
Professor
Materials Engineering Department
Universidad Politécnica de Madrid
Madrid School of Mines
Madrid, Spain

D. Van Zyl
Professor of Mine Life Cycle Systems
Norman B. Keevil Institute of Mining Engineering
University of British Columbia
Vancouver, British Columbia, Canada

A. C. Zomosa-Signoret
Manager, Sustainable Development
Servicios Industriales Peñoles
Torreon, Mexico

Foreword

Editor J. A. Botin, advisory board members, co-authors, and industry experts who contributed information for *Sustainable Management of Mining Operations* have come together to address a critical issue of our time. As worldwide economic growth continues to drive demand for mineral resources, our future will be defined by our ability to meet this demand in an environmentally and socially responsible manner.

The teamwork involved in the publication of this book exemplifies the multi-disciplinary and multi-organizational approach needed to achieve sustainable resource development. We know that no single branch of academic research holds a singular, transformative breakthrough. Instead, the complexity of the problem must be matched by the expertise with which many disciplines are blended to create technologically incremental solutions. We also know this is not just an academic or technological problem but one that is deeply ingrained in society and our way of life, in industry and entrepreneurship, in globalization, and in public policy.

It is abundantly clear that we must harness the greatest minds in the world to confront the challenges implied by civilization's quest for a sustainable, comfortable, and material standard of living. At the same time, we must develop the great minds essential to working through these issues in the decades ahead. To fail to do so would squander a critical—perhaps irrecoverable—opportunity.

It is also worth noting that this project was hosted by our two prestigious and long-established schools of mines, Escuela de Minas de Madrid and Colorado School of Mines, both focused on the global and sustainable character of today's mining. Together, they have served the mining industry in Europe and America from the 18th century to the 21st century. Over many years, both institutions have benefited greatly from cooperative agreements that promote exchanges of students and professors. This book exemplifies our shared commitment and will serve as an important contribution to the body of knowledge about sustainable resource development.

M. W. Scoggins, President, Colorado School of Mines
B. Calvo, Director, Escuela de Minas de Madrid

Preface

Mining is among humankind's oldest industries, having been essential to improving our quality of life since the Stone Age. Today, mineral resources are at the core of many human activities, from housing, household goods, industrial equipment, and energy to high technology and space exploration. Our quality of life will continue to depend on the availability and extraction of minerals for the foreseeable future.

The minerals industry has been entrusted with the responsibility of meeting this need for minerals without compromising the interests of future generations. Today, mining companies' charters include acting with environmental and social—as well as commercial—responsibility. This presents unique challenges for both today's mining engineers and tomorrow's.

The title, *Sustainable Management of Mining Operations*, refers to the vision and process that effectively integrate economic, environmental, and social considerations aimed at creating sustainable mine operations in the 21st century and beyond.

The book focuses on sustainable management at the operations level and how to integrate sustainability into the organization at all levels. It deals with three management functions that are key for sustainable management: corporate strategy, human resources management, and operations management. We have sought an international perspective from a global array of authors to address these functions.

The book represents a team effort from chapter editors, authors, and mining company representatives and others contributed information, and the guidance of our Project Advisory Board. Any plaudits this work may earn belong to those who have shared their time and knowledge so generously. Special thanks go to my friend and colleague, Dr. Thomas Davis, professor of geophysics at Colorado School of Mines, whose leadership and support made the project possible.

The project has been cohosted by Colorado School of Mines and Universidad Politécnica de Madrid (Escuela de Minas de Madrid, or Madrid School of Mines), my two "alma mater" institutions, which have served the minerals industry for more 150 years.

J.A. Botin

CHAPTER ONE

Introduction

J. A. Botin

The use of mineral resources has been fundamental to human activity: from housing to household goods, from industrial equipment to energy, and from high technology to space exploration, mining has provided the basics of life to the human race. The mining industry produces energy, metals, and minerals that are essential to economic prosperity and a better quality of life. As important as these benefits are, mining activity produces social and environmental impacts on communities, and requires a more responsible mining practice—it requires *sustainable mine management*.

Today, mining faces unparalleled challenges brought by globalization and increased social and environmental awareness. Many mining companies have acknowledged the challenge and have stated their commitment to the values of sustainability. However, the public perception of the environmental and social performance of the minerals industry remains poor.

Furthermore, significant driving forces are acting on the market, whereby corporate performance in sustainable development issues is becoming increasingly related to measurable economic returns and increased value to shareholders. A better reputation is becoming a competitive advantage through improved control of business risks and increased business opportunities. In this context, mining managers are expected not only to comply with but also to lead in the development of increasingly demanding corporate policies and regulations for environmental control, safety, and social responsibility.

This book characterizes the concept of sustainable management as the management approach that integrates sustainability throughout the organization of the company. Each section of the book focuses on sustainable management from a different perspective, management level, or stage of the mine life cycle.

MINING AND SUSTAINABLE DEVELOPMENT

A discussion of the conceptual meaning of sustainable development and the multiple views and emphasis found in the literature is beyond the scope of this book. A brief description of this concept and what it means in the context of mining is introduced in this section and briefly discussed in Chapter 3. For more detailed analysis on the different meanings of sustainable development, the reader is referred to "The Meaning of Sustainable Development" by Michael Redclift,[1] "Capital Theory and the Measurement of Sustainable Development"[2] by Pearce and Atkinson, and *Sustainable Mineral Resource Management and Indicators: Case Study Slovenia*[3] by Shields and Solar.

An early reference to sustainable development was made in 1980 in a report on renewable resource management by the International Union for Conservation of Nature, where sustainable development was defined as "a strategy framework by which economic development can progress, whilst simultaneously enhancing human development and ensuring the long-term viability of those natural systems on which that development depends."[4] Later, in 1987, the World

Commission on Environment and Development (the Brundtland Commission) issued a report in which sustainable development means "to meet the needs of the present generation without undermining the capacity of future generations to meet their needs."[5]

In most definitions, sustainable development integrates three separate strands of thought about the management of human activities. The economic dimension focuses on the economic needs of society, such as adequate livelihood, productive assets, and systems. The social dimension refers to social and cultural needs, for example, health, education, shelter, cultural institutions, and norms. The third dimension deals with the maintenance of ecosystems and the natural resource base.

In the context of mining, one of the more notable references to sustainable development was made in 2002 by the United Nations Environment Programme in a publication titled *Berlin II Guidelines for Mining and Sustainable Development*. *Berlin II* stated: "If sustainable development is defined as the integration of social, economic and environmental considerations, then a mining project that is developed, operated and closed in an environmentally and socially acceptable manner could be seen as contributing to sustainable development. Critical to this goal is ensuring that benefits of the project are employed to develop the region in a way that will survive long after the mine is closed."[6]

More recently, mining associations, nongovernmental organizations, and minerals industry groups have attempted to create sustainable development standards or frameworks outlining performance requirements on environmental, human rights, and social issues associated with the minerals industry. Some of these *frameworks* are described in Chapters 3 and 4 and are referred to throughout this book.

SUSTAINABLE MANAGEMENT: THE CHALLENGE

The concept of sustainable management of mining operations or sustainable mine management, as characterized in this book, refers to a management approach that efficiently integrates economic, environmental, and social issues into the mining operations; aims to create long-term benefits to all stakeholders; and tries to secure the support, cooperation, and trust of the local community in which a mine operates (that is, a continuing social license to operate). Among many other issues, sustainable mine management deals with strategy, responsible exploration and project feasibility decisions, and managing for operational efficiency, improved risk management, enhanced stakeholder relationships, and corporate reputation. Overall, it deals with seeking long-term competitive advantages through responsible management of environmental and social issues.

An essential requirement for sustainable mine management is the corporate commitment to the values of sustainability, but this is not sufficient. Also essential is the development of a business culture where sustainability is a high professional and business value. Furthermore, an organizational structure with specific roles and integration mechanisms and adequate management systems is also required.

Regarding business culture, a well-established business code is a necessary but insufficient condition. Sustainable management relies on individual ethical conduct and trust to foster full participation of stakeholders and encourage commitment among them. It allows decision making at appropriate levels in the organization and encourages individual risk-taking for continuous improvement. Without trust, earning and maintaining social license is not achievable.

Therefore, sustainable mine management is a major challenge. This is evidenced by the strong management focus on the "Sustainable Development Framework" published by multi-participant initiatives offering policy guidance on sustainable development to the minerals

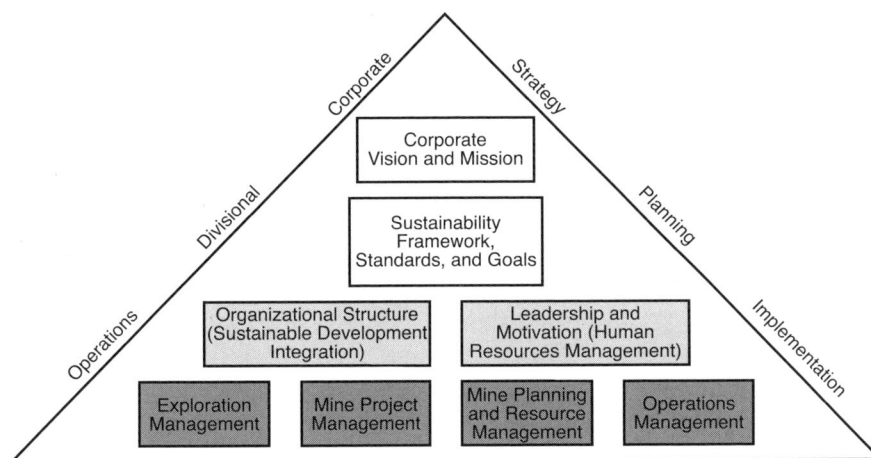

FIGURE 1.1 Sustainable management of mining operations—an integrated model

industry. For example, the International Council on Mining and Metals' principles of sustainable development[7] make references to culture, structural integration, strategy implementation, continual improvement, and other challenging management tasks.

The author's concept of sustainable mine management is that of a management approach that integrates a business culture, strong leadership, and an organizational structure that strives for long-term economics benefits through sustainability. This concept is represented in the graphical model in Figure 1.1. In this model, sustainability must be vertically integrated at three organizational levels (corporate, divisional, and operations) and three functional levels (strategy, planning, and implementation). In addition, the implementation of sustainability goals at the different stages of the mine life cycle (exploration, project, mine planning, and production) requires an organizational structure with adequate integration mechanisms[8] and a business culture in which sustainability is a high professional value.

THE BUSINESS CASE ON SUSTAINABILITY

There is ample evidence of the link between environmental and social performance and company financial performance.[9] However, the question of whether there is a business case for sustainability performance remains open to debate. Ultimately, any business case for sustainability requires that sustainable management practice leads to improved profitability and added value to owners; in this regard, the potential benefits of sustainability need to be quantified.

The quantitative evaluation of the benefits described here and cost concepts may or may not provide a business case. If a business case exists, sustainable management will be driven by business considerations and become fully integrated in the business cycle.

Sustainable management offers a variety of potential benefits:

- Enhanced corporate reputation and lower risk profile
- Higher operational efficiency, derived from sustainable management of safety and health, energy usage, ore resources, and the production processes

- Improved planning and control, derived from the implementation of management systems (e.g., ISO 14001[10], ISO 9001[11]) and the continuous improvement philosophy associated with sustainable management

- Greater advantage in access to mineral resources, resulting in lower resource acquisition cost and reduced project failure rates

- Greater advantage in recruiting and retaining human resources, resulting in improved leadership and motivation, initiative, and decision making at lower levels

- Easier and more economical project financing, derived from investors' perception of the positive financial consequences of social license in financing new mining projects

- Lower project development costs through improved stakeholder relations and a faster permitting process

Achieving these benefits has an associated cost related to a larger company structure, the partnership or sponsorship in community development projects, and, in general, the costs of earning and maintaining the social license.

My personal perception is that, at present, many mining companies are already in the position to substantiate a business case from sustainability performance and are probably achieving the competitive advantages of sustainable management. Looking to the future, the generalization of a "business case" for sustainability[12] will be enhanced by the increased social and political sensitivity on environmental and social issues and the corresponding increase on the prices and the production costs of mineral commodities.

BOOK OVERVIEW

Mining and Mine Management: Historical Background (Chapter 2)

This chapter is intended to provide nonengineers with both a broad and a selective look at a number of historical aspects of mining. These aspects have had an impact not only on management in the industry but on civilizations around the world and on the millions of miners who have made the industry possible. Chapter 2 also looks at two schools that have played a large role in educating mining engineers over the years and that have provided considerable support throughout the research stages of our book project: the Escuela de Minas de Madrid (Madrid School of Mines) and the Colorado School of Mines.

What Sustainability and Sustainable Development Mean for Mining (Chapter 3)

This chapter reviews the ideas underlying sustainability and sustainable development, independent of mineral development and mining. It discusses the implications of these ideas for the minerals sector and for the responsibilities of mining companies. Finally, it reviews what companies have done to put these principles into practice and the project-level tools that have been developed to guide mine managers.

Strategic Issues in the Mining and Metals Industries (Chapter 4)

Sustainability is becoming a fundamental pillar of mining companies' strategy. Not infrequently, operational managers lack sufficient understanding of the rationale for corporate strategic decisions. To address this lack and to help managers comprehend companies' strategic choices, this chapter provides an overview of the main concepts and issues related to the corporate strategy of mining companies. The chapter presents the theoretical basis of the corporate strategy in a simple, comprehensive way, as well as the nature and influence of the external and internal factors affecting mining firms' competitiveness.

Integrating Sustainability into the Organization (Chapter 5)

This chapter discusses the organizational structure of a mining company, focusing on the challenge of integrating the values of sustainable development and sustainability down to the operational levels of mining companies, as well as the organizational structures and the management roles, systems, and tools required for integration. It describes corporate vision and strategy on sustainability and on how to integrate the use of environmental management systems, the management of social stakeholders' expectations, and the concepts and application of the "community partnerships" approach in integrating the expectations of community and local stakeholders at operating levels.

Human Resources Management (Chapter 6)

This discussion of human resources management focuses on the integration of sustainability values in the business culture and ethics in the organization. It describes the processes of recruiting, training, leadership, and recognition as the keys to employee commitment and trust, thus allowing for the integration of decision making down to the lowest possible levels in the organization, a necessary condition for sustainable management.

Management of Exploration (Chapter 7)

Mineral exploration is the technical and management process applied to the search for mineral deposits in the earth's crust. At least in its early stages, exploration does not require changing the use of the land where it is undertaken, so its social and environmental footprint is rather limited. However, the manner in which exploration is conducted is critical to the social license; therefore, it must be conducted in an environmentally responsible manner and in close communications with local stakeholders. Chapter 7 presents an overview of corporate exploration strategy and highlights a sustainable approach to exploration management based on conducting exploration activities in a responsible and transparent manner, aiming to achieve the license to operate at early stages of the mine life cycle.

Managing Project Feasibility and Construction (Chapter 8)

Chapter 8 deals with the sustainable management challenges at the preproduction stages of mining projects. It presents an approach to project management in which the focus is placed on the sustainable feasibility evaluation of economic performance, safety and health, and environmental and social responsibility as the key management issues involved in ensuring long-term benefits throughout the mine life cycle. The chapter compiles the authors' personal experiences as project managers. It aims to share the authors' perspectives without attempting a systematic approach to project management systems and techniques.

Mine Planning and Production Management (Chapter 9)

This chapter outlines the main management issues concerning responsible management in mine production activities. The focus is on various units of operation such as planning, excavations, processing, ground control, material handling, maintenance, waste disposal, and waste management. These topics will be discussed in relation to responsible management for a sustainable mining operation.

INFORMATION SOURCES

Some material in this book has been drawn from publications, sustainable development institutions, mining companies' Web sites, and personal communications. In all cases, these contributions are acknowledged and referenced in endnotes, tables, or figures. We gratefully acknowledge the permissions given to us by these companies, individuals, publishers, and organizations.

NOTES

1. M. Redclift, "The Meaning of Sustainable Development," *Geoforum* 25, no. 3 (1992): 395–403.

2. D. W. Pearce and G. Atkinson, "Capital Theory and the Measurement of Sustainable Development: An Indicator of Weak Sustainability," *Ecological Economics* 8, no. 2 (1993): 103–108.

3. D. J. Shields and S. V. Solar, *Sustainable Mineral Resource Management and Indicators: Case Study Slovenia* (Ljubljana, Slovenia: Geoloski zavod Slovenije, 2004).

4. IUCN/UNEP/WWF, *World Conservation Strategy: Living Resource Conservation for Sustainable Development* (Gland, Switzerland: IUCN, UNEP, and WWF, 1980).

5. World Commission on Environment and Development, "Our Common Future" (Report of the World Commission on Environment and Development. 1987).

6. United Nations Conference on Environment and Development, *Berlin II Guidelines for Mining and Sustainable Development* (New York: United Nations, 2002).

7. International Council on Mining and Metals, "ICMM Sustainable Development Framework," 2003.

8. J. R. Galbraith, *Strategy Implementation: The Role of Structure and Process* (St. Paul: West Publishing, 1978).

9. M. Grieg-Gran, "Financial Incentives for Improved Sustainability Performance: The Business Case and the Sustainability Dividend," MMSD project of IIED and WBCSD, 2001.

10. International Organization for Standardization, ISO 14001, Geneva, Switzerland. Available from: www.iso.org/iso/iso_catalogue/management_standards/iso_9000_iso_14000.htm.

11. International Organization for Standardization, ISO 9001, Geneva, Switzerland. Available from: www.iso.org/iso/iso_catalogue/management_standards/iso_9000_iso_14000.htm.

12. L. Horowitz, "Improving Environmental, Economic and Ethical Performance in the Mining Industry," *Journal of Cleaner Production. Part 1. Environmental Management and Sustainable Development* 14, no. 3 and 4 (2006): 307–308.

CHAPTER TWO

Mining and Mine Management: Historical Background

W. Eckley

INTRODUCTION

I swear there is no greatness or power that does not emulate those of the earth.

—Walt Whitman, *Leaves of Grass*

When the poet Walt Whitman penned that line, he was no doubt thinking of both the physical and metaphysical relationship between Man and Nature. Moreover, in a more practical sense, he was surely aware of the greatness and power of the mineral resources lying beneath the earth's surface, resources used over the ages by human curiosity and ingenuity in the growth and development of the world's civilizations. For early miners of such minerals as gold, copper, tin, and iron, the relationship with the earth was clear enough: they had to find the minerals, take them from the earth, and put them to use. Although these miners are lost to past ages, their legacy lives on in today's dynamic mining industries, which provide the world with comfort, convenience, and achievement.

Primitive though their activities were, the early miners had to be concerned with the tensions of management. Over the millennia, these tensions became more complex to recognize and more difficult to orchestrate. Like management in any of today's business ventures, mining management has itself become a science—to be studied, developed, and followed. In short, it has become a profession. Economic theories, political conflicts, human relations, and environmental questions—just to name a few—are constant concerns of modern mining management. Moreover, the future offers little if any diminution in the plethora of management problems, which get more complicated and challenging as time passes.

While the past is seldom considered relevant in a world primarily concerned with the present, a glance at history can occasionally prove valuable. This chapter of *Sustainable Management of Mining Operations* offers nonengineers an informative, detailed introduction to some of the many challenges facing the mining industry in a rapidly changing world. It looks both broadly and selectively at a number of historical aspects of mining that have not only had an impact on mining industry management but also on the civilizations of the world and the millions of miners who have made it all possible.

FROM THE BEGINNINGS

Early man, nomad that he was, relied on hunting, fishing, and gathering for his subsistence, a way of life that was time-consuming and precarious. Even in those primitive times, however, recognizable forms of cooperation were evolving to maintain and increase longevity. With no writing skills and little more than grunts and child-like pictures scraped on cave walls for communication, this early man was establishing simple forms of management that, through the millennia, would ultimately lead to the settling of lands suitable for planting and harvesting crops. Agriculture and the permanency it brought gave impetus to the first civilizations, particularly in the fertile valleys of the Nile and Euphrates rivers. This early period, in which man was using various stones and flints for weapons as well as for crude farming implements, set the stage for the Age of Metals, a new and creative chapter in human development, with hammering giving way to melting.

Gold was found in many areas of the world, and, because of its peculiar qualities of malleability and longevity, was used extensively for decorative jewelry and coins. In the search for gold, other metals were discovered and used, among them silver, copper, tin, and iron. Melting led to smelting, and smelting led to new dimensions and uses of these basic minerals. Copper and tin became bronze, while zinc and copper became brass. The curtain was indeed being raised on what was to become the modern age of minerals. The Ages of Copper and Bronze were succeeded by the Iron Age. Abundant and relatively cheap to mine, iron was the strongest and most magnetic of the heavy metals. Iron's ubiquitous properties eventually made it the most indispensable of metals, particularly with the emergence of wrought iron, cast iron, and the many types of steel. Iron was widely used by the Assyrians and the Egyptians, as well as by those who followed them in biblical times.

Because history is a continuum with overlapping timelines, it is difficult to label any period or endeavor accurately. This is certainly true when trying to reach a workable division of ages that cover thousands of years. The Age of Metals, for example, covers only 1% of the five hundred thousand years of human existence. So how does one pinpoint when mining became modern with any accuracy? The only way is with a broad brush. T. A. Rickard, in his *Man and Metals*,[1] points to the medieval years (450–1450) as a time when the recognition of the true value of metals brought considerable attention to the search for them and to the improvement of mining technology—the beginnings, as he put it, of the real art of mining. Mining for the ancients had been little more than a grubbing in the ground, carried out mostly by slaves or near slaves with little incentive to improve their craft. Because of limited space, these miners often had to work on their backs, using hammers, picks, and shovels, with the broken ore passed from man to man in bags weighing 100 to 200 pounds. Plutarch complained that "one cannot much approve of gaining riches by working mines, the greatest part of which is done by malefactors and barbarians."[2] These often-brutal working conditions in mines were not to be ameliorated for some time.

Columbus's voyages of discovery unleashed a host of adventurers seeking their fortunes in what was aptly called New Spain. In the relatively short period of 60 years, a myriad of gold, silver, copper, tin, and mercury mines appeared in the areas that would one day become Mexico, Peru, Chile, Columbia, and Bolivia—mines that served the industrial and coinage needs of a booming Europe unable to provide its own resources. Spain alone imported more than 200 metric tons of gold and almost 20,000 metric tons of silver from New Spain between 1520 and 1560. Carlos Prieto, in *Mining in the New World*,[3] points out that, in addition to the welcome economic impetus in Europe, the mines of New Spain led to the development of agriculture and trade, as well as the building of roads and ports, the formation of cities, and the establishment of educational institutions throughout the area. New Spain was indeed on the road to becoming the Latin American nations that exist today. The economic and political roles of the Conquistadors were accompanied by the efforts of the Catholic missionaries who did their part in spreading Christianity to the natives and turning them into a valuable and obedient workforce in the mines.

With the value of metals increasing and the need to find new sources of ore, efforts to improve mining technology increased to the point that mining itself was becoming a legitimate industry in both the new and the old worlds. The areas of Cornwall in Britain and Saxony on the Continent, for example, did much to bring mining out of the dark ages.[4,5] Not only did they exchange miners and mining techniques, they also organized miners' associations that did much to raise the status of individual miners, even if it was minimally. It would not, moreover, be long before mining schools were founded throughout New Spain. Then, too, the mining industry was attracting men of superior ability and of sufficient capital, who were willing to assume the risks that could lead them to fortune or misfortune. One such risk-taker was Jacob Fugger, the grandson of a German weaver, who moved the family's assets into mining interests in 1473. By 1495, the Fuggers had a fortune based on copper and silver mines, which enabled them to become the bankers for many European kingdoms. Fugger also had a positive role, according to Rickard, in turning mining in Germany into a respected industry, firmly established as a prime factor in the conquest of nature and an essential agent in the advancement of humankind.

Another key figure of the period, though not in a monetary sense, was George Bauer, better known as Georgius Agricola, the author of *De Re Metallica*,[6] a work that in the view of its translators, Herbert and Lou Hoover, speaks for itself as a masterful, systematic treatment of the sciences of mining and metallurgy. It also stands as a work of art for the artistic woodcuts that illustrate various aspects of mining. Born in 1494 in Glauchau, Saxony, Agricola was swept up in the Protestant Reformation and the age of exploration that was part of it. Graduating from the University of Leipzig in 1518, he went to Italy to taste the beginnings of the Renaissance, studying philosophy, medicine, and the natural sciences. Although his career combined that of physician, university professor, and holder of several public offices, his primary fascination lay in observing and studying mineralogy and mining. Interestingly enough, it was his investment in a mine, appropriately named God's Gift, that made him a wealthy man. At this time, the arguments over advances in technology, particularly in the use of minerals, and what they might mean to humankind had been common over the centuries.

Although Agricola certainly knew there was no going back, he nevertheless defends the use of iron, copper, and lead in weapons. If such were done away with, he points out, men would quickly turn to hand-to-hand combat, poison, starvation, and all manner of horrible means of torture to do away with an enemy. Moreover, he argues, wars are not caused by products of mines in themselves, but from man's own vices—anger, cruelty, discord, passion for power, avarice, and lust. "For my part," he says, "I see no reason why anything that is in itself of use should not be placed in the class of good things." To the argument that agriculture was more stable than mining, Agricola responds that in the long run "mining would be more productive and profitable than that land which, if you sow, it does not yield crops, but if you dig, it nourishes many more than if it had borne fruit." He admits that none of the arts is older than agriculture, but that of the metals is not less ancient. In fact, they are at least equal and coeval, he argues, "for no mortal man ever tilled a field without implements."

In answer to the view that mining is a dangerous occupation for the miners on its cutting edge, Agricola notes simply that occurrences that killed or injured miners "are of exceeding gravity, and moreover fraught with terror and peril, so that I should consider that the metals should not be dug up at all, if such things are to happen very frequently to the miners, or if they could not safely guard against such risks by any means." He also puts much of the blame for mining accidents on the carelessness of the miners themselves. Yet, he had great respect for the miner: "For, trained to vigilance and work by night and day, he has great powers of endurance when occasion demands, and easily sustains the fatigues and duties of a soldier."

The footnotes that Hoover, the translator, supplied are valuable by-products of Agricola's monumental work. In one of the more interesting notes, Hoover discusses how from early periods to modern times, mining laws have favored four groups with certain proprietary rights regarding mining claims and the actual mining carried out on them. These were the overlord, the state, the landowner, and the mine operator and discoverer. Time and place, of course, made for variations in the power and structure of these agencies. During the Middle Ages, for example, the stronger of the potentates were quick to interpret early Roman mining law in affirming their right to dispossess the weaker. The growth of individualism in the 19th century gave landlords and miners wider rights at the expense of the state. Sentiment since then, however, has been to grant greater restrictions on mineral ownership in favor of the state.

MINING AND THE INDUSTRIAL REVOLUTION

With the dawn of the 19th century, Europe was on the threshold of a new kind of revolution—the Industrial Revolution. While it had been on its way for some years, it arrived with a force that was to be felt first in Europe and eventually throughout the civilized world. Using iron, coal, coke, and steam, this revolution was marked by steam locomotives, steam ships, rolling mills, mass-production factories, urbanized industrial centers, and some rather horrifying working conditions for the common laborer. For two centuries, European nations, particularly those with overseas empires, had been practicing mercantilism, the general goal of which was to export more products than were imported and to accumulate as much bullion as possible. Adam Smith in his widely read *Wealth of Nations*[7] argues that mercantilism had essentially run its economic course and should be replaced with increased industry undergirded by free trade. Britain was the first to realize the implications of steam power and mass production in redefining economic strength. As William Willcox and Walter Arnstein point out in *The Age of Aristocracy*,[8] Britain had the prerequisites to meet this new economic challenge: a supply of natural resources, agriculture to feed a population that doubled between 1750 and 1821, a large-scale division of labor, sufficient capital to finance new industries, a demand for consumer goods, and technological innovation. Coal, coke, and iron were the key resources needed, and Britain had sizeable deposits of all three, as did central Europe for that matter. Britain led the way in coal mining, followed by Belgium, France, and Germany. Steam took over from water to run machines that were constantly growing larger and more complex, and the flap of belts that ran them became a common sound in British and European factories and would remain so until the middle of the 20th century.

The nations of Europe found that it was cheaper to get base metals from New Spain than from their own mines, which in some areas were showing signs of depletion. The Cerro de Paso mine in Peru, for example, produced ore containing copper, silver, and gold from a single mine. Mexico, moreover, struck a bonanza in 1762 that produced 31 million dollars worth of silver from a single mine. Mines of Cerro de Potosi in Bolivia produced silver valued at more than a billion dollars between 1505 and 1801. So productive were these mines that Potosi was at once a luxurious and splendid city, notes Prieto, that imported not only European merchandise, but silks and carpets from Africa; perfumes from Arabia; diamonds from Ceylon; crystal, ivory, and precious stones from East India; and spices, aromatics, and porcelain from Ternate, Malacca, and Goa. Indeed, Potosi was so famous that in every language the phrase "worth a Potosi" became a symbol for wealth.[9]

It is interesting to note that as the Industrial Revolution was moving toward its zenith, gold was discovered in California and other western areas of America, setting off the rush to riches that thousands hoped they would find at the end of a rainbow of gold and silver. For one brief instant, the miner was truly his own boss. Equipped with little more than a pick and pan, he made or lost, usually the latter, his fortune. If this was the miner's apotheosis, it was a short one to be sure. As Rickard laments, "The free miner, the heroic youth of the Golden Age, was succeeded by the hired mechanic; the increasing complexity of the civilization that he himself had helped so largely to build by his gift of metals, proved his undoing. The democratic ideal was submerged beneath the industrial complex."[10] As romantic as this view might seem, it is basically true. The Industrial Revolution spawned considerable social and political change in both Europe and the Americas. If the miner, or any worker for that matter, was ever his own boss, mass production pretty much made him a cog in a machine that he himself did not fully understand. What he came quickly to learn, however, was that in widely developed unity there could be significant strength. Thus was unionism born, and thus was the strike to become the basic weapon used in the perennial conflict between labor and management.

As the Industrial Revolution progressed, pressures on mining grew rapidly. Production of coal in Britain, for example, grew from 10 million metric tons in 1810 to 200 million metric tons in 1875. Miners employed increased from 69,000 in 1801 to more than a million and a half in 1914 in some three thousand mines. While mining procedures were improving to some degree, basically they were not much different from what they were in the 16th century. According to Bernard Cook, in Victorian Britain, between 1850 and 1914, approximately one thousand miners were killed annually. He also notes that work in the mines was still strenuous, disagreeable, and ultimately debilitating, with miners still having to work on their knees in coal seams half a meter high. There was some effort to improve conditions in the mines with Shaftesbury's Mines Act of 1842, which outlawed work by women and children under the age of 10 in the mines. During the first half of the19th century, it was common for miners to be paid monthly and that any advances in pay had to be spent in the Tommy shop, a store run by the mine, whose prices were 20% more than those of the average market. This was a pattern that soon found its way into company mining towns of North America.

Accompanying the Industrial Revolution were insurgent movements in Spanish and Portuguese America. As Prieto notes, independence in Spanish America was an inevitable consequence of the maturing of the people who made up the empire.[11] The American and French revolutions were also having considerable influence. Between 1810 and 1814, such insurgent movements were carried out in the Spanish kingdoms and provinces of Argentina, Chile, Ecuador, Paraguay, Peru, and Venezuela, as well as in Portuguese Brazil. Of these, Mexico and Peru were the most valuable to Spain. The independence achieved did not signal any significant democratic overtones in governmental struggles that would plague most of Latin and South America for almost two centuries. Whatever the forms of government, however, these areas were, and still are, major players in the mining of minerals needed by the industrial world.

A fascinating story that actually begins long before the Industrial Revolution and continues to this day is that of Rio Tinto, a sulfur and copper mine in the area of Huelva, some 30 kilometers inland from the gulf of Cadiz. In June 1556, at the order of King Philip II, a group of Spaniards were searching for some ancient mining diggings in this area. First used by the Phoenicians and later by the Romans, these mines had faded from history for some centuries. Because of Spain's financial problems, Philip was eager to increase the country's mining interests. The searchers found a great pile of slag, from which they garnered some ore samples that assayed positively for copper. Because of the expense involved in building a seaport and a railroad to get minerals to the coast, the area was not developed significantly until 1723, when the Rio Tinto mines were reborn to become one of the greatest, if not the greatest, copper mining centers of the world. Rio Tinto Company Limited, an international consortium led by the British, purchased the mine from Spain for four million pounds and succeeded in bringing the mine on line.

The story of Rio Tinto during this period reads very much like a movie script. The mine grew rapidly by the 1880s with more than eight thousand workers from Spain and Portugal laboring in the mines. The British managers lived in their rather posh compound, while the miners lived in more primitive dwellings—not an unusual division in mining towns of the world. It was not long before social tensions arose between miners and managers. The former, discontented with wages, crowded living conditions, and rules limiting the drinking of alcohol, were mobilized in late 1887 by Maximiliano Tornet, who organized them into secret groups resembling the Molly Maguires, who were operating in the anthracite mining areas of Pennsylvania at about the same time. Civil guards, called the blacklegs, came on the scene to quell demonstrations by striking miners, with the result that on February 4, 1888, shots were fired that killed or wounded 48 demonstrators.

David Avery,[12] in his excellent history of Rio Tinto, *Not on Queen Victoria's Birthday*, describes the impact of this tragedy on both the British and the Spaniards. For the former, it "represented a crystallization of all the fears and all the uneasiness which lay beneath the surface life at Rio Tinto." For the latter, it was a marked break with the past and a symbol of the present, "in which they were subjected without any right of participation, to the decrees of authority which could in the last resort depend on armed might to enforce its rule." In short, this violent episode brought both sides to a reality that neither wanted nor could escape. The idea that opposites generate each other held true for each side. The British were the demons for the Spaniards, and the Spaniards were the demons for the British. Simply put, the British cared more for a profitable mine while the miners cared more for better living and working conditions.

The years of World War I and World War II were difficult for Spain. Rio Tinto, even though it needed new blood in management and considerable repair, played a significant role in helping the nation get back to normal. Since 1954, Rio Tinto has been reorganized on the international economic scene a number of times. Indeed, at this writing, Chinese steelmakers are competing with BHP Billiton, the world's largest mining company, with headquarters in Melbourne, Australia, to take over London-based Rio Tinto Limited. Thus does a fascinating mining story continue.

DIGGING FOR KING COAL

Underground mining, whatever the mineral in question, has many common characteristics, and it has always been of significant interest to the general public. Yet, it may well be that the underground coal miner has captured more of the imagination of that general public than any other. Arguably, his job has always been the dirtiest, the hardest, and the most dangerous in the occupation of mining. Moreover, it is carried out deep in an unforgiving netherworld that in itself seems almost supernatural. It is not surprising that many miners, hard rock or coal, still hold to the folklore belief that mines of any kind are the home of the little people—ghosts, gremlins, gnomes, or what have you—that can have both good and evil motives. Cedric Gregory, in *A Concise History of Mining*,[13] points to one of the more interesting of these beliefs, held by the early coal miners of Cornwall, in creatures called the Tommy Knockers who teased the miners and mimicked the sound of their picks. The miners believed that the Knockers were the souls of departed spirits drifting between heaven and hell and wandering about the mines. There are, however, more realistic problems facing coal miners of today.

At the beginning of the 20th century, 344 thousand of the 30 million workers in the United States were coal miners, most in the anthracite mines of eastern Pennsylvania and the bituminous mines of southern West Virginia.[14,15] Once it was learned in the mid-1800s that there was coal in these areas, speculators grabbed whatever land they could and began seeking labor to mine coal.

Because of the scarcity of the native Pennsylvania Dutch (German background) and the native Scotch-Irish laborers of West Virginia, mine owners sought and welcomed immigrants from such places as Cornwall, Italy, Serbia, Slovenia, Slovakia, and Hungary. These immigrants, along with a number of southern African Americans, joined the native labor to provide a varied but effective workforce, even if its members did not always get along well with each other.

The larger coal companies began construction of towns around their mines to house miners and their families. It has been said that miners belonged to the company because they were treated by the company doctor, lived in a company house that might be taken from them without notice, shopped at the company store where prices were high, and were eventually buried in the company graveyard. While there were advantages for both company and miner, it was without doubt an economic boon for the company to keep the miner in debt to the store. Some companies built churches and hired ministers to provide "acceptable" religious services. David Corbin, in his excellent study, *Life, Work, and Rebellion in the Coal Fields,*[16] tells of a mine owner advising his superintendent "to never lose sight of the fact that the sole purpose of the company was to make money for the stockholders, and that matters of conduct tending to produce a contradictory result should be promptly squelched with a heavy hand."

As it is with all miners, the coal miner's basic concern is the number of loads that he produces, while that of the manager is the overseeing of an operation that will produce a positive bottom line. Each of these concerns touches on the very core of human needs and values of the individual holder. While they no doubt bring inherently different life views in terms of culture and education, miner and manager must fashion a working relationship that will have a chance at solving the conflicts that have been instrumental in the shaping of the mining industry.

The underground mine itself is in many respects the miner's world, the one where he makes his living and risks his life in the process. It is in fact a world that the mining manager, who spends the majority of his time on the top, may not fully understand. In a newspaper interview,[17] one West Virginia miner described a telling incident regarding his uncle's first day at the No. 9 mine: "When he got to the bottom of the elevator and the doors opened up and he looked back in the heading (tunnel), he went right back up and said, 'I quit.'" Yet, most miners do not quit as long as the job is there. In his private, almost secret world, he lives for the moment or for the tenor of the whistle that ends the shift and leads to a double shot and a beer.

Though difficult to explain, the lure of the mine that is a theme for a number of folk ballads and country songs is real enough. It is highly doubtful that anyone has made up a song about mine managers. Miners across the world have a pride and a respect regarding their jobs, perhaps even a touch of arrogance. After all, it is the miner who goes back into a mine to rescue his mates in time of disaster. The manager who does not understand and appreciate this "other" culture of the mines will never fully succeed in his own job.

Advocates of social Darwinism, mine owners, and other industrialists saw unionism as the forerunner of socialism and a threat to the very core of business, and they were prepared to fight it as often and as hard as necessary. Anxious to enlarge its influence, the United Mine Workers of America sent labor organizers into the coal-mining areas to awaken miners to the value of a union. John Mitchell, who had worked in a mine at the age of 12, eventually became president of the United Mine Workers in 1898. He led two strikes in the anthracite region of Pennsylvania, one in 1900 and one in 1902. The latter lasted 23 weeks and garnered a 10% increase in pay, a 9-hour work day, the right of miners to name overseers of the weighing of coal produced daily by a miner, and a conciliation board to hear grievances.

Two other strikes in American coalfields reflect the vehemence and violence that could result when mine owner and striking miner faced each other. The first occurred in 1914 in the mining towns of southern Colorado. The strikers wanted recognition of the United Mine Workers union and improved working conditions. Realizing that they were going to be evicted from their company houses, they moved into tents. On April 20, after some 7 months of the strike, 200 company guards attacked and burned the main miner camp at Ludlow, killing 21, including 11 children, and wounding 100. The second major strike occurred in the coalfields of southern West Virginia in 1920. Suffice it to say that West Virginia was a battlefield in which thousands of miners fought against scabs, Baldwin-Felts detectives, state police, and National Guard troops. Lon Savage, in his penetrating study, *Thunder in the Mountains*, describes one day's battle (May 12, 1920) at the town of Matewan, across the river from Kentucky:

> As if on schedule, fighting erupted in the rainy morning hours . . . when the bullets peppered down from the West Virginia mountains on a half dozen mining towns near Matewan. Immediately, nonunion miners returned fire from the Kentucky mountains. By mid-morning, fighting was general, and shots rang out from every mountain. No one could tell who was shooting at whom. Nonunion miners in the valley shot blindly back, and deputies opened fire with machine guns. Hundreds, then thousands of shots boomed and echoed until in places they became a continuous roar.[18]

Only after the federal government declared martial law and sent in the Army did the fighting stop. The mines of southern West Virginia were not unionized until the 1930s.

The coal-mining industry of Britain, too, had its labor problems, including strikes and political conflicts that helped bring down the Heath government in 1974. The industry had been nationalized in 1947 in an attempt resuscitate it. Between that year and 1982, however, the number of British coal miners declined from 700 thousand to 200 thousand. The fate of British coal mining was essentially sealed. One rather interesting example of the impact of the Industrial Revolution and the later reduction in coal mining in Britain are the Rhondda Valleys, which until the 1850s were thinly populated. With the discovery of vast amounts of coal, migration to the area resulted in population growth from less than a thousand in 1851 to 169,999 in 1924. The valleys were packed with some 53 collieries and hundreds of houses. Upwards of five thousand men were employed in the five pits of the Lewis Merthyr colliery, producing nearly a million metric tons of coal each year. Cheap coal from overseas eventually made mining in the valleys a losing proposition. Lewis Merthyr stopped mining in 1983, and the last of the valley pits closed in 1990. The same fate awaited Wylam, a former collier village whose mining days began in the 11th century when the monks of Tynemouth mined both coal and iron. At one time, the village supported five coal mines and two iron works. Following the closure of the mines in the mid-20th century, the area became rough grassland with scattered trees and is now the Wylam Haughs Nature Reserve, a fitting grave for the miners who toiled beneath it.

A TALE OF TWO SCHOOLS

It is appropriate here to look for a moment at two schools that have played a large role in educating mining engineers over the years and have provided considerable support throughout the research stages of our book project: the Escuela de Minas de Madrid (Madrid School of Mines) and the Colorado School of Mines.

Having both established international reputations for excellence as institutions of higher learning in mining and engineering, the Madrid School of Mines and the Colorado School of Mines have not only prepared students in the practical aspects of mining but have also carried out significant research projects. These projects have opened new avenues in mining, mineralogy, stewardship of the earth, and mine management.

The Academy of Mining in Almadén was founded in 1777 by Charles III, King of Spain. Several years earlier, the strategic mercury mine in Almadén had suffered a long-burning fire that necessitated closing it for a considerable time. Two other mercury mines were unable to produce the amount of mercury needed to process the silver being mined not only in Spain, but also in New Spain (Peru and Mexico), so the Spanish economy was suffering. To answer the need, the king set up the academy and brought a number of professors from Freiberg to aid in its organization. The Underground Geometry and Mineralogy Chair, under the direction of Don Enrique Cristobal Storr, was the first to be established and followed the Germanic way of teaching.

The academy was moved to Madrid in 1836 and became the Madrid School of Mines (Figure 2.1), and its educational mission began to grow dramatically. The degree of mining engineer was established in 1857 with a program based on scientific and experimental education replacing the older practical approach. This 5-year degree program encompassed 2 years of basic sciences, 2 years of applied sciences, and a fifth year of specialization.

In 1896, the School of Mines moved to its present campus in Madrid, a handsome new main building designed by Velazquez Bosco and decorated by Daniel de Zuloaga. Subjects added to the curriculum have included steel production, microscopy, petrography, mineral micrography, liquid fuels, groundwater geology, geophysics, industrial engineering, computer sciences, and management and environmental engineering. Two new degrees include geological engineering and energy, fuels, and explosive engineering.

The Colorado School of Mines (Figure 2.2), established roughly a century after the Madrid School, was a creation of the developing American frontier as it moved across the West. Unofficially founded by an Episcopalian missionary in Golden, Colorado, it was one of a trio of schools: a preparatory school, a divinity school, and a mining school. The embryonic mining school that would eventually become the Colorado School of Mines was the only survivor of the three. From the beginning, its role was obvious: to educate and train engineers, who were sorely needed to guide and develop the hard-rock mining that was such a large part of the American West. So important was it that the State of Colorado took control of the fledgling school in 1874. Thanks to the academic leadership of such early presidents as Regis Chauvenet, Victor Alderson, and Melville Coolbaugh, the Colorado School of Mines was able to continue with some confidence on its educational journey.

FIGURE 2.1   Escuela de Minas de Madrid (Madrid School of Mines)

FIGURE 2.2   Colorado School of Mines

Through the last half of the 19th century and the early part of the 20th in America, mining was key to the development of the lands west of the Platte River. The discovery of gold in California in 1848 drew thousands of Americans and foreigners alike to try their luck at becoming rich overnight. Indeed, mining became the largest nonagricultural source of jobs in the West. Ten years later, gold was found in Clear Creek Canyon near Golden, Colorado. Although Colorado did not experience the same frenzy that California did, it was surely on the map for the gold and silver mining, just as it would be later for coal and uranium mining.

Mining was caught in the process of change. The romanticized days of the prospector were coming to a close if they had ever existed. Going underground and demanding more technology and capital, mining would soon be a bonanza of big business and millionaires who would build opera houses and shoe their horses in silver.

The Colorado School of Mines today is more than a mining school. Its roots have spurred the growth of a number of disciplines in addition to mining: geology, metallurgy, chemistry, chemical engineering and petroleum refining, petroleum engineering, geophysics, physics, economics and business, liberal arts and international studies, electrical engineering, mechanical engineering, civil engineering, environmental engineering, mathematics, and computer science. All of these degree programs at the Colorado School of Mines, with the exception of liberal arts and international studies, offer PhD programs.

THE TOUCHSTONE OF MINING

Just as the Greeks had their misgivings about minerals and their uses, so too do various groups today. The latter, of course, have no illusions that the mining of minerals or the drilling for oil can, or should be, eliminated or even diminished. Each age has offered its costs and its gifts, its problems and its solutions regarding the advances of technology and the minerals required for those advances. In a world that's, ironically, growing smaller in terms of communication and interaction as it grows larger in terms of population, competition has become the insidious by-product of that conundrum. When the economic factors of cost and profit are taken into consideration, the level of complexity rises dramatically. For mining, the line between these two factors is fine indeed and requires that management come into play early and strongly. In addition to the broader concerns mentioned earlier, the physical activities of mining involve risk and capital. Finding, evaluating, digging, loading, and hauling of minerals and coal are the physical activities of mining, whether it's done underground or at the surface. These activities all require various levels of skilled labor, as well as capable managers to plan and oversee them.

Since prehistoric times, management of one fashion or another has run the world's affairs, from household to religion, from government to the largest of corporations. And it has not been an easy task at any level. Confucius had it right when he said that if you would set government in order, you must first set your family in order; if you would set your family in order, you must set yourself in order; if you would set yourself in order, you must first call things by their right names. The last, of course, is the most difficult to do. Like all management, mining management at any level is never easy. Successful mining, then, needs a touchstone, and that touchstone is capable and imaginative management carried out by individuals and teams who recognize and accept their professional, moral, and ethical responsibilities.

Just as technology has more often than not outpaced the making of war, so has it often outpaced the organizing of business and industry, with the result that managers face constant growth and change. In this race, as it were, Frederick Turner's rules of efficiency[19] and William Whyte's *Organization Man*[20] have been left in the dust of that growth and change. Mining certainly has had to change its own methods of operation and management over the years. Profit, of course, still remains the hoped-for bottom line, but the challenges and conflicts inherent in that quest are numerous and show no evidence of riding off into the sunset. It is impossible here to rank these challenges and conflicts, but undoubtedly, labor relations, mine safety, and environmental questions are of perennial importance. In an ideal mine, of course, the mineral would be plentiful and easily accessible; the relationship between miner and manager would be mutually rewarding; and there would be no environmental problems to muddy the waters. Reality, however, more often presents a different scenario.

Mine safety has been a constant concern throughout the history of mining, particularly in recent times. Although Agricola is right when he says that many mining accidents are the result of carelessness on the part of the miner, many more are the result of insufficient regulations or the failure to follow established ones. While this chapter was in development, three tragic coal-mine accidents occurred: one in the state of Utah in which nine miners were killed, another in Ukraine in which scores of miners were killed, and a third in China in which 96 were reported killed. During the 45 years of communist control in Russia and Eastern Europe, many mine disasters were not reported to the rest of the world. Both Ukraine and China presently rank high in mining accidents. Since the fall of communism in 1991, 4,700 Ukrainian miners have been killed, the worst single accident occurring in 2000, when 80 miners died in a dust explosion. China averages approximately five thousand deaths from mining accidents each year. One of the worst mining disasters in the United States occurred in the anthracite fields in 1869, when a fire-damp explosion killed 108 men and boys. This, of course, is not the place simply to point the finger of responsibility. Suffice it to say that of all mining, coal mining in particular, has to be examined more carefully regarding safety; moreover, it is the mining manager who must play a significant role in that effort.

Environmental damage is a worldwide problem that is probably even more significant and more difficult to solve than mine safety problems. Although mining has contributed its share of this damage, directly or indirectly, it is certainly not alone in polluting the environment or in contributing to the warming of the earth. China, with its dynamic population growth and industrial expansion, again finds itself as a leader in damaging the environment. As one might expect, however, China points to other nations that should bear some of the guilt. An article in the *New*

York Times of August 26, 2007, notes that Britain, the United States, and Japan polluted their way to prosperity and worried about environmental damage only after their economies had matured and their citizens were demanding blue skies and safe drinking water. One Chinese climate expert complains that China has to deal with environmental problems while it is still poor and with no model to follow.

At this writing, the Colorado legislature is debating a bill that would impose new restrictions on mining in Colorado, especially uranium mining. Hard-rock miners have expressed concern that such restrictions would force them to curtail existing operations at the very time that they want to reopen mines. Some Colorado mountain communities that suffered environmental damage as a result of earlier mining operations and whose economies have suffered because of mine closures are now interested in having mining resumed.

The mining manager, to be sure, has much to worry about. Along with concern for the productivity and safety of miners, he or she must satisfy owners, stockholders, governmental regulators, banks, accountants, environmentalists, politicians, and the miners digging and loading the tonnage. He or she must, moreover, keep up with technological advances that sometimes must seem like the mechanical rabbit at a dogtrack: they never get caught. The mining manager must wear many hats and bear heavy responsibility but with few absolute answers. Fortunately, the greatness and power of the earth remains as a beacon.

NOTES

1. T. A. Rickard, *Man and Metals* (New York: Whittlesey House, 1932).

2. Plutarch, "Comparison of Crassus with Nicias," vol. 3, *Plutarch's Lives*, ed. A. H. Clough, trans. John Dryden (Boston: Little Brown, and Co., 1895), 376.

3. Carlos Prieto, *Mining in the New World* (New York: McGraw-Hill, 1973).

4. Eugenia W. Herbert, Bernard Knapp, and Vincent C. Pigott, *Social Approaches to an Industrial Past* (London: Routledge, 1998).

5. Martin Lynch, *Mining in World History* (London: Reaktion Books, 2002).

6. Georgius Agricola, *De Re Metallica* (New York: Dover, 1950).

7. Adam Smith, *Wealth of Nations*, Vol. X, The Harvard Classics, ed. C. J. Bullock (New York: P.F. Collier & Son, 1909-14; Bartleby.com, 2001) www.bartleby.com/10/ (Accessed: November 15, 2008).

8. William Willcox and Walter Arnstein, *The Age of Aristocracy* (Boston: Houghton Mifflin, 2001).

9. Prieto, *Mining in the New World*.

10. Rickard, *Man and Metals*.

11. Prieto, *Mining in the New World*.

12. David Avery, *Not on Queen Victoria's Birthday* (London: William Collins and Sons, 1974).

13. Cedric Gregory, *A Concise History of Mining* (Exton, PA: A.A. Balkema, 1980).

14. Joan Champion, *Smokestacks and Black Diamonds* (Easton, PA: Canal History and Technology Press, 1997).

15. Lon Savage, *Thunder in the Mountains* (Pittsburgh: University of Pittsburgh Press, 1990).

16. David Alan Corbin, *Life, Work, and Rebellion in the Coal Fields* (Chicago: University of Chicago Press, 1981).

17. John Henderson, "Eating a Mile Underground," *Denver Post*, December 12, 2007.

18. Savage, *Thunder in the Mountains*.

19. F. J. Turner, *The Frontier in American History* (New York: Henry Holt and Company, 1935).

20. W. H. Whyte, *Organization Man* (New York: Simon & Schuster, 1956).

CHAPTER THREE

What Sustainability and Sustainable Development Mean for Mining

R. G. Eggert

INTRODUCTION

The index of the two-volume, 2,400-page *SME Mining Engineering Handbook*[1] does not contain the words *sustainability* and *sustainable development*. The world certainly has changed. Today, one would expect these words and the underlying concerns about mining's environmental and social challenges, and how they relate to economic and commercial considerations, to be prominent in any new edition of this landmark publication. Sustainability and sustainable development have taken on heightened—some would say paramount—importance since publication of this book in the early 1990s.

Despite the prominence of sustainability and sustainable development in current discussions about mining, these terms are prone to hyperbole, confusion, and disagreement over what they mean and imply for mining.

This chapter seeks to clarify the concepts of sustainability and sustainable development and what they imply for mine management. It begins by reviewing the ideas behind sustainability and sustainable development, independent of mineral development and mining, then defines four implications (or principles) for the mineral sector and discusses what they mean for the responsibilities of mining companies. And finally, it reviews what companies have done to put these principles into practice and what project-level tools have been developed to guide mine managers.

BROAD CONCEPTS

Sustainability and sustainable development emerged as key social concepts out of concern that many current, commercial activities are *unsustainable*—that is, these activities may result in such significant environmental damage and social disruption that future generations will be worse off than the current generation. There is general agreement—and who could disagree with the goal or ideal?—that decisions today not make future generations worse off than the present generation and, more specifically, that commercial decisions not only reflect the quest for profits but also result in appropriate protection of the natural environment and in social justice.

The concept of sustainability originated in the field of renewable resource management. The idea is that a forest, fishery, or other renewable resource be managed such that the rate of harvest does not exceed the rate of regeneration—thus sustaining indefinitely the stock of the natural resource. This original idea can be applied to other environmental and natural resource issues, such as managing an ecosystem so that the stock of natural resources that make up an ecosystem is sustained, maintaining a desired degree of biodiversity, or maintaining a desired level of air or water quality.

Think of sustainability, in this sense, as *environmental sustainability*. It also can be viewed as physical sustainability in that the unit of measure is physical (e.g., number of species, or parts per billion of a specific air pollutant). Environmental sustainability emphasizes maintaining the ability of the natural environment to provide the life-sustaining services and the aesthetic qualities it provides to humans (e.g., clean air and water, or scenic vistas). This view of sustainability also embodies the belief that the natural environment should be maintained or sustained for its own sake, independent of how human beings interact with and use the environment.

Economic sustainability, a second form of sustainability, emphasizes sustaining or enhancing human living standards. A starting point for assessing economic sustainability is a measure of well-being at present, such as per capita income. Broader starting points incorporate other, less purely economic determinants of human well-being, such as education levels, life expectancy, and income distribution. The United Nations Human Development Index (http://hdr.undp.org/statistics) is one example. Sustainability requires that these indicators of well-being be at least sustained if not enhanced. At a national level and more conceptually, the concept of green accounting goes a step further and attempts to assess whether a given level of well-being today is, in fact, sustainable in the future.

Traditional gross domestic product (GDP) is the estimated value of goods and services produced by an economy, which at a national level is equivalent to income, in turn, a first approximation of human well-being. Traditional GDP, however, will not be sustainable if today's economic activities diminish our ability create goods and services (i.e., human well-being) in the future—as would happen if today's activities completely wore out existing manufacturing and other productive equipment and facilities. The solution is to adjust traditional GDP estimates to account for net investment in productive facilities (investment purchases minus depreciation), yielding a measure of the level of production (and analogously income) that is sustainable because it leaves the capital stock intact. Similarly, our overall level of human well-being will not be sustainable if current activities "wear out" natural resources and the environment. So-called green GDP takes traditional GDP as a starting point but includes the following adjustments: (a) net investment in human-made productive facilities as noted previously (which can be a positive or negative adjustment, depending on the balance between investment and depreciation), (b) net changes to the stock of mineral and energy resources (which can be subtracted from or added to GDP, depending on the net effect of reserve depletion from ongoing mineral and energy production and reserve additions during the same period), and (c) the estimated value of the net effect of improvements to environmental quality due to pollution-control expenditures and of environmental damage caused by current economic activities. To be sure, valuation is not simple, but considerable work has been done investigating how to assign economic value to natural resources and the environment.[2]

Social and cultural sustainability is the third major form of sustainability. It focuses on social justice. It emphasizes how the benefits and burdens of economic activities are distributed; is a specific distribution fair? Most commercial activities yield benefits and burdens that are not shared equally across society. For example, a new manufacturing facility may bring net benefits to a national or provincial economy but may leave the local, host community worse off on balance if the facility brings in workers from outside the local economy and leaves local unemployment unchanged, creates significant environmental damage for which the local community is uncompensated, or leads to social problems because the influx of workers strains the local infrastructure of housing, schools, and medical facilities. Concerns such as these raise the question of the extent to which parties affected by a new commercial development have or should have a role in deciding whether the development occurs and under what terms. In what circumstances is an affected party entitled to compensation? To what extent are affected parties entitled to share in the net benefits or profits of an activity? The process through which issues such as these are resolved is an important aspect of social and cultural sustainability.

Note that each form of sustainability—environmental, economic, sociocultural—can be thought of as one-dimensional in that the objective is to maintain or sustain something that is either environmental (e.g., maximum allowable rate of pollution), economic (e.g., per capita income or some other measure of human living standards), or social (e.g., a fair distribution of wealth). Each form of sustainability gives priority—either explicitly or implicitly—to an objective in one dimension (e.g., environmental quality) at the potential expense of alternative objectives in the other dimensions (e.g., economic and social justice).

Sustainable development, in contrast, is inherently multi-dimensional. Development that is sustainable will appropriately balance and integrate environmental, economic, and social aspects of an activity. Sustainable development, in other words, strives to *simultaneously* sustain or even enhance environmental quality, human (economic) living standards, and the fairness of the distribution of the burdens and benefits of development (social justice). Sustainable development represents economic development that is consistent with society's preferences for environmental quality and social justice. There is no single measure of sustainable development; rather, progress toward this goal is less prone to quantification and instead is indicated only by looking at a variety of environmental, economic, and sociocultural indicators. Given that pursuing goals in one dimension (say, environmental) may come at the expense of goals in another dimension, progress toward sustainable development requires careful attention to the social institutions and processes that facilitate the resolution of conflicts and the simultaneous pursuit of multiple objectives.

In translating these broad concepts into actual decisions, the issue of scale is important. Is our priority the sustainability or sustainable development of a local community? A province or subnational region of a country? A nation? Or the world as a whole? What is sustainable for a local community may not be sustainable or optimal for a nation or the world as a whole. In other words, there may be and are conflicts between what sustainability and sustainable development mean for different people, organizations, and social and political groups.

Four Implications or Principles for Mining

What do these broad concepts of sustainability and sustainable development mean for the mining sector? What they certainly do *not* mean is that we should strive to sustain forever the life of an individual mine, which of course is impossible—although the life of an individual mine can be sustained to some degree through development of additional reserves adjacent to existing reserves and through technological innovation that makes it technically and commercially feasible to extract previously uneconomic rock. Furthermore, these concepts do not mean that we should strive to sustain forever and at all costs the life of an individual community; there are some mines in sufficiently remote locations that there are no viable (commercial) opportunities that do not revolve around mining.

Instead, this author's view is that the concepts of sustainability and sustainable development suggest the following principles or social goals associated with mining:

1. Facilitate the creation of mineral wealth;
2. Ensure that mineral development occurs in an economically (socially) efficient manner;
3. Distribute the surpluses from mining fairly; and
4. Sustain the benefits of mining even after a mine closes.

Facilitate the Creation of Mineral Wealth

Before minerals can contribute to sustainability or sustainable development, mineral wealth needs to be created. To be sure, minerals themselves will not be created in the earth's crust over time periods relevant for human beings. The creation of mineral wealth, in contrast, requires human activities to acquire sufficient knowledge of a mineral occurrence (or the chance of occurrence) that someone is willing to purchase exploration, development, or mining rights.

Ensure That Mineral Development Occurs in an Economically (Socially) Efficient Manner

The key idea or goal of economic efficiency is to maximize net benefits, in this case from mining. The definitions of benefits and costs need to be broad enough to include full social benefits and costs. Starting points for estimating benefits and costs are the revenues received and costs incurred by profit-seeking mining companies. In addition, and significantly when evaluating sustainability or sustainable development, one needs to consider the possible spillover benefits and costs associated with mining. Spillover benefits could include local purchases of inputs (e.g., food, fuel, spare tires) by a mining company that otherwise would not occur, local spending of mine-worker wages that otherwise would not occur, health improvements experienced by local communities from clinics a mining company funds, and roads or other infrastructure a mining company builds for a mine that also are useful for a local community. Spillover costs could include environmental damage, cultural impacts on indigenous peoples, and social disruptions often associated with mining development, such as increased rates of alcoholism or prostitution. Economic efficiency requires that decision-makers acknowledge and recognize the unpaid nature of spillover costs—in other words, unpaid costs need to be paid. More specifically, local communities should be compensated for spillover costs they bear from mining. Overall, economic efficiency, defined here as maximizing the net social benefits of an activity, incorporates at least conceptually the environmental and social problems that lead to concern that these activities are unsustainable.

Distribute the Surpluses from Mining Fairly

Mining often generates surpluses (economic profits or, as they sometimes are called, rents) even after mining companies pay the spillover costs referred to previously, either by compensating local communities or others for bearing these costs or through activities such as expenditures on environmental protection that reduce or eliminate the creation of these costs. How these surpluses are distributed—among the owners and workers of private mining companies, national governments, local governments, communities, and other organizations—is a key issue in social and cultural sustainability and sustainable development. This issue is a critical aspect of the quest for social justice.

The core conceptual question is: What is fair? As noted by Young,[3] the Aristotelian approach emphasizes proportionality—for a fair distribution, allocate surpluses according to each party's contribution to creation of the surplus. For mining, proportionality suggests that business partners should share in profits in proportion to their financial contributions to a project. Other allocations based on proportionality are less easily quantifiable—government's fair share when it funds infrastructure used by the mine (but usually also by other members of the community) or society's share if it is argued that a mineral deposit is a gift of nature and thus contributed by society at large to the project. A second approach to fairness, associated with the philosopher Jeremy Bentham, emphasizes utility—distribute surpluses such that their use leads to the greatest good. For mining, it is difficult to know how to put this into practice. Which party has the ability to put mining's surpluses to best use? A mining company, through reinvestment in mineral development? A local government, though investment in schools and public infrastructure? A national government, which might take surpluses from the region of a mine and invest them in public infrastructure in a poorer region of the country? A third approach, attributed to the philosopher John Rawls, emphasizes helping the least well-off groups in society. For mining, such an approach might emphasize directing surpluses to activities that alleviate poverty.

Sustain the Benefits of Mining Even After a Mine Closes

Investing an appropriate portion of the revenues from mining in sustainable assets can, in effect, make the benefits from mining permanent. This principle is perhaps the key lesson from the concept of economic sustainability. The idea is simple: save and invest a portion of current income each period, spend at most the investment's income each period and in so doing sustain the ability of the investment's corpus to fund spending indefinitely into the future. This principle can be implemented in various ways. As Hanneson[4] notes, there are a number of important investment issues to consider:

- *How much to save and invest:* The answer depends on what a society wants to sustain— the current level of well-being, growth in the level of well-being, or some other goal? The higher the goal, the higher the savings rate must be.
- *By whom:* Governments on behalf of a local community or national society? The mining companies themselves, which presumably would emphasize investments in sustaining a mine's life or in developing other mineral deposits?
- *In what:* Financial assets? Or public investment in roads, schools, electric power, schools, hospitals, scientific and technical research and development?
- *Where:* In the mining community, region, or nation? Wherever the potential investment returns are highest given the riskiness of an investment?

Implications for Corporate Responsibility and Mine Management

The previous section, articulating principles for the mineral sector, ignored the important issue of who is responsible for pursuing the objectives these principles represent. Just what the appropriate roles and responsibilities of governments, private companies, and other nongovernmental organizations are in mineral development and in pursuit of these principles depends, of course, on one's view of how social activities should be organized. To most economists and in market economies, production of most goods and services—including minerals and mineral products— is left to private enterprises because, in well-functioning markets, private enterprises have the incentive to create wealth in an economically efficient manner (satisfying the environmental and social requirements for sustainable development described previously). In the process of striving to maximize profits for their owners, mining enterprises provide employment for workers, purchase inputs from suppliers, and so forth. As part of the process, these enterprises have the incentive to minimize costs for a given level of output, maximizing the creation of surplus or net benefits (profits from the perspective of private enterprises).

In an ideal world, governments—acting on behalf of society at large—would establish appropriate legal and regulatory frameworks that facilitate achievement of the four principles of sustainable mineral development identified previously. In particular, most economists argue that government intervention in specific markets, including mining, be limited to activities that facilitate market activity, promote fairness in the distribution of burdens and benefits, and promote economic efficiency by correcting imperfections in actual markets that would prevent them from being "well functioning" or economically efficient. In mining, governments would facilitate mineral exploration, mine development, and mining through policies that, among other things, define property rights, create processes for private companies to obtain approvals for exploration and development, and outline and describe the rights and responsibilities of private companies during extractive activities and after mining ceases. Government would define mechanisms to define what a fair distribution of benefits and burdens is and then to achieve this fair distribution. Governments would intervene more directly in mineral activities to correct the inefficiencies that otherwise would occur when environmental damages and social disruptions receive inadequate consideration in decisions about mineral exploration, mine development, and mining.

In such an ideal world, private mining companies would strive to maximize profits based on their evaluations of the commercial attractiveness of the various investment and operational options they face in mineral exploration, mine development, and mining. Private companies would not need to consider their "responsibility" to local communities beyond striving to maximize profits in ways consistent and compliant with social norms as codified in laws and regulations. Companies would *not* find it advantageous or appropriate to go beyond complying with existing rules for environmental protection, worker health and safety, and investments in the communities in which they operate.

This ideal or conventional view of a company's responsibilities can be questioned on at least three grounds. First, mining companies might find it more profitable over the longer term to go beyond compliance if such activities develop goodwill with workers and in the community in which they operate—even if existing laws and regulations clearly define the framework for mining, and government institutions are capable of implementation and enforcement. In this case, companies would consider whether an investment in goodwill today fosters higher profitability in the future, even if it sacrifices current profits. Second, companies may find it more profitable over the longer term than otherwise to go beyond what is minimally required when comprehensive laws and regulations do not exist or when government and social institutions for implementing rules are weak, leading to inconsistent or spotty enforcement or when the political, legal, and regulatory environment is in a state of flux. In this case, mining companies would go beyond what is minimally required to avoid or minimize future conflicts. Companies in effect would become surrogate governments or development agencies in choosing how much environmental protection to build into a mine, how much engagement to have with local communities in ensuring fairness in the distribution of benefits and burdens, and how much attention they should give to developing mechanisms to sustain the benefits of mining after mining ceases through trust funds, investment in other economic sectors to promote diversification, and other activities. The third way the ideal or conventional view of a company's responsibilities can be questioned is more controversial—that companies have a moral responsibility to go beyond mere compliance with rules regarding the natural environment, worker health and safety, and how they interact with and support the communities in which they operate—even if going beyond mere compliance leads to lower profits.

In the first two cases, companies would act in their own self-interest, undertaking activities that reduce profits today in exchange for expected higher profits in the future. In this sense, the first two situations are variations on the ideal or conventional view of a company's social responsibilities rather than entirely different perspectives. The third case, however, when companies would sacrifice profits over the longer term in the public interest, is controversial. A detailed assessment of whether companies *should* undertake profit-reducing activities for moral reasons is beyond the scope of this paper. For such an assessment and review of the important issues, see Vogel,[5] Hay, Stavins, and Vietor,[6] and Yakovleva.[7] Suffice it to say that private companies do undertake activities in the first two categories out of necessity.

PUTTING CONCEPTS INTO PRACTICE IN MINING

Whether motivated by requirements of public policy or company decisions to go beyond simple compliance, most private mining companies undertake a variety of activities aimed at one aspect or another of sustainability or sustainable development. Yakovleva[8] groups these activities into three categories: the natural environment, worker health and safety, and community development and stakeholder engagement. Yakovleva reviewed annual reports and other information published by the ten largest gold mining companies during the period 1998–2001 and found that the majority of companies made public commitments and disclosures in all three categories. For the environment, the majority of companies commit to meet established rules and regulations, to improve environmental performance, to meet the standards of sustainable development, and to minimize environmental impacts and effects; they have environmental management plans and systems for, for instance, environmental monitoring, tailings management, water management, air quality, revegetation, assessment of ecological risk and environmental impacts, independent environmental audits, and reporting of environmental performance.

Similarly, in the area of worker health and safety, the majority of large gold mining companies have formal health and safety policies, a commitment to improve performance in this area, dedicated health and safety offices, training programs, monitoring of all workers, HIV/AIDS policies or programs, and health and safety audits and reports. Finally, in the category of community development and stakeholder engagement, the majority of large gold mining companies have formal social and community policies and consult with stakeholders; offer education and training to members of the local community; contribute to community activities such as schools and clinics; cooperate with nongovernmental organizations, industry associations, and government on social programs; and have established philanthropic foundations. To be sure, some of these activities are required by existing government policies and regulations. But many go beyond mere compliance. Key approaches are disclosure of policies, independent audits of performance, and public reporting.

To this point, this chapter has discussed general principles and concepts. The rest of the chapter describes several specific initiatives aimed at putting sustainability and sustainable development into practice in the mining sector.

Broad, Multi-Participant Initiatives

Between 2000 and 2002, a group of large mining companies undertook the Global Mining Initiative (GMI), dedicated to understanding the relationship between sustainable development, on the one hand, and mining, on the other. GMI had two lasting outputs. The first is the final report of the Mining, Minerals and Sustainable Development (MMSD) project, an independent review of the key issues, initiated through the World Business Council for Sustainable Development and then performed by the International Institute for Environment and Development. The review involved regional partnerships focusing on issues specific to southern Africa, South America, Australia, and North America; 23 global workshops and expert-group meetings; and more than 150 commissioned studies. The final report appeared as the book *Breaking New Ground* (International Institute for Environment and Development, 2002).[9] It identifies nine key challenges, which broadly speaking encompass the four principles articulated earlier in this chapter. The second lasting output is the creation of the International Council on Mining and Metals (ICMM)[10] based in London. ICMM is an industry association charged with carrying out the work investigated by MMSD and more generally to promote mining that is both viable for companies and that contributes to broader sustainable development.

Between 2001 and 2003, the Extractive Industries Review was performed. The review, commissioned by the World Bank and conducted independently, examined the role of the World

Bank in the extractive industries, including oil, natural gas, and mining. It was prompted by strong external criticism of the bank's role in financing and promoting these sectors, particularly that the extractive industries were *not* consistent with sustainable development and poverty alleviation precisely because of the environmental damage and social disruption caused by extraction. The review found that the World Bank does have a role to play in the extractive sectors but only if its activities promote sustainable development and poverty alleviation, which in turn require that three enabling conditions be met: that pro-poor public and corporate governance exists, that extractive projects have stronger environmental and social requirement than at present, and that there be respect for human rights as verified by third parties. The detailed findings of the review are available at the World Bank Web site,[11] and the World Bank's response can be viewed at the International Finance Corporation Web site.[12]

Three other international initiatives relate to mining, even though they are not targeted directly at sustainability and sustainable development. First, the Equator Principles are a voluntary set of guidelines for banks and other financial institutions to use in managing environmental and social issues that arise in lending to investment projects. The principles[13] were developed in 2003 and revised in 2006. As of early 2008, 59 institutions had become signatories. Second, the Global Compact,[14] initiated in 2000, is a voluntary network of businesses and other organizations committing to adhere to practices consistent with appropriate corporate behavior in the areas of human rights, labor practices, the environment, and corruption reduction and prevention. Third, the Extractive Industries Transparency Initiative[15] promotes full disclosure and independent verification of company payments and government revenues from oil, gas, and mining. Launched in 2003, the initiative aims to improve private and public governance in resource-rich nations by making it more difficult for companies and governments to undertake activities that the public at large finds inappropriate.

Project-Level Tools

At the scale of an individual mine, community, or region, implementing any of the principles described to this point in the chapter typically is not simple or straightforward. Developing appropriate processes to consider and, when necessary, balance economic, environmental, and social considerations is essential.

The Seven Questions Framework

One approach is the Seven Questions framework[16] for assessing sustainability (Figure 3.1). The Seven Questions help answer the overall question: Is a project or operation and its net contributions to sustainability positive over the longer term?[17] Developed as part of the MMSD project described previously, the questions are the product of two workshops, one in Canada and the other in the United States, involving some 30 participants and experts from academia, First Nations/Native Americans, government, industry, organized labor, and other nongovernmental organizations. The Seven Questions are broad and serve as starting points for more detailed examinations of specific issues:

1. *Engagement:* Are engagement processes in place and working effectively?
2. *People:* Will people's well-being be maintained or improved?
3. *Environment:* Is the integrity of the environment assured over the longer term?
4. *Economy:* Is the economic viability of the project or operation assured, and will the economy of the community and beyond be better off as a result?

Assessing for
Sustainability

1.
Engagement.
Are engagement processes
in place and working effectively?

2.
People.
Will people's well-being
be maintained or improved?

3.
Environment.
Is the integrity of the environment
assured over the long term?

4.
Economy.
Is the economic viability of the project
or operation assured, and will the
economy of the community and
beyond be better off as a result?

5.
**Traditional and
Non-Market Activities.**
Are traditional and non-market activities
in the community and surrounding area
accounted for in a way that is acceptable
to the local people?

6.
**Institutional Arrangements
and Governance.**
Are rules, incentives, programs, and
capacities in place to address project
or operational consequences?

7.
**Synthesis and
Continuous Learning.**
Does a full synthesis show that the
net result will be positive or negative
in the long term, and will there be
periodic reassessments?

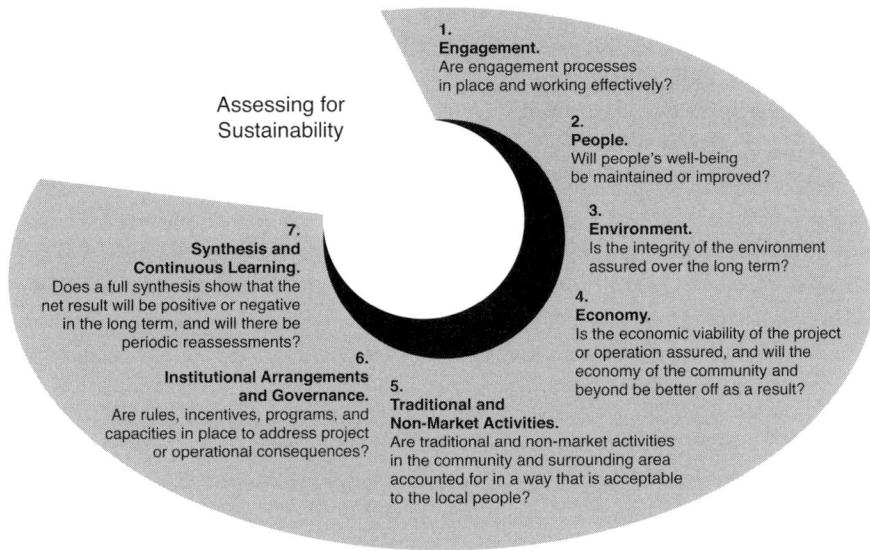

Source: International Institute for Environment and Development, 2002

FIGURE 3.1   Seven Questions to Sustainability

5. *Traditional and Non-Market Activities:* Are traditional and non-market activities in the community and surrounding area accounted for in a way that is acceptable to the local people?

6. *Institutional Arrangements and Governance:* Are rules, incentives, programs, and capacities in place to address project or operational consequences?

7. *Synthesis and Continuous Learning:* Does a full synthesis show that the net result will be positive or negative in the long term, and will there be periodic reassessments?

Each broad question is answered through assessment of a hierarchy of more specific questions, indicators, and measures. For example, an affirmative answer to the first question (engagement) requires that all interested parties have the opportunity to participate in decisions that affect their futures and that engagement processes are understood and agreed upon by the interested parties, consistent with legal, institutional, and cultural characteristics of the community and country. Appropriate processes need to have dispute resolution mechanisms, reporting and verification, adequate resources, and informed and voluntary consent. There are similar requirements for affirmative answers to the other questions.

Seven Questions is a framework for discussion, aimed at helping identify issues important to the various interested parties. It is a mechanism for evaluating these issues and the trade-offs that sometimes are inevitable among economic, environmental, and sociocultural considerations and objectives associated with mining. It is a mechanism for finding solutions to conflict that are mutually beneficial to all parties.

The Seven Questions framework was applied to the experience of the Tahltan people with mining.[18] The Tahltan people are a First Nation in northwestern British Columbia, Canada. Over the last 50 years, a number of gold mines have operated in Tahltan traditional territory. Since the middle 1990s, Barrick Gold's Eskay Creek gold mine has operated there.

The goals of the Seven Questions review were to evaluate the experiences of the Tahltan people and develop a strategy to guide future relationships between the Tahltan people and the mining industry. The review was carried out through a 3-day workshop in 2003, involving twenty-eight Tahltan people, five industry participants, four government representatives, and a facilitator. The Tahltan assessment consisted of separate evaluations of mineral exploration, operating mines, and mine closure. The assessments contributed to a mining strategy that the Tahltan people developed for themselves, with three objectives: first, to communicate that the Tahltan people support mining on their lands if it is done "right"; second, to facilitate Tahltan participation in mining and mineral activities; and third, to ensure that community concerns—especially those related to health, society, and culture—receive adequate attention. The strategy identified a number of potential, specific actions that the Tahltan people could take to achieve these objectives and that the mining industry and government could take, too. One would like to write that the Seven Questions framework has been an unqualified success for the Tahltan people and the mining companies active on Tahltan lands. Although the framework helped facilitate interactions among the various interested parties in the Tahltan region, it has not eliminated all tensions. In 2005, 35 Tahltan elders occupied the office of the elected Tahltan chief for more than 3 weeks and demanded that he step down for giving inadequate attention to consultation with communities in negotiating on behalf of the Tahltan people for a number of new mining projects.[19]

The Community Development Toolkit

The World Bank and ICMM jointly developed another similar but more detailed framework to guide interactions between mining companies and communities, the *Community Development Toolkit*.[20] This toolkit is aimed primarily at mining companies, although it can be used also by communities, governments, and other nongovernmental organizations. The existence of such a toolkit might suggest that mining companies in the past have not paid attention to communities. Such is not the case. As the toolkit notes, companies frequently funded schools, clinics, and other facilities in communities in which they operated. Unfortunately, in many cases these projects—although well intentioned—were not as successful as they might have been and were not long lasting (i.e., sustainable). The reasons were many: the companies or local elites chose the projects without broad community involvement; outsiders built or ran the project; only the affluent community members, and not the poor, had access to the project; or the local community did not have the capacity to manage the project, especially after the mine closed. It was against this backdrop that the toolkit was developed. It consists of 17 tools to facilitate community development over the mining-project life cycle, including exploration, feasibility, construction, operations, decommissioning and closure, and postclosure (Figure 3.2):

Assessment:

1. Stakeholder identification: to identify all the people with an interest or who might be affected by a project

2. Social baseline study: to profile a community and its regional and national setting

3. Social impact and opportunities assessment: to identify and assess the impacts of a project, positive and negative, and to evaluate how to manage them

4. Competencies assessment: to determine the adequacy of your team's skills, knowledge, and understanding and, if deficient, evaluate what other cometencies may be necessary

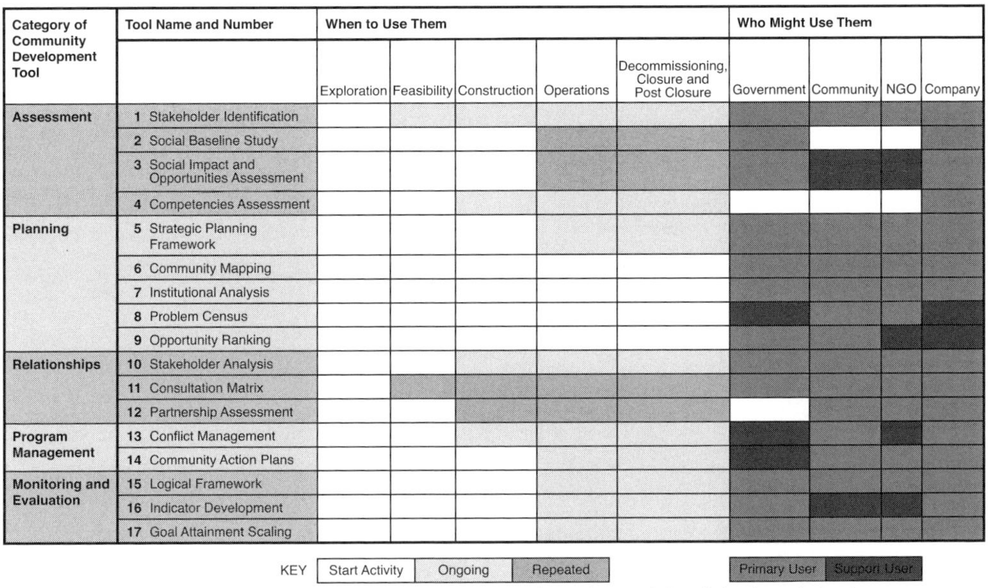

Category of Community Development Tool	Tool Name and Number	When to Use Them					Who Might Use Them			
		Exploration	Feasibility	Construction	Operations	Decommissioning, Closure and Post Closure	Government	Community	NGO	Company
Assessment	1　Stakeholder Identification									
	2　Social Baseline Study									
	3　Social Impact and Opportunities Assessment									
	4　Competencies Assessment									
Planning	5　Strategic Planning Framework									
	6　Community Mapping									
	7　Institutional Analysis									
	8　Problem Census									
	9　Opportunity Ranking									
Relationships	10　Stakeholder Analysis									
	11　Consultation Matrix									
	12　Partnership Assessment									
Program Management	13　Conflict Management									
	14　Community Action Plans									
Monitoring and Evaluation	15　Logical Framework									
	16　Indicator Development									
	17　Goal Attainment Scaling									

KEY　　Start Activity　Ongoing　Repeated　　　　Primary User　Support User

NOTE: This matrix provides a general guide to the tools including who might use them and when during the project cycle.

Source: ESMAP, World Bank, and ICMM.

FIGURE 3.2　Overview of ICMM's Community Development Toolkit

Planning tools:

5. Strategic planning framework: to assess why you want to contribute to community development, what your development objectives are, how you will achieve these objectives, and how you will know when you have succeeded

6. Community map: to facilitate the local people's planning of their community's physical layout

7. Institutional analysis: to evaluate the variety, strength, and linkages among institutions in the community

8. Problem census: to facilitate identification and prioritization of development issues in the community, involving all interested parties

9. Opportunity ranking: to help decide what projects to start first by considering both priority and feasibility

Relationships tools:

10. Stakeholder analysis: to evaluate the level of interest of the various stakeholders in the project

11. Consultation matrix: to develop a system for consulting with stakeholders that is appropriate both in frequency and detail

12. Partnership assessment: to analyze potential partners in community development

Program management tools:

13. **Conflict management:** to identify, understand, and manage conflicts through to resolution so that these conflicts do not disrupt activities of any of the stakeholders

14. **Community action plans:** to implement solutions to problems identified in the planning process

Monitoring and evaluation tools:

15. **Logical framework:** to develop clear outputs and outcomes using verifiable measures of progress toward goals

16. **Indicator development:** to choose indicators that are transparent and can stand up to external scrutiny

17. **Goal attainment strategy:** to measure the degree to which outputs and outcomes are being met

The toolkit contains detailed questions and templates for each tool.

Finally, at the stage of mineral exploration, Environmental Excellence in Exploration publishes an online toolkit of good practices and examples of environmental and social corporate responsibility in topics such as community engagement, land acquisition and access, camps, and drilling. It also has checklists and case histories. It is a consortium of companies and coordinated by the Prospectors and Developers Association of Canada.[21]

CONCLUSIONS

So, what do sustainability and sustainable development mean for mineral development and mine management? At a minimum, they mean that mining companies and mine managers will continue to work with local communities and local and national governments to integrate environmental and social issues into technical and commercial decisions about mining. They mean that processes will continue to be important—the processes through which companies and all other interested parties to mine development identify, discuss, and resolve issues that require integrating the technical, commercial, environmental, and social aspects of mining.

One hopes, more broadly, that the growing experience of companies and communities interacting with one another will lead to broader agreement about what is to be sustained, for whom, and through what process. To date, there has been little consensus on how to put the broad concepts of sustainability and sustainable development into practice. Nevertheless, there are limits to the extent that we should expect worldwide agreement to emerge. What is sustainable to a community in one part of the world may not be sustainable in another community in another part of the world. Sustainability and sustainable development are all about social preferences for the appropriate balance among the economic, environmental, and sociocultural dimensions of human activity.

Suggested Readings

This chapter draws significantly on four earlier writings by the author.[22–25] In addition to the publications cited in the text, the following papers and books are recommended. For more on the broad concepts of sustainability and sustainable development, see *Our Common Future*,[26] referred to as the Brundtland report and credited with initiating discussion of sustainable development, Jamieson,[27] Pezzey and Toman,[28] National Research Council,[29] and World Bank.[30]

For more on how sustainability and sustainable development relate to the mining sector, see Auty and Mikesell[31] and Otto and Cordes.[32]

For more on managing the range of sustainability issues at the mine site, see Rajaram, Dutta, and Parameswaran[33] and the various publications prepared by the International Council on Mining and Metals,[34] the Minerals Council of Australia,[35] and other industry professional associations.

NOTES

1. H. L. Hartman, ed., *SME Mining Engineering Handbook*, 2nd ed. (Littleton, CO: Society for Mining, Metallurgy and Exploration, 1992).

2. William D. Nordhaus and Edward C. Kokkelenberg, eds., *Nature's Numbers: Expanding the National Economic Accounts to Include the Environment* (Washington, DC: National Academy Press, 1999), 250.

3. H. P. Young, *Equity: In Theory and Practice* (Princeton, NJ: Princeton University Press, 1994).

4. R. Hanneson, *Investing for Sustainability: The Management of Mineral Wealth* (Boston: Kluwer Academic, 2001).

5. David Vogel, *The Market For Virtue: The Potential and Limits of Corporate Social Responsibility* (Washington, DC: Brookings Institution Press, 2005).

6. Bruce L. Hay, Robert N. Stavins, and Richard H. K. Vietor, eds., *Environmental Protection and the Social Responsibility of Firms: Perspectives from Law, Economics, and Business* (Washington, DC: Resources for the Future, 2005).

7. N. Yakovleva, *Corporate Social Responsibility in the Mining Industries* (Aldershot, Hampshire, England: Ashgate Publishing, 2005).

8. Ibid.

9. International Institute for Environment and Development, "Breaking New Ground," MMSD Final Report (London: IIED, 2002), www.iied.org/mmsd/finalreport/index.html (accessed February 7, 2008).

10. International Council on Mining and Metals, www.icmm.com.

11. Extractive Industries Review, http://go.worldbank.org/PMSHHP27M0 (accessed February 7, 2008).

12. International Finance Corporation, "World Bank Group Management Responses," www.ifc.org/eir (accessed February 7, 2008).

13. "The Equator Principles," www.equator-principles.com (accessed February 7, 2008).

14. United Nations Global Compact, www.unglobalcompact.org (accessed February 7, 2008).

15. Extractive Industries Transparency Initiative, http://eitransparency.org (accessed February 7, 2008).

16. International Institute for Sustainable Development, *Seven Questions to Sustainability: How to Assess the Contributions of Mining and Mineral Activities* (Winnipeg: International Institute for Sustainable Development, 2002), www.iied.org/mmsd/mmsdpdfs/145_mmsdnamerica.pdf (accessed February 8, 2008).

17. Ibid.

18. *Out of Respect: The Tahltan, Mining, and the Seven Questions to Sustainability*, Report of the Tahltan Mining Symposium, April 4–6, 2003 (Winnipeg: International Institute for Sustainable Development, and Dease Lake, BC: Tahltan Central Council, 2004).

19. Ron Collins, "Tahltan Chief Jerry Asp Removed from Office by Elders," www.minesand communities.org/Action/press535.htm (accessed February 8, 2008).

20. ESMAP, the World Bank, and ICMM, *Community Development Toolkit* (Washington, DC: ESMAP and the World Bank and London: ICMM, 2005), www.icmm.com (accessed February 8, 2008).

21. Environmental Excellence in Exploration online toolkit, www.e3mining.com (accessed February 8, 2008).

22. R. G. Eggert, "Sustainable Development and the Mineral Industry," in *Sustainable Development and the Future of Mineral Investment*, eds. J. M. Otto and J. Cordes (Paris: United Nations Environment Programme and Metal Mining Agency of Japan, 2000).

23. R. G. Eggert, *Mining and Economic Sustainability: National Economies and Local Communities*, monograph commissioned by the Mining, Minerals and Sustainable Development Project. Available on the CD-ROM accompanying *Breaking New Ground: Mining, Minerals and Sustainable Development*. Earthscan, 2002.

24. R. G. Eggert, "The Mineral Economies: Performance, Potential Problems and Policy Challenges," in *Managing Mineral Wealth* (Addis Ababa: United Nations Economic Commission for Africa, 2004), 7–48.

25. Roderick Eggert, "Mining, Sustainability and Sustainable Development," in *Australian Mineral Economics*, monograph 24, ed. Philip Maxwell (Carlton, Victoria: Australasian Institute of Mining and Metallurgy, 2006) 187–194.

26. World Commission on Environment and Development, *Our Common Future* (Oxford: Oxford University Press, 1987).

27. D. Jamieson, "Sustainability and Beyond," *Ecological Economics* 24, no. 2-3 (1998): 183–192.

28. J. C. V. Pezzey and M. A. Toman, "Sustainability and Its Economic Interpretations" in *Scarcity and Growth Revisited: Natural Resources and the Environment*, ed. R. D. Simpson, M. A. Toman, and R. U. Ayres (Washington, DC: Resources for the Future, 2005).

29. National Research Council, *Our Common Journey: A Transition Toward Sustainability* (Washington, DC: National Academy Press, 1999).

30. World Bank, *World Development Report 2003: Sustainable Development in a Dynamic World* (Washington, DC: World Bank and Oxford: Oxford University Press, 2003).

31. R. M. Auty and R. F. Mikesell, *Sustainable Development in Mineral Economies* (Oxford: Clarendon Press, 1998).

32. J. M. Otto and J. Cordes, eds., *Sustainable Development and the Future of Mineral Development* (Paris: United Nations Environment Programme, 2000).

33. V. Rajaram, S. Dutta, and K. Parameswaran, *Sustainable Mining Practices: A Global Perspective* (London: Taylor and Francis, 2005).

34. International Council on Mining and Metals, www.icmm.com.

35. Minerals Council of Australia, www.minerals.org.au/.

CHAPTER FOUR

Strategic Issues in the Mining and Metals Industries

J. L. Rebollo

INTRODUCTION

Sustainability is becoming a fundamental pillar of any mining or metals company's strategy. That's because today's society has a much better understanding of the consequences of the interaction between human activity and the ecosystem, and people have become increasingly conscious of the importance of this issue for both present and future generations. As a consequence, the social license to operate is becoming more difficult to obtain, forcing companies to better integrate the concept and the values of sustainability into their strategy.

The dictionary[1] defines *strategy* as "the science and art of using all the forces of a nation to execute approved plans as effectively as possible during peace or war." If this definition were translated into business terms, it would be "the science and art of using all the forces of a company to create and execute plans as effectively as possible in order to reach its objectives." The adjective *strategic* is defined as "highly important to an intended objective." As a corollary, the strategy of a company is the science of dealing with its more important issues.

Regardless of the skills of its managers, no company can conduct business by improvising decisions. Mining is no exception; managing a mining business requires exhaustive internal and competitive analysis, which results in a set of guiding principles that constitute the strategic plan.

The conception of the company's strategy is the highest-level task that managers can perform. Although the broad objectives are fixed by the shareholders, management is in charge of identifying the actions and means to allow the corporation to get as close to these objectives as possible.

Designing a company's strategy includes a permanent redefinition of its business geometry: its size, markets, geographical implementation, technologies, and human resources, as well as social and political attitude. Strategy is intrinsically dynamic; it can be said that an excessively static strategy is always a poor one.

All too frequently, operational managers lack sufficient understanding of the rationale for strategic corporate decisions. The goal of this chapter is to aid in the comprehension of these strategic choices, presenting the theoretical basis of the corporate strategy in a simple, comprehensive way, as well as the nature and influence of the external and internal factors affecting the competitiveness of mining companies.

The work of many authors has contributed to the creation of the modern methodology for corporate strategy and strategic planning. However, Michael E. Porter[2] and Alfred Humphrey merit special mention because of the relevance of their contributions; their ideas have inspired some elements of this chapter.

BASICS OF BUSINESS STRATEGY: SPECIAL CHARACTERISTICS OF THE MINERALS INDUSTRY

This section presents the basic principles governing the most important function of management: the strategy. It starts with some basic ideas and concepts, then continues with a description of the main external and internal factors and their roles in a company's life. It focuses specifically on the more relevant issues for the mining and metals industries.

The Corporation and Its Interfaces

Corporations are open systems interfacing actively with their environments. There is a continuous flow of information between corporations and their shareholders, employees, clients, suppliers, banks, financial analysts, media, and authorities, as well as with their political and social neighborhoods. The quality and intensity of these flows are not the same for every sector or for every company within each sector.

A corporation's development of its strategic plan must take into consideration the dissimilar interactions with these stakeholders. The importance to the company of the different interfaces differs with the kind of business and its environment. For example, the mining sector will give more importance to environmental issues than will the banking or retailing sectors, which will put more emphasis on suppliers' issues.

An essential condition for successful strategies, in all companies and all sectors, is their elaboration as a complex, multivariable exercise, integrating the potential influence of each one of the factors that may affect a company's business.

The mining and metals business, because of its specificity, requires a particular kind of relationship with its stakeholders. Any mining company needs what is called "social license to operate." Although the feeling of mistrust that has often obstructed the relationship between mining companies and some of their stakeholders is less present today, a certain degree of caution still remains. It is evident that mining activities are still regarded by the public as threatening and hazardous. Few citizens will object to the installation of a research center in their neighborhood, but it will be a hard task to convince them to accept a mine near their homes. It is obvious that a strategic plan for developing such a mine can't be limited to the geological, operational, market, and financial considerations. It must include an exhaustive analysis of the social, political, and environmental implications that could be crucial for the feasibility of the project.

The management structure of a company must reflect these external interfaces, so it is increasingly common in mining companies for the senior executive in charge of sustainability matters to occupy a seat on the board of directors, along with the technical, marketing, and financial executives. This was seldom the case in the past.

The interaction between a corporation and its stakeholders must respond to a well-established, coherent, and consensual plan aligned with the corporate strategy. In addition to the marketing communication that aims to improve the commercialization of products or services, corporate communication must build an appropriate image of the company itself, creating the best possible reputation among all stakeholders. It must be professionally and efficiently designed and executed, because operational and functional managers do not necessarily have the appropriate skills to perform such a task.

Crisis communication is a particularly important element of this communication function for the minerals industry. Because of the nature of its activities, the minerals industry is especially exposed to sudden unplanned events that can have serious repercussions on a company's image and threaten its equilibrium. Industrial accidents, casualties, effluent spills, and similar incidents need to be handled and communicated in a very professional manner. Although these events are, by their nature, unpredictable, the reaction of the company can and must be prearranged. A crisis cell, perfectly trained for this task, must be put in place immediately. It must be in charge, not only of taking the first measures to cope with the crisis, but also of arranging and executing the internal and external communication to minimize the negative repercussions of the event.

Another important element, although less specific to the mining sector, is financial communication. Apart from the institutional reports (e.g., annual financial report, sustainability reports), the company should institute a set of contacts with its financial stakeholders, analysts, and shareholders. Data supplied must be truthful, clearly expressed, and consistent with previous communications. The key expression of this consistency is the coherence of the units and references used. There is a strong temptation to change the time references or the summarizing schemes in order to hide unsatisfactory performance from view and make time series analysis difficult. This would be a mistake; in the long run, it will surely provoke a negative perception of the entire business, as well as a reaction of mistrust that will have a serious effect on the credibility of the company.

Reputation is an important intangible asset in mining companies. It can be improved in the framework of an intelligent strategy or degraded by wrong policies and poor communication. It is not only a matter of prestige. It allows companies to be better accepted by their environment and, as a consequence, increase their competitiveness in many fields.

Methodology of Strategic Planning

It was stated previously that strategy is the science and art of dealing proactively with the important issues of a company. It is basically an empirical science, built more on practical experience than on theory. Years of practicing this science by thousands of corporations, together with the analysis of the results of the application of diverse principles, have consolidated a methodology that can be used to develop successful strategies. Numerous authors have worked on compiling and synthesizing these experiences, but special mention should be given to Michael E. Porter,[3] one of the most innovative and pertinent authors in this field.

The strategic planning process sets the principles and guidelines that have to be applied to achieve the company's objectives and to determine how these objectives will be reached.

The first step consists of analyzing the present situation: "Where are we now?" This asks for an unbiased analysis of today's situation, identifying the strengths and weaknesses of all the components of the company: markets, technologies, human resources, political background, environmental issues, financial situation, and so forth. It is in reality a very classical exercise. In the 1970s, business theorist Alfred Humphrey developed the SWOT (strengths, weaknesses, opportunities, and threats) method at Stanford Research Institute; this method can be very helpful in systematizing this analysis.

The second step is a bit more complicated: "Where do we want to go?" Do we want to be the biggest company in the sector? Do we want to maximize the return to our shareholders? Do we want to help the society through clean, sustainable operations? Or, perhaps, do we want a weighted mix of all of these? We must define how we ideally want the company to be in the future. The answers to these questions often lie with the shareholders, but management can and must provide quantified options. It is extremely important, although it is not always the case in practice, to integrate macroeconomic information into strategy design,[4] especially in the commodities industry.

The third step defines the strategic plan: "How do we get there and with what timing?" Depending on the nature of the business, companies can take either of two approaches to strategic management. The first approach is the industrial organization approach, which is based on economic principles like competitiveness and costs and resources allocation. The second, the sociological approach, is based on human interactions, like behavior or customer satisfaction. The industrial organization approach is suitable for companies with activities that are more material, like the mining, automobile, food, machinery, or electronics industries. The sociological approach is appropriate for companies producing and selling more immaterial goods, like software, medical services, and media.

Generally, both approaches are used together, with a preponderance of one or the other. In the case of the mining and metals industries, the industrial organization approach predominates.

Dimensions of the Strategy

The strategic planning domain has two dimensions: time and degree of generality. On the time dimension, the strategy can be short or long term.

The short-term strategy portrays the measures needed to run the business in a time horizon of less than 1 year. Thus, it manages the present. Its main goal is maximizing value by establishing decisions tailored to the observed situation and conditioning factors. It is set up within the framework of the long-term strategy and is generally driven by opportunistic considerations. It is usually set up within the framework of a task force involving a number of specific competences: marketing, operations, human resources, financial, legal, and so forth.

The long-term strategy prepares the future, is essentially structural, and seeks to maximize the long-term value of the company. It is oriented to the development of sustainable competitive advantages. It usually results from a deep analysis carried out by a multidisciplinary team led by the company's top executives and the chief executive officer (CEO). It often involves profound changes in the company's structure, spread over several years, and has low reversibility. As Gary Hamel expressed acutely, "a company that cannot imagine its industry's future may not survive to see it."[5] In practice, changes to the long-term strategy of a company need the approval of the board and must be submitted to the shareholders' assembly.

On the other dimension, the "degree of generality," strategic planning involves different levels, from high to low: corporate, business, and functional. These levels are correlated with the organizational structure:

- The corporate-level plan is the most general, conclusive, and essential plan a company can establish. It provides the necessary guidance for all the activities of the company and defines the financial and geographical geometry of the corporation, through mergers, acquisitions, divestitures, and spin-offs.

- The business-level plan defines the technological and commercial strategy, product policy, and human resources development.

- The functional-level plan deals with the most operational issues of the company: work organization, production planning, maintenance, and so forth.

Figure 4.1 illustrates the two conceptual dimensions of the strategy. The sizes and positions of the defined areas depend on the specific circumstances of each company.

The strategic plan must extend over all domains of the company—functional, business, and corporate—with the overall goal of "being better than competitors" for every option adopted. All options shall be generated through a process of creative thinking,[6] getting away from routine, and traditional, well-known alternatives. They must be challenged, checking that all assumptions are realistic and feasible, answering all "what if . . ." questions, and cautiously taking into consideration the probabilities and consequences of the worst-case scenario.

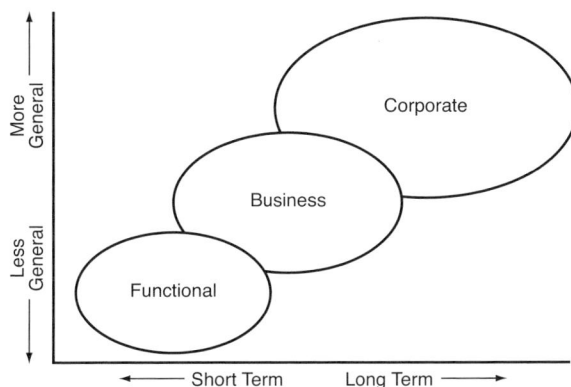

FIGURE 4.1   Two conceptual dimensions of strategy

Operational Effectiveness Versus Strategy

It is important to differentiate between operational effectiveness and strategy. A company can reach high levels of operational effectiveness through the application of modern management tools (reengineering, benchmarking, outsourcing, total quality), but this does not necessarily lead to value creation and sustainable profits, unless these actions are well inserted in the framework of an adequate strategy.

Because of the importance that costs have in the minerals industry's competitiveness, management frequently immerses itself in an intensive exercise of improving operational efficiency and costs, neglecting the strategic agenda. Management techniques often take the place of sound strategy. Operational effectiveness is necessary but not sufficient. In the early 1900s, although some lead producers made substantial investments in order to increase the production of bismuth (a by-product of lead concentrates) as a way to increase their competitiveness, the U.S. Food and Drug Administration was considering banning its use in pharmaceutical products, then its main use. The result was an overproduction that was not at all in line with market demand. Once again, productivity considerations took the place of strategic thinking.

The search for competitive advantages as a way to create value is the main goal of the corporate strategy. These advantages must be sustainable and long lasting, without degrading the existing ones, and developed in the framework of a sound long-term strategy. For example, reducing the premiums over the London Metal Exchange (LME) prices of a metal will allow increasing sales and reducing stocks in a tough market scenario. But this advantage is only temporary because competitors will quickly imitate the move with equivalent results. The final consequence is shrinking margins for the entire industry. This strategy does not create a permanent, sustainable advantage, so it is not a good one. A much better alternative would be to reduce the amount of impurities in the metal ingots, through research and investments, differentiating the product from the competitor's. The dominance of benchmarking over real strategies leads to imitation and ultimately to strategic convergence.

The maximum possible operational efficiency that a company can ideally reach using the best technologies, human capital, and external resources has been called the "productivity frontier" (Figure 4.2).[7]

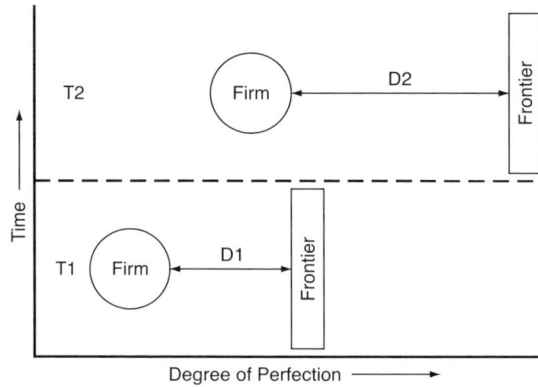

FIGURE 4.2   Productivity frontier

Some companies try to create competitive advantages by moving forward their operational efficiency toward the productivity frontier (Figure 4.2). As the production factors (education, technologies, external supplies, etc.) progress continuously, the frontier moves forward, neutralizing part of the progress. Except for the ones that manage to create differentiation, the result is an exhausting race that does not create competitive advantage for anyone. The search for operational excellence is necessary but not sufficient. It must be performed in the framework of a competitive, value-creating strategy.

Generic Strategies

Among the multiple strategies that a company can implement, Michael Porter[8] has identified three generic ones as common factors for most successful businesses: cost, differentiation, and focus.

Cost Strategy

The first generic strategy is reducing costs. It is obvious that if a company is able to produce an equivalent good or service at a lower cost than its competitors, it gains a competitive advantage because its margins will be higher for a given market price. This generic strategy is by far the most important for the mining and metals industries. Reducing costs is sometimes contradictory with other functions that may be essential to a good competitive situation. It is easy to reduce research, development, or training costs, but the consequences over the long term can be catastrophic. However, there is an optimal equilibrium to be found between cost reductions and long-term competitiveness.

The cumulative cost curve (Figure 4.3) is a strategic analysis tool that represents the unit cost by company in a given sector, related to the cumulative production. The cost usually represented is the *net cash cost*, defined as the total production cash cost credited with the value of the by-products produced per ton. The width of the columns is proportional to the amount of units produced per year. If the market price of the metal is Q_y, the margin of company D will be positive (P). For this price, all companies in the sector except J and K have positive margins. If the market price is Q_x, then company D will have a negative margin (L). At this market price, only companies A and B have positive margins.

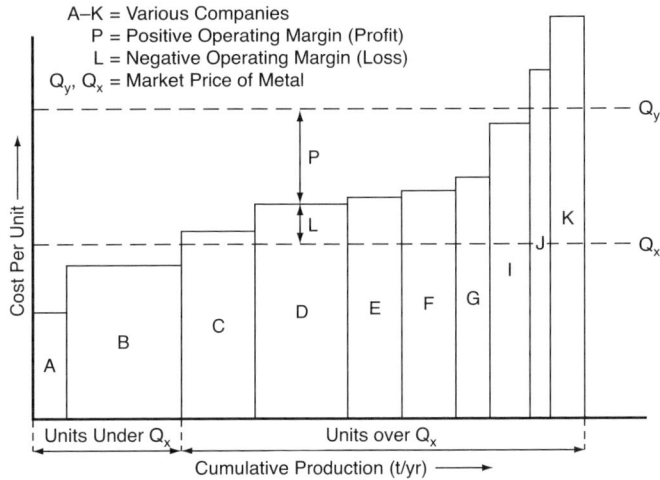

FIGURE 4.3   Cumulative cost curve

The curve allows the companies to evaluate their cost position versus that of their competitors. It also allows for estimating the total amount of production in the sector with associated costs lower than those of a given company. The shape of the curve fluctuates with time. It is then possible to assess changes in competitiveness for the different actors.

These curves are largely used in the minerals industry to carry out competitive analyses. They are published and made available to the industry by several specialized consulting firms for the most important metals.[9]

Another instrument commonly used in prospective cost analysis is the "experience curve" (Figure 4.4). It has been demonstrated that the average manufacturing cost per unit of a product decreases regularly with the cumulated number of units produced. This is due to the "learning experience," which makes the production process more efficient as a consequence of the improvements introduced over time. It has been suggested that the relative reduction in cost each time that the cumulated production doubles is constant and specific for each industry, product, or process, going from 10% to 30%. In the late 1970s, the Boston Consulting Group (www.bcg.com) stressed the link between this fact and the strategy formulation: companies should capitalize on the situation within the life cycle of a product to gain the competitive advantage arising from lower costs. The curves can be drawn empirically for each product or process.

Differentiation Strategy

The second generic strategy is differentiation. Although more important in other businesses, it has a certain weight in the minerals industry. Differentiation consists of doing business in a way that's distinct from that of the competitors. Although some metal producers would choose to specialize in low-cost mass production of a few basic metals, others may decide to transform some impurities contained in the concentrates in minor or precious metals. Differentiation can be a way to overcome the negative effect of some external factors, like high labor costs, by taking advantage of other favorable ones, like a highly qualified labor force. Because of the traditional interchange of technical and organizational information between mining companies, best practices are rapidly imitated. In addition to the data interchanged by bilateral benchmarking exercises,

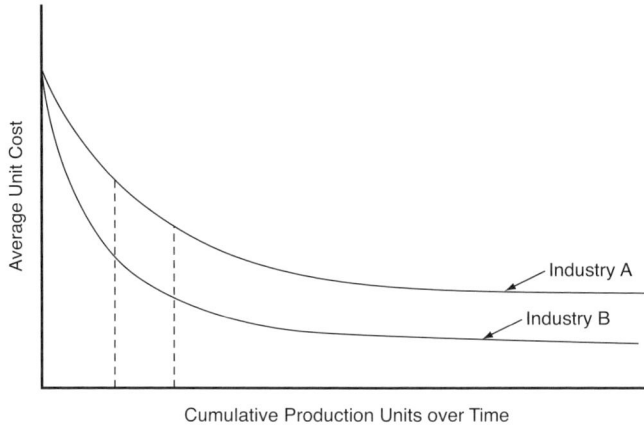

FIGURE 4.4   Experience curve

strategic knowledge may be diffused through visits, institutional communication, or even informal exchanges. Competitors tend to copy successful strategies, and finally most players end by having the same ones. A good strategic analysis will recognize the competitors carrying out similar strategies and identify different paths that may be the key for success. The paths to be followed for the companies to get competitive advantages are not always obvious. The fight to create value is sometimes fierce; therefore, it is not surprising that a frequent motivation for mergers and acquisitions is the opportunity to surpass a competitor, following this strategy: "if you can not beat it, buy it."

Focus Strategies

The third generic strategy is focus, which means concentrating the activity on a particular market segment, product group, or geographical area. After the analysis is done and the decisions are made, management should not be distracted with secondary options. Consistency is an essential quality necessary to implement any strategy. All the resources of the company must be put at the service of its main objectives. The cost of "management distraction" is not only the amount of the labor costs but also the loss of value creation over the main strategic objectives. Consistency is a great virtue for a corporate strategy. It is, most of the time, the consequence of strong and intelligent leadership.

Mining companies often face this question: "Where do we position the business in the value chain?" Because it can only mine ores and sell metal concentrates, more questions develop: "Or should we get the smelting process integrated, producing ingots and alloys? Or go further and fabricate derivate products like sheets, pipes, die casting pieces, packaging items, or chemical by-products? How large should the spectrum of activities be? Should it be reduced to a few metals? Fabricated in a broader variety? Should we enlarge its mining activities to products like bricks, clay, kaolin, refractory products? Include some energy commodities like uranium or coal and so forth? Why not oil, tar sands, natural gas...?"

Rather frequently, companies implement strategies based mainly on growth. Growth by itself is not a good strategy. The growth approach may lead to fundamental errors that may damage the value-creating function. On the other hand, a sound strategy can also lead to growth when used in addition to value creation.

Dynamics of the Strategy Formulations

Although an efficient strategy must be consistent, it should not be rigid. As the factors affecting a company's performance change with time, the strategy needs to be adapted to these changes. This flexibility ought to be higher for the short-term (functional) strategy and lower for the long-term (corporate) strategy. Middle-term strategy must be adapted to the market and environmental changes (e.g., variation in production output as a function of the prices, raw materials supply, competitor's commercial policy).

Only drastic and permanent changes in the fundamental parameters should lead to modifications of the long-term strategy. For the minerals industry, the main factors that would justify this would be substantial changes in metals' end uses and markets, environmental regulations, social and political changes, or significant adjustments of competitors' roles.

Consistency is a great virtue of winning strategies in the long term. It is a common factor among the most successful mining companies that have constantly created shareholder value and survived crises. It is also an indicator of good strategic analysis. Long-term equilibrium may appear to be challenged temporarily in the middle and short term, which sometimes induces unnecessary strategy changes in companies whose business analysis lacks rigor and pertinence.

Strategy and Leadership

Strategy is not an exact science. Although some of the elements of good strategic planning come from a scientific analysis of data, other variables, like sociological and psychological elements, are more difficult to quantify. Forecasting market data, technology evolution, or sociopolitical trends and constraints is not always an easy exercise. For the fuzziest areas of the subject, intuition becomes a useful tool. With such a complex set of forces intervening, strong personalities with solid intellectual frameworks are necessary.

The percentage of time that managers actually devote to looking at the company's future, on average, has been estimated at 1% to 3%.[10] This figure is probably lower for the minerals industry. Building the future in mining means knowing how the markets will be and understanding the trends in metal uses, the threats of product substitution, and the evolution of the social license for its activities. It also means comprehending the progression of the energy and labor markets, the main components of its costs. Producing 1 t of aluminum from bauxite requires 12,000 kW·h of electricity. An ingot of aluminum is in fact a concentrate of energy. It is obvious that a long-term strategy for this metal (and many others) can't be disconnected from the energy facts.

Managers need to have a vision and be able to transmit it to the organization. Each company has its cultural "genes," the evolution of which is neither easy nor fast. Long-term strategies need to be followed by "genetic mutations" that can only be forced by strong leadership.

Someone has to do the job of building the future concept of the corporation. This task should be an essential element of the top executive's job description. A CEO must "see the future," create the necessary competences (human, technological, financial), and build the way to it. Most of the abrupt changes presently seen in companies' top management are not motivated by a lack of capacity for managing the present but because of insufficiencies in building the future of the corporation. Strong, intelligent chiefs are a rare species, expensive and not easy to find, but indispensable. They should be able not only to design the future but to decide at what pace the company should reach it.

Corporations are complex, dynamic systems in which the cause/effect relationship is not always evident or intuitive. Massachusetts Institute of Technology professor Jay Forrester[11] has developed models that attempt to quantify the effect on performance of the multiple internal and external factors. But in practice its application is not simple and most of the time managers must compensate for the lack of a rigorous approach by pragmatic experience/intuition-based decisions, built on team-developed foundations, as well as on experience and a deep knowledge of markets and competitors. The results are often surprisingly good, which proves that, once again, a manager's personality and leadership are essential conditions for winning strategies.

The same qualities are necessary for managers to obtain the loyalty of their employees and to implement the strategy successfully. When the turnaround of the company is evident, the employees find themselves happy and motivated.

Special Characteristics of the Minerals Industry

The concepts exposed are applicable to a broad range of activities, although the ones that are more significant for the minerals industry have been emphasized. Some characteristics of the minerals industry are special and not shared with many other sectors.

Contrary to most activities, a mine's geographical location is imposed. A mine cannot be moved as a factory or a commercial unit can. When the orebody has been discovered, some of the parameters of the future business are already locked: geographical location, political environment, availability of energy and water, transportation conditions, climate, and so forth. Little can be done to change these parameters.

Considering that labor, energy, and transportation are usually the most important components of the cost, it is easy to understand that the location of an orebody is a major element in the strategy of the business. The comparative net present value of two ideally equal mines depends on where they are located.

The mining activity strongly interacts with its social and ecological environments. Probably no government will oppose the existence of a computer research center, but it will have something to say in the case of a new mine or smelter. Even if the apparent economic returns for the host country (or region) constitute a powerful incentive, the drawbacks of the mining activity are considerable. The conditions imposed will surely be constraining, sometimes not counterbalancing the benefits expected from the operation.

Another particularity of the minerals industry is the relatively small number of actors. Mining production involves relatively few companies. The consolidations that have taken place lately reduced this number even more. Smelter activity is even more concentrated, with a smelter typically having sufficient capacity to process the production of several mines.

The products of mines and smelters (concentrates, materials, metals, alloys) are not always sold to end users. This particularity, together with the long cycle between the extraction of the ore from the mine and the use of the metal by the manufacturing companies, introduces a hysteresis or lag in effect in the transmission of the market signals from the end user to the miner. The miners know their clients, the smelters, well but have little knowledge about the manufacturers' activity and even less of the end users'. This circumstance places additional difficulty on the task of building the strategy of mining companies. The more a mining company is integrated downstream in the value chain, the better its vision will be on the future uses and prices of its products.

Prices for some important mining products are determined daily by organized markets (LME, Commodity Exchange [COMEX], New York Mercantile Exchange [NYMEX], and others) and are applicable to most transactions. These markets provide instruments (derivatives) that allow the sellers and the buyers to operate with future prices, fix prices over a certain period, or limit the range of prices applicable during a given interval of time. The LME, COMEX, and NYMEX markets provide services like pricing, hedging, and physical delivery for most commercial metals. The LME holds a large network of warehouses throughout the world where producers can sell metals at prevailing daily prices and customers can buy them at published, transparent prices.

Although it has other virtues, the existence of physical metal stocks at the LME warehouses brings up delay in companies' reactions, complicating their strategic thinking. A mining company can increase production over the level demanded by its clients and sell the surplus to the LME. Increasing capacity reduces the unit cost immediately, and the products will not augment the companies' stocks because they can be sold to the LME. But in the long term, prices will go down under the pressure of the increase of LME stocks, and the total industry will see decreasing profits (and not only the companies that originated the stocks).

Not all prices of mining products are determined by commodity exchange pricing. Some, like coal, bauxite, iron ore, or minor metals, are negotiated directly between sellers and buyers. These specificities make the strategic planning exercise of mining companies (and, more generally, commodities companies) particularly complex.

EXTERNAL FACTORS INFLUENCING THE STRATEGY

External factors are referred to as those business factors that are imposed on a company by its external interface and which it can do little to change. The legal framework, market structure, financial partners, politics, society, and competitors are external factors, given that they exist and function independently of the will of the company. Although the company can have some leverage on some of them (e.g., lobbying and publicity), they cannot be considered as variable parameters in the strategic equation. Nevertheless, they are essential in the construction of the company strategy.

The Legal Framework

The *legal framework* is probably the external factor that most constrains the strategy of a mining company. Mining operations are linked to a specific geographical region (i.e., country, state, province, and town), and the applicable law within the region establishes the legal framework of the activity, the boundaries of what a company can or cannot do with the different components of the business.

During the exploration and development stages, a mining company must follow a specific regulatory framework (and pay taxes and royalties as required by law) in order to obtain and retain the exploration claims and mining concessions. In this process, the company must interact and subscribe agreements with the appropriate government agencies, evaluate risks and opportunities, and make strategic decisions accordingly. Also, mining companies must negotiate and agree with landowners to gain access to the land required for the exploration and mining activities.

Mining companies require government permits during all stages of the mine life cycle. For example, permits are required during the prefeasibility and construction stages (e.g., project, roads, power lines, water supply, environmental), mining operations (e.g., mine plan, blasting, and ore transportation), and mine closure (e.g., reclamation, decommissioning, postclosure).

Corporate, environmental, labor, commercial taxation, and many other business issues are subject to laws and regulations that are specific to each country or region. Expatriate company executives managing mining operations abroad frequently make the mistake of trying to extrapolate situations that were analyzed under the prism of their country's legislation. In this context, the company, as well as its executives and staff, may incur liabilities in relation to the mining activities. Proper consideration must be given to this, as the associated costs could change the substance of the strategy.

As several chapters in this book discuss, environmental regulations are becoming an increasingly important strategic factor for the mining sector, so they need to be carefully considered before any decision is made.

The main target of any strategy is to maximize the shareholder value. In this regard, the applicable fiscal laws must be considered cautiously because this may become an impediment to reaching the financial goals. It may be advantageous to establish alliances and get the help of local companies or individuals. This is a pragmatic way of coping with country and local regulations.

If the standards and governance rules of the company ask for a universal application of some principles, it shall be necessary to evaluate their impact on the project carefully.

None of this is banal. The feasibility of a mining project often depends on the strategic trade-offs management will take during negotiation with the local authorities. The economic and financial analysis of mining projects must include a realistic estimate of all costs associated with legal issues during all stages of the mine life cycle and consider how far the company can go, integrating the margins of error, and keeping in mind that other competitors may also be candidates for the business.

Shareholders' Role and Influence

Shareholders are the ultimate drivers of a company's strategy. In addition to the ownership of shares of publicly held corporations, shareholders have other rights, among them the right of information and the right to influence and approve certain management decisions. These rights are exercised mainly through voting at the general shareholder meetings.

Principal shareholders are those who own a substantial stake in the company and have significantly more power than small shareholders. Even if their stake is lower than the legal voting majority, principal shareholders may have real power over management, and often, the dispersion of ownership makes it easier for them to achieve formal control of voting and thus decisions.

Small shareholders do not usually have significant influence on the company. However, they can reinforce their influence by creating *shareholder associations* or *shareholder committees*. The shareholder associations work outside the framework of the corporation and are often led by corporate law professionals. Shareholder committees are internal bodies resulting from the cooperation between management and shareholders, aiming to improve communication and transparency.

The by-laws of each company set out the procedures by which certain management decisions (like changes in strategy) must be submitted to the shareholders for approval. Even if it is not legally required, it is a sound management practice to submit the matters concerning long-term strategies for shareholder approval. It is also important that political, social, and environmental issues are included with this information because shareholders are increasingly concerned about the sustainability performance of the company.

From an economic viewpoint, the mining business has two distinctive characteristics: (1) it follows a cyclical pattern, and (2) it requires substantial capital investments with long-term maturity. Evidently, a business with these characteristics calls for solid shareholders, that is, investors willing to commit a considerable amount of capital in volatile ventures and to wait years for a return. For example, the initial capital investments in a mining project must not only pay for the exploration costs but also for development of the mine and plant facilities, site development infrastructure, and overburden waste removal; there may be years of negative operational cash flow. A frequent joke among mining professionals says that the ideal shareholder must have "deep pockets and large shoulders." In this context and in the case of stock or bond issues and other financial operations, a major strategic task for top management is to seek the right investors who will became future shareholders.

Markets

Metals and minerals markets are singular commodity markets. They are global, and the producers (the mining companies) are somewhat disconnected from the end users. Metals and minerals are homogeneous, nonperishable products, traded mainly on the basis of price with a comparatively small number of suppliers and clients. In this context, decisions concerning markets are of strategic importance. Quantities, qualities, costs, and destination are issues that can make the difference between successful and unsuccessful mining companies.

Metal concentrates are nonstandard products and each mine has its own concentrate signature (i.e., differing grades, fluxes, impurities, and by-products' content, grain size, flow moisture point, self-oxidizing behavior). Even concentrates produced at different moments in a mine's life can be fairly dissimilar. Smelters may require a specific flow sheet to process a specific metal concentrate. Therefore, one of the first strategic decisions to be made during the preproduction stages of a mining project is the choice of smelters that may be technically and economically suitable for processing the concentrates. Changes in the flow sheet and capacities are possible, although expensive and technically complex for both mines and smelters. Given the capital investments involved, mistakes at this stage can be fatal. On the other hand, the geographical location plays an essential role, as it affects the transportation and distribution cost of the products.

Another factor requiring careful consideration during the feasibility stage is the effect on the market of the new capacities brought on stream by the project. Critical to this effect are the analysis of consumption trends, the spare capacity of the competitor's facilities, and the impact on prices and markets of the new production. Hence, the price scenario used for the final feasibility study should be established after due consideration of the effect on prices of the increase in supply. This effect will be higher if the production capacity added by the project is higher relative to the total market of the related commodity.

Another strategic issue is the possible convenience of the mine/smelter integration. If production capacity is large enough and local conditions are adequate (i.e., energy cost and supply, transportation infrastructure, labor availability and cost, access to harbors or railways), the construction of an on-site smelter may be a good option. Vertical integration can also be realized when the concentrates are processed in smelters owned by the same company if they are close enough to the mine. Other strategic options for the concentrates are swapping with competitors or selling them in the concentrate markets. In all cases, the factor that weighs most heavily in this strategic decision is cost. The structure to be adopted is often the one that minimizes the total cost of the chain from mine to clients.

If concentrates are to be sold to third parties, an important strategic issue is deciding the number and structure of the agreements with customers (e.g., amounts sold through long-term agreements with smelters versus short-term sales to concentrate traders, swapping conditions for third parties).

The market strategy should integrate the risk of product substitution (which is linked with prices), the quality requirements of potential buyers, and the environmental conditions. There are many examples of product substitution in the metals sector. Aluminum cans are progressively replacing tin-coated steel cans. Plastics can often be substituted for steel in the automobile industry (at least partially). Organic molecules can replace bismuth in pharmaceutical applications. Clay or cement shingles can replace zinc sheets in roofing. Polypropylene plumbing pipes take the place of lead and copper. Lead could be replaced, at least partially, by nickel or cadmium in the battery industry. Metal oxide pigments can be replaced by organic pigments. A prolonged shortage of a specific metal can cause sharp increases of prices, thereby inducing goods manufacturers to substitute more economical raw materials. Negative environmental effects can dramatically reduce the use of some metals, such as is the case with mercury and cadmium.

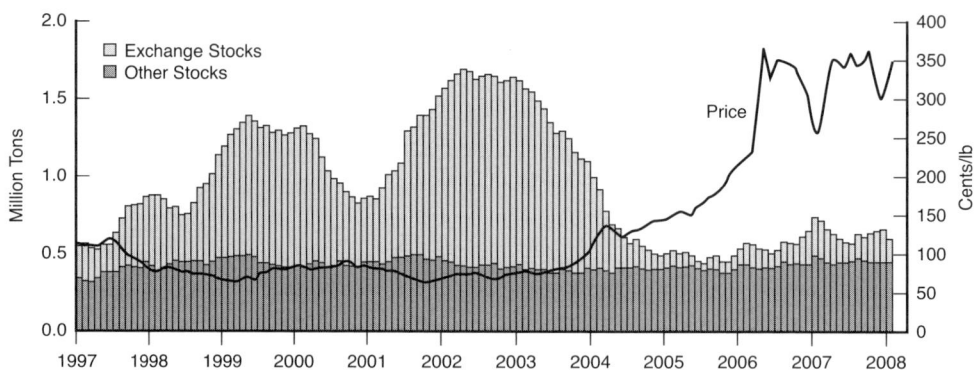

Source: Rio Tinto 2008 Chart Book

FIGURE 4.5   LME stocks and metal prices

Price Mechanisms

The main characteristic of the metals supply/demand balance is low supply elasticity. Smelters generally operate at close to 100% of nominal capacity and have very little flexibility in changing production rate. This is a requirement to reduce costs: the bigger the production is, the lower the fixed cost per ton. Mines are designed for a given production capacity, and a production increase requires capital investments and time for implementation. It is difficult for a mine operation to cut back on production because unit prices would increase and competitiveness would erode. As a consequence, supply is unable to react rapidly to changes in demand, thus generating stress and price volatility in the market.

Changes in demand generate variations in stocks (very visible through the LME warehouses' stocks) and stresses in supply (Figure 4.5). From the customers' perspective, adjusting to price fluctuations is not easy because product substitutions require changes in technology that only make sense in the long term.

Mines and smelters are also subject to sudden incidents that can cut production (e.g., accidents, breakdowns, strikes, or unforeseen environmental problems). Depending on the amount of production loss, these incidents may have a significant impact on metal prices.

Metal prices are global, established daily at organized market platforms like the LME or the New York–based NYMEX or COMEX exchanges. LME is the world's premier nonferrous metals market.[12] In addition to pricing, LME provides services like hedging future price risks, physical trading, storage, and delivery. Prices are fixed daily as the result of a high volume of trading on the "ring." Producers and consumers can hedge at the LME as well. Hedging allows for managing the risk of price changes by offsetting that risk in the futures markets. Through hedging, the industry can decide on the level of risk it is prepared to accept. The LME can physically buy and deliver approved brands of metals from its authorized warehouses. The level of stocks in these warehouses is published daily, along with the prevailing prices.

The United Nations study groups (copper, lead, zinc, and nickel) are intergovernmental organizations that periodically provide a source of industry statistics about production, consumption, new project status, and other essential information on the metals industries and markets. Their role is to contribute to the transparency and equilibrium of these markets. They also provide forums in which industry and government representatives can exchange information.

A number of professional organizations are financed and steered by the industry. They work on subjects of common interest for all members with respect to the laws of competition. Among other tasks, they develop statistics that help the industry in understanding the market situation.

All these data and facts together help industry strategists gain a better understanding of the future evolution of markets and prices.

LME physical stocks and metal prices are inversely correlated. Figure 4.5 presents the evolution of copper prices and LME stocks of this metal[13] from 1997 to 2007. Although the correlation is not perfect, it can be seen that, in general, higher prices correspond to lower stocks and vice versa. The evolution of stocks (sum of LME "visible" stocks and the ones in the hands of producers, consumers, and traders) is a good indicator of future price trends, which is broadly used by tacticians to forecast short- and mid-term price evolution.

The study of the variations in the LME physical stocks is one of the tools that the industry uses to forecast future prices. Strategic hedging decisions are an important consequence of this.

Regarding risk management policies, there are basically two doctrines: one that maintains that price fluctuation is a normal market phenomenon and companies must live with it and assume the economic consequences, and the other that defends the idea that the intelligent use of derivatives limits the risks and maximizes the margins of the companies. There are companies adopting each of these doctrines and it is difficult to argue for or against them. What is clear is that, given its importance, the issue of price risk management is strategic and companies should publicly state their hedging strategies to allow stakeholders to act accordingly. In fact, the real and difficult issue is the balance between cost and benefits as a consequence of these two strategies.

It is therefore important that management seeks shareholders' approval of their policies on fixing and/or hedging metal prices, exchange rates, or interest rates. Price hedging is generally considered as a conservative strategy, allowing smelters to buy concentrates and sell the metal contained at the same price paid a few months before (independent of the price fluctuations that may have occurred). It limits losses if prices go down but also prevents profits if prices go up. On the other hand, fixing the future prices or exchange rates implies a bet on unknown events. All players, including the shareholders, can have an opinion about this. If things turn out wrong, the financial consequences can be enormous. Shareholders often say, "Managers are paid to run the company, not to gamble with our money."

Finally, regarding the utilization of derivatives, one must consider that their use requires highly specialized knowledge and tight control of these activities by top management. History is full of cases of the misuse of these financial instruments and its catastrophic consequences. Also, the operating cost of the group in charge of these operations within the company's organization can be substantial and may offset potential benefits.

Financial Strategy and Corporate Finance

Financial aspects are among the most important external factors affecting a company's strategy. Financial resources are scarce and often impose limits on the activities of a mining company. Therefore, as the company develops its strategy, the issue of how to finance its activities often arises. Three financing mechanisms are possible: (1) use of internally generated funds, (2) use of debt, and (3) financing by equity. Each one has its own advantages and drawbacks.[14]

A first element of financial strategy is of a fiscal nature. Interest and dividends are both financial costs, but interest is deductible from taxable income, while dividends on common stock are paid out of after-tax income. Therefore, a company pays lower taxes when financed by debt.

Sources on Percentage of the Total Expenditures

	A	B
• Internally generated cash*	72%	52%
• Financial deficit	28%	48%
• Covered by:		
- Net stock issues	4%	34%
- Net increase in long-term debt	20%	10%
- Net increase in short-term debt	4%	4%

*Cash flow from operations less cash dividends paid to stockholders.

FIGURE 4.6 Two examples of sources and uses of funds in mining companies

Issuing securities implies a set of formalities that consumes time and money. It also involves an exercise in transparency and publicity that some companies may not want to go through. On the other hand, a loan implies a deal between the company and its banker; therefore, the facts and strategies of the company are disclosed to a lesser extent.

Those in the financial world (analysts, banks, investors) carefully watch the debt/equity ratio of mining companies. This parameter is considered as one of the indicators of financial health. Too much debt versus equity can be considered as a weakness and might affect the share value. The structure of funds of most mining companies shows that the internally generated cash is preponderant over the stock issues and the short- and long-term debt. In fact, this structure is not exclusive to the minerals industry. The adoption of a scheme for financing the activities is a strategic decision and is part of strategic planning.

Figure 4.6 shows two case examples of sources and uses of funds. Case A refers to a company with high internal cash flow, which implies a low financial deficit, covered mostly by long-term debt. Short-term debt is used mainly to finance current commercial activities and small investments. Case B presents a company with lower internally generated cash flow, which leads to a higher financial deficit. Investments and working capital are covered by stock issues and long-term debt. Companies represented by Case A have a strong actual cash flow and will use it to finance new projects. Companies represented by Case B, although they may have high potential, have low cash flow, so they need to seek investors to finance their projects.

The minerals industry is a cyclical one. There are endless discussions about the structure and interval of these cycles, but it is well known that a period of high prices is always followed by one of low prices and vice versa. The length of these cycles can be tens of years. Managers must know how to deal with this phenomenon, which affects the ability of the company to invest and compete. Intuitively, one might gain the impression that the time after a high activity period is when the company is in the best position to undertake new investments and projects. What happens in reality (Figure 4.7) is quite to the contrary.

When pulling out of a recession, mining companies are not operating at 100% of capacity. They sell from existing stocks, which releases cash, and they do not invest heavily because they have spare capacity. After a peak in the cycle of economic activity, companies are immersed in substantial investments and new projects, so they may have difficulties in adjusting their expenditures to the recession. At the same time, their sales decline and the need for funds increases.

An important component of the strategy of mining companies consists of anticipating the effects of the cycles on their financial equilibrium. The deep ups and downs of prices can shake a company's financial system and force managers to adapt their decisions accordingly.

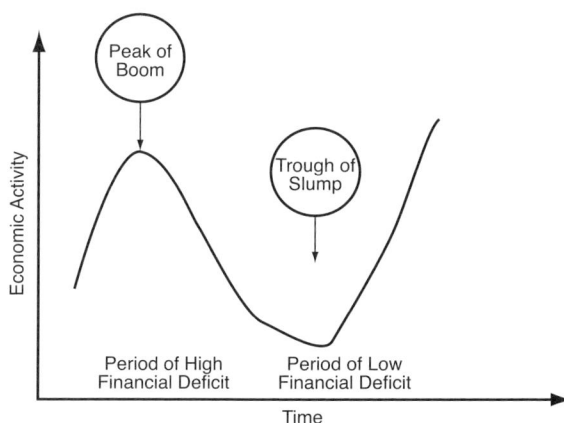

FIGURE 4.7   **Financial deficit versus economic activity**

Politics, Society, and the Corporation

Today, mining and smelting are controversial activities. The intensity and importance of the political and social interactions are higher than for most other industries. The latent value of a mining asset can only be achieved if the company is able to obtain what has been called "social license" and runs its operations in compliance with the specific conditions of the host country and local communities.

"Know-how" on social and political management represents an important competitive advantage today. Some mining and metals companies have developed specialized skills to operate in specific social and political environments. This is the case with Teck Cominco in the Arctic mines or Umicore in the heart of the most densely populated regions of Europe.

Some years ago, mining companies' activities mainly consisted of finding the best possible orebody, extracting the richest ore as quickly as possible, and moving to another place to start the process all over again. However, as social and environmental sensitivity grew, mining companies became conscious of the environmental and social impacts of their activities. Governments progressively tightened environmental regulations as society became increasingly aware of ecological matters. The notion of sustainability is rapidly gaining importance, but the speed of this process is different among countries. It may take decades for environmental regulations in certain developing countries to reach the level of severity of those already in force in some industrial countries (Sweden, France, Germany, the United States, and Japan). Environmental organizations are gaining importance worldwide. Today, their membership, financial capacity, and influence have reached a considerable level.

This increasing ecological awareness has manifested itself in many countries as a surge of "green parties"—political parties whose programs are for the most part ecological. These parties are gaining strength in many countries, especially in those of the European Union. Their presence in the European Parliament (as of the most recent elections in 2004) is assured by two parliamentary groups, the European Greens/European Free Alliance and the European United Left/Nordic Green Left, which represent 83 members out of a total of 732, for 11% of the total membership. In fact, given that public opinion today is sensitive to sound ecological principles, most of the other political parties include some "green" ideas and projects in their programs. Obviously, these circumstances have a strong influence on national environmental policies and regulations. Given the representation of the dominant parties (European People's Party/ European Democrats and Party of European Socialists), the "green" parties have a certain role of arbitration, which gives them more effective power than their numerical representation might otherwise indicate.

The laws reflect equilibrium among social forces that defend their own interests. In the specific case of the environmental and sustainability matters, two opposing forces try to influence legislators. They are what Paul B. Downing calls the "emitters lobby" and the "receivers lobby,"[15] where each lobby represents a sector of the society. The former defends the potentially polluting activities while the latter defends the environment. Although the emitters have the strength of monetary resources, the receivers have the strength of the votes. In the context of the minerals industry, the emitters are represented by international professional associations such as the International Council on Mining and Metals (ICMM),[16] European Association of Metals (Euro-Metaux),[17] Independent Petroleum Association of America,[18] American Petroleum Institute,[19] and European Petroleum Industry Association,[20] among others. The receivers are represented by international nongovernmental organizations (NGOs) like the International World Union for Conservation of Nature,[21] World Wide Fund for Nature,[22] Friends of the Earth International,[23] and Greenpeace International,[24] among others.

The organizations representing the industry, like ICMM or EUROMETAUX, strive to convey to the public and authorities the image of an industry that cares about the environmental and social impact of its activities, playing the role of interface between corporations, political authorities, and society. They make available to the companies updated information about the legislative and social trends of the countries in which they operate, gathering scientific evidence on the effects of mining and metals activities on the environment or human health. On the other hand, the ecological NGOs defend the public and the ecosystem in general from potential aggressions from the industry and put pressure on the legislators to implement more stringent environmental rules. Both sides are well organized and well financed. They have traditionally had a difficult dialog, although since 2003, the world forum organized by the Global Mining Initiative in Toronto has resulted in an improvement in their relationship and exchange of ideas.

Today, it would be unconceivable that a corporation could build its strategy while ignoring its social environment and without developing an adequate social interface. The success of mining and smelting activities is greatly dependent on the ability of managers to interact and understand the requirements of the society. Basic rules must be respected: for the industrialized countries, "be a good neighbor"; for developing countries, "be a good partner"; for indigenous people, "be respectful."

The way in which the mining rent is distributed between mining companies and host countries is still a highly controversial subject. If the total tax[25] (sum of royalties and other fiscal taxes) is too high, the companies would not be encouraged to explore and invest and, at the limit, would discontinue existing operations with the corresponding permanent loss for the host country. If the tax level is too low, more companies will be encouraged to invest, but their activity will not benefit the local economy sufficiently. An optimal level exists between these two extremes and should be sought. Experience demonstrates that a fluid dialog between the two parties is best. A last consideration is the effect of price fluctuations on the government's attitude: when the prices are high, the temptation to increase taxes is high, too. If the prices drop, it will be difficult to return to the previous tax levels. Again, a transparent attitude on the part of the companies, explaining the effects of price volatility over the long term, would help in finding the right balance.

Another strategic factor to be considered is the economic policy of the host country. It has been observed by some researchers[26,27] that countries with abundant natural resources show a lower gross domestic product growth over the long term than those whose economies are based on transformation and manufacturing industries. It is what has been called the "resources curse." The reasons for this will not be analyzed in this chapter, but it is important to highlight that the strategy of the mining companies in developing countries must integrate the forecasted values of its macroeconomic parameters. The distribution of the mining rent between companies and the host state can change through time and through price levels, modifying the financial returns of the companies.

The Environment as a Strategic Factor

Mining companies have traditionally built their strategies by integrating the macro and micro-economic parameters, as well as the market, technological, and geopolitical ones. The environmental parameter was incorporated only when it had an impact on operating costs. Circumstances have changed and now the environmental factor has a determining role in the strategies of corporations.

The first reason for this change was that increased social and environmental awareness fostered the emergence of rules and injunctions, and complying with these rules forced mining companies to undertake expensive remediation measures. The impact of these measures can, in certain cases, compromise the financial viability of the activity.

A new legal framework can substantially modify strategic equilibriums. For example, the redefinition of the technical requirements for a tailings dam can increase costs to a level at which the entire activity may no longer be viable.

On the positive side, environmental pressure stimulates creativity and induces improvements in technology and productivity. It can be observed that, as a general rule, industries located in countries with tough environmental regulations have better productivity than those in more permissive countries. For example, pressures to improve working conditions in an unsafe process can drive the company to replace workers with robots, in a revolutionary move that can change the overall conception of the activity and reduce costs dramatically. A positive impact in quality is also observed where quality improves with improved environmental performance. The opportunity for environmental investments must be measured by their overall effect on the activity, including not only the direct costs/benefits effects, but also an assessment of the value of an improved public image for the company, as well as the consequences of an improvement in employees' working conditions and satisfaction levels.

Corporations' progressive internalization of the costs of sustainability will modify the long-term trend of commodity prices. Product substitution must then be carefully analyzed in order to reorient the company's product strategy eventually.

The environmental factor also plays an important role in mergers and acquisitions, mainly in developed countries. Some attempts have failed because of the difficulty of getting a common view of the partners' environmental liabilities. Although it is relatively unproblematic to evaluate the assets, allocating a net present value to environmental liabilities may be a much more difficult exercise. The reasons for this difficulty are diverse. By taking over the environmental liabilities of a potential partner, a company also assumes their future consequences, such as when, for example, the applicable legislation changes. Both parties may have a common view on it if the assets are in the same country but may disagree substantially if they are spread around the world because of their different levels of awareness on the countries' environmental outlook. The merger itself may be compromised or killed, given the magnitude of the financial liabilities linked to environmental impacts and rehabilitation.

Competition: Rivalry or Alliance

Competitors represent the final strategic external factor to be considered. Companies do not operate alone within their sectors. Other companies are trying to reach similar targets using similar means. Stated informally, to compete in business is to strive with other companies in order to get the biggest piece of a financial cake that is, by nature, limited in size. A company is financially more competitive than another if its shareholder return is consistently higher.

The main goal of a company's management team is to maximize the return on investment for its shareholders. The best way to achieve this is to create competitive advantages and the method to do so is to develop and implement a good strategic plan.

John E. Tilton[28] differentiates between two kinds of competitiveness in the minerals industry. The first one is the natural competitiveness, originating in the endowments of the national economy: high-grade/low-cost orebodies, cheap and abundant energy, capable and well-educated employees, and so forth. Identifying these advantages and putting them at the service of the company is a strategic exercise that asks for specific management skills. The second type of competitiveness is the policy-induced competitiveness, originated by nations' or companies' actions and developed from inside by implementing appropriate policies or generated by summing up the qualities of other companies with one's own through mergers and acquisitions.

Natural competitiveness is not always revealed; it must be identified by the companies or nations. The presence of massive concentrations of natural resources, like nickel in New Caledonia, zinc in northern Alaska, or diamonds in Botswana, have been unveiled by geological exploration that was not always sponsored by the host country. Furthermore, natural competitveness (e.g., skilled workers, a good communications network, energy availability) can't be implemented by short-term political decisions.

Policy-induced competitiveness must be built up internally through intelligent management initiatives oriented toward developing competitive advantages. It belongs to the category of strategic internal factors that will be analyzed later.

Mergers and Acquisitions

One of the ways that companies reinforce competitiveness and increase shareholder value is by creating alliances. This is a strategic move aiming to develop synergies in order to reduce costs, expand markets, and improve the competitive position of the company.

Two or more companies, after a careful analysis of their business structures, can organize the merger of their activities under a sole management. The net present value of the merged company should be higher than the sum of the components separately. Growth or size alone does not justify mergers.

A merger can take place at the initiative of one of the parties, which buys the other without the agreement of the target company, or by mutual consent. In the first case, if the merger takes place against the will of the management of the seller, the shares must be purchased in the market from their actual holders, usually paying a premium (a price higher than the publicly listed price) in order to encourage the holders to sell. To oppose this, the seller's management can, and often does, claim that the share value proposed undervalues the company. It can also convince its shareholders to build some barriers that make the transaction difficult, such as a "poison pill," and so forth. The announcement of a hostile takeover is usually followed by a transitory decrease in the share value of the buyer and an increase of the share price of the seller. This is mainly due to the effect of the premium paid.

The strategy of the merged company has to be redefined. It can't simply be an aggregation of the strategies of the merged companies. This constitutes a very constructive and fundamental exercise for the new company. It is an exceptional occasion for team-building and values-setting. Management must go through numerous hurdles: cultural differences, organizational schemes, ethical principles, accounting methods, human resources systems, language, and so forth.

The practical process to execute the merger formally is complex, particularly when the participants are publicly listed companies. A complete and detailed memo on the objectives, means, and forecasted results of the merged organization must be published. Both parties must go through a process of "due diligence," and the entire operation must be approved by the authorities of the markets in which the participants' stock prices are quoted. The management is legally liable for the consequences of any missed or incorrect information supplied.

The synergies generated by a merger come from cost savings and revenue increases. They are estimated and included in the resulting business plan.

The postmerger phase is critical for the new company, as the success of a merger is not guaranteed. A high percentage of them never meet the initial targets. A recent study conducted by the consulting firm McKinsey[29] shows that the seller typically captures most of the shareholder value created. The synergies are often overestimated: 70% of the mergers studied (170 cases) failed to achieve revenues synergies and one quarter of them overestimated cost synergies by at least 25%.

Mergers often take place not as the result of mutual convictions about the opportunities they present, but rather based on the personal will of the buyer's CEO. It is always appealing to manage a bigger company and have more power, but this is not a good reason for forcing a merger. The excessive ego of the CEO, as well as the biased interest of the buyer's management, is a common factor in many failed mergers.

INTERNAL FACTORS INFLUENCING THE STRATEGY

Among the variables influencing companies' strategies, internal factors are the ones over which a company has better control. Governance, personnel, financial doctrine, technology and research, organization, and communication are issues that can be directly managed to serve the company's strategy. Unlike external factors, internal factors are under the direct control of management so they can be implemented in the way that maximizes the chances of achieving strategic objectives.

Corporate Governance

The Organisation for Economic Co-operation and Development (OECD)[30] defines corporate governance as "the system by which business corporations are directed and controlled." There is no universal set of standard principles for corporate governance, but those published by OECD are widely shared. OECD governance rules specify the distribution of rights and responsibilities among different participants in the corporation, such as the board, managers, shareholders, and other stakeholders, and highlight the rules and procedures for making decisions on corporate affairs. By doing this, it also provides the structure through which company objectives are set and the means of attaining those objectives and monitoring performance.

Corporate governance rules are issues by companies to ensure that shareholders and other stakeholders are properly treated, to guarantee transparent communication with the company's interfaces, and to define the rights and obligations of the board of directors. They should address four issues:

- The rights of the shareholders;
- The role of the stockholders;
- The need for transparency; and
- The responsibilities and codes of action for the board.

In terms of corporate ethics and social and environmental issues, the governance rules of a modern corporation normally go beyond legal obligations. For example, some multinational mining companies voluntarily implement the same standards of health, safety, environment, and ethics for their subsidiaries in developing countries as they do in the developed countries in which they operate. Though voluntary, corporate governance rules are a public unilateral commitment, the violation of which may incur legal liabilities.

Corporate governance rules must be clearly stated, both internally and outside the company. The adoption of these rules is not only a matter of ethics but also a key element in improving economic efficiency and growth. Corporate governance rules create a trustful environment that encourages investment. They must be a reference for all stakeholders and provide the necessary confidence for the correct functioning of a market economy.

The definition of the corporate governance rules has a high strategic content. Although they impose restrictions on the autonomy of the company, they consolidate its reputation as a well-founded partner for business. Both aspects—autonomy and reputation—are extremely important for the corporation, and experience shows that the balance of both is always positive in the long term.

Corporate Culture and Internal Policies

In every company, the management and employees develop a specific culture that becomes a distinctive brand of the organization over time. Each company unconsciously develops an idiosyncratic corporate culture. The values of the top management strongly influence the corporate culture. Because it develops over a long period of time, corporate culture is difficult to change as it involves a complex set of people's powerful beliefs. Companies frequently publicize their culture as assets (e.g., openness, innovation, respect, teamwork, commitment), which not only contribute to the reputation of the company but also to the improvement of its value-creating process.

In our rapidly changing world, rigid cultural values may become an obstacle. It is quite common in mining companies for the technical culture to overpower other factors. Although technical excellence is indispensable, the excessive dominance of technical considerations over financial or commercial issues can diminish competitiveness and put the future at risk. The company's culture may restrict its ability to change strategy, which is essential in case of a merger. Years of building an identity can't be easily erased if the shareholders decide on a major change in strategy. In some cases, the human factor is simply incapable of keeping up the pace, so the postmerger implementation may be seriously jeopardized. A certain degree of cultural compatibility is crucial in mergers and acquisitions.

A poor strategic situation is often a symptom of the lack of a distinctive corporate culture yielding to confused and conflicting organizational arrangements. Companies in this situation are often incapable of defining their position among the generic strategies. The minerals industry sector leaves little room for focus or differentiation strategies because most companies tend to privilege low-cost strategies. The concept of "cost strategy" is much broader than the simple elimination of superfluous costs. It requires the implementation of a set of functional policies, as well as the development of a cost-reduction-oriented culture. Volume of production, the quality of the mining assets, the optimization of the flow of products, and intelligent choices of suppliers and clients are the main ingredients for successful results.

Efficiently managed companies should adopt and publicize clear internal policies (in line with their corporate values) on matters like personal conduct, responsibilities, complaints, equal opportunities, integrity, external communications, and so forth. When these rules have been adopted, they become an element of the company's strategy.

Human Resources—The Human Capital

Human resources are means of production, expandable and transferable. The human capital is composed of creative, social beings. The phrase "a company is worth what its men and women are worth" conveys well the importance that human capital has for the company.

Human capital represents tangible assets for any company although it is not accounted for in the balance sheet. Companies pay a salary and benefits to its employees and, by doing so, it "rents" the services of its human resources. Therefore, unlike other company assets, human resources are free to leave the company and bring their human capital to another company, maybe even a competitor. The contribution of employees to the company can and should be enhanced through training and education. Additional investment in education yields additional output.

An effective recruiting strategy is essential for the success of the company and vice versa. Unsuccessful companies do not attract talent. The quality of the employees has a multiplicative effect on the results of the company: the higher their quality is, the better the results will be, which in turn will attract even better people, and so on. Building up a high-quality management team is essential for the success of corporations.

The human resources (HR) management function is one of the most important factors in a company's strategy. Employees are recruited to accomplish the task of driving the corporation as defined by a strategic plan. Because of this, the HR function is more often becoming integrated in the top management of mining companies. Although it has an administrative component, it is a real strategic function.

HR management calls for special skills (e.g., psychology or sociology) that are not always present among business leaders. Recruiting is a complex function that doesn't consist of hiring the best brains but the ones that best fit the company's strategy and culture. Achieving this requires an in-depth knowledge of the company's strategy, good communication skills, and psychological skills.

Employees must qualify for their jobs, not more nor less. Hiring overqualified employees for a given job is not a positive achievement. It creates stress and frustration within the organization. On the other hand, underqualified employees do not contribute adequately to the optimal functioning of the company, reducing its competitiveness, although intelligent individuals can be trained for specific jobs if they are endowed with the appropriate qualities. This is always a good investment for the company.

Any employee possesses two types of qualities: those endowed—like intelligence, memory, and personality—and those acquired by training and education. Endowed qualities are inherent to the individual, permanent, and not modifiable. The acquired qualities can be achieved through education and training. Each job in the company requires a specific balance between these two categories of human qualities. Figure 4.8 shows the relationship between dominant qualities and the degree of complexity/uncertainty of the function for different categories of jobs.

In general, jobs involving making decisions with a higher degree of uncertainty require individuals with more endowed qualities and vice versa. It is interesting to remark that employees'

Human Resources
Personal Qualities—The Right Fit

More Uncertainty

Less Uncertainty

CEO

Division Manager

Plant Manager

Operations Worker

More Trainable Qualities

More Endowed Qualities

Dominant Qualities

FIGURE 4.8   Dominant qualities versus degree of complexity/uncertainty of the function

compensation follows the same pattern. Jobs involving higher complexity and uncertainty in their business environment usually have higher compensation.

Continuous progress of the human capital is critical to corporate success. Companies must provide their employees with appropriate training and opportunities to increase their experience. Leading personnel through an intelligent career path and exposing them to rich and stimulating practices is the best way to create and retain a motivated and competitive team. Inversely, routine, lack of challenge, and inadequately matched skills and interests always lead to poor collective performance.

A common feature of highly competitive organizations is the existence of a well-structured and transparent system for the evaluation of employees' performance, based on which the management can draw the career paths and determine the appropriate compensations. This system must be competitive, so the employee will not be tempted to leave for another company that pays better. The "Talent Toolbox" section in Chapter 6 addresses this issue in greater detail.

For a company, its organization is the cement that keeps its structure solid. It has to be designed to maximize employees' psychological energy. To this end, the best tool is motivation, which is the state of mind that moves individuals to action. Many factors affect the motivation of an individual to fulfill a job. Figure 4.9 represents the most important motivating factors in the context of business.

The three vertices of this triangle are satisfaction at work and recognition, financial incentives and working conditions, and power and influence. Basically, an individual is motivated at work by a combination of the three factors, but the relative weight of each factor varies for each person as well as with the stage of his or her career. Some people are more motivated by money, others by satisfaction or power and influence. A good HR manager must be able to determine the position of each employee in the motivation diagram in Figure 4.9 and provide him or her with the optimal conditions for maximum motivation. HR management includes the search for the best equilibrium among these three factors—the one that better uses the psychological energy of the people.

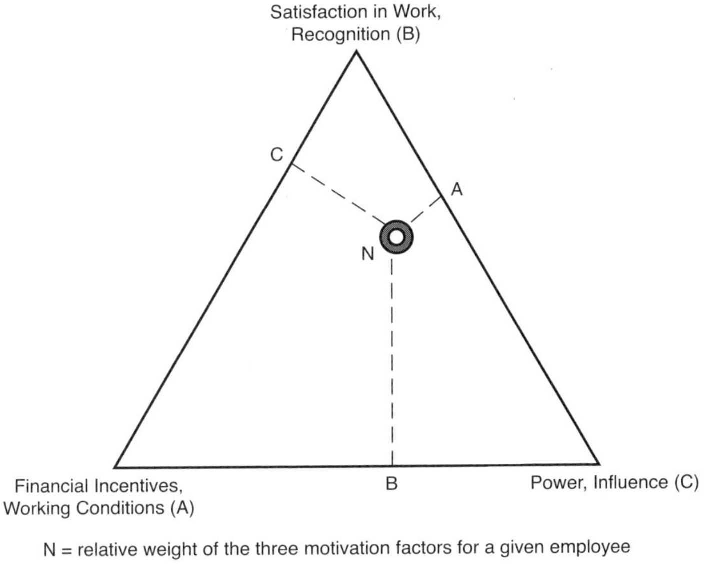

N = relative weight of the three motivation factors for a given employee

FIGURE 4.9   The most important factor of human resources motivation

The strategic plan dictates the kind of personnel the company needs. A company that has adopted a "cost and focus" strategy would privilege "doers," pragmatic people capable of building and implementing strong cost-cutting measures, while a "differentiation" strategy needs more "intellectuals," creative people able to invent new products, processes, and strategies.

Financial Doctrine

With the exception of the issue of how to finance its activities, most other internal financial decisions correspond to short-term issues. However, some of them belong to the strategic category, specifically those concerned with investment analysis, management of financial risks, and the use of derivatives.

Investment Analysis

Companies invest with the purpose of generating a cash gain from buying an asset at a price lower than its real value. Given the amount of investments in the minerals industry, the matter is of strategic importance.

It is necessary to distinguish between investment decisions and financial decisions. The former is related to investing in projects of any nature; the latter refers to how to obtain the funds for the investment. Investment decisions correspond to business managers and involve multidisciplinary analysis of projects within the framework of the strategic plan. Financial decisions are more a matter of financial specialists and are in fact a much less strategic issue.

Focusing on investments, two categories may be considered: (1) compulsory investments, and (2) optional investments. Compulsory investments are those that the company is obliged to make given the circumstances (e.g., installing a filter on a stream of gases following an injunction of the authorities). Optional investments are freely decided upon by the company (e.g., to increase the production of a mine or modernize a section in a smelter). The company's decision may be motivated by strategic, market, financial, political, technical, or environmental considerations.

From the study of an investment opportunity, managers obtain a stream of forecasted cash flows over the life of the project. Together with qualitative factors, many decision-making tools and techniques may be applied to the analysis of the opportunity of the project. The most common tool is the net present value (NPV), which may or may not be complemented by other financial indicators like the internal rate of return, the payback, the average return on book value, or the profitability index.[31]

The NPV method has the advantage of integrating the notion of risk, because the rate of interest adopted depends on the level of uncertainty that the project involves. Another advantage is that it gives an absolute value of the project (sum of the discounted cash flows over the life of the project), which allows prioritizing different projects, as well as optimizing the allocation of financial resources among them.

While considering the influence of different factors (technical performance, costs, raw material prices, metal and energy prices, markets, etc.) on the cash flows and on the NPV, the effect of the experience curve (Figure 4.4) is frequently undervalued. In other words, the time and pace needed for the project to reach 100% of its potential is often underestimated. As the cash flows of the first years are the most greatly affected by this undervaluation, the impact on the real NPV can be dramatic.

The probabilistic approach of investment analysis allows measuring the risk of a project. The most common algorithm is the Monte Carlo method, described in the following section.

For companies with large organizations and broad activities, the process of deciding on investments is, for practical reasons, structured in an organic way. This allows the different levels of the organization to share the power to decide within specific categories and sizes of projects. These procedures are part of the internal policies established by the management and are subject to internal auditing.

Measuring the Risk on Investment Decisions

The quality of an investment decision depends on the quality of the parameters and data used for its analysis. If the estimation of metal prices or costs is poor, the quality of the results obtained will also be poor ("trash in, trash out"), because the reliability of the NPV rests on the statistical variance of the parameters used.

The Monte Carlo method[32] consists of a set of simulations of the NPV, adopting probabilistic values for the different variables instead of fixed ones. Simulations are performed using random sampling on the statistical distributions of the different parameters (Gaussian or non-Gaussian). These are determined by expert systems built by specialists in the different areas. The result is a statistical distribution of NPVs (or any other financial indicator) instead of a unique value, which allows associating a margin of error with the calculated indicator (Figure 4.10).

The advantage of this numerical method is that it gives a statistical distribution of the NPV, allowing the assignment of a bracket of probabilities to the median value.

Financial resources are generally scarce, so the problem companies face is to prioritize capital expenditures. The decision-making process is facilitated when a measure of risk is associated with the NPV of each project. A simple way to systematize these decisions is to represent the projects in a two-dimensional diagram (Figure 4.11). Each point represents a project with the value of its NPV on the horizontal axis and the measure of its risks on the vertical axis. Projects situated in the lower right quadrant must have preferential attention because they present together high NPV and low risk. Projects on the upper right rectangle have high NPV but also high risk. Part of this risk can be reduced or eliminated by investing in improving the quality of the estimations (e.g., investments in lab testing designed to gain knowledge of the flotation behavior of the ore or market research to reduce the risk associated with estimating the prices). After that, these projects could be moved to the first category. Projects A and B, having low risk but also low NPV, should be discarded (in principle) because they may involve a certain dose of "management distraction," the cost of which is difficult to estimate at the time of calculating the stream of cash flows. Finally, project C has no interest at all, because of its low NPV and high risk.

Obviously, the criteria used by mining companies when deciding on investments are not exclusively financial. They are used in conjunction with their strategic guidelines, rejecting ventures that do not fit with it. For example, a company that has adopted the strategy of only operating world-class mines will not invest in a small mining project, even if its financial picture looks

Project "A"	Year 1	Year 2	Year 3
Earnings	X1	X2	X3
Costs	Y1	Y2	Y3
Income	X1 – Y1	X2 – Y2	X3 – Y3
Depreciation	Z1	Z2	Z3
	CF1	CF2	CF3

Probability Distributions for Three Project Variables:

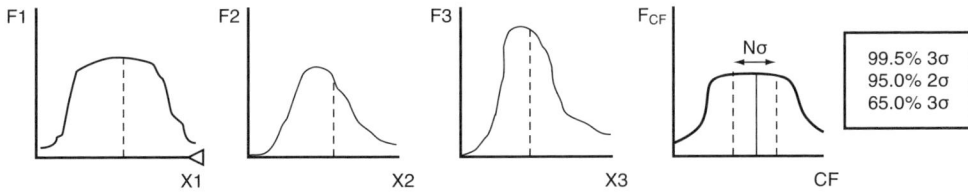

FIGURE 4.10   Measuring risk in investment decisions using the Monte Carlo method

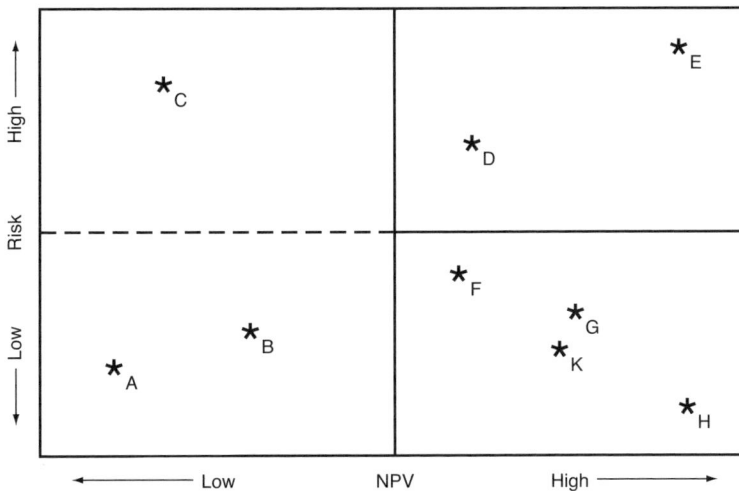

FIGURE 4.11   A two-dimensional investment decision model: profitability versus risk

good. Similarly, a company strategically focused on precious metals will not consider investing in base metals, and a company producing downstream metal products, like oxides, sheets, or die casting parts, will not invest in massive primary production.

The sustainability of new activities is becoming a significant factor in many companies' investment philosophies. For example, some companies forbid themselves from buying land in industrialized countries. Instead, they look for ways to rent, lease, or find other similar methods that keep them out of hidden potential pollution liabilities that could appear over the life of the activity. Mining projects located in developing countries always undergo a study of their sustainability in parallel with pure financial analysis.

Some investments, especially those of social or environmental character, cannot be analyzed under the focal point of finance because their return is difficult to quantify. These investments are sometimes imposed (legal injunctions) and at times decided upon voluntarily. Companies always have good reasons to make these investment decisions, but these reasons are qualitative most of the time. The benefits coming from them are in the field of the social license, public image, or reputation. A typical example of this would be investments focused on emission sources, designed to keep their environmental impact at a lower value than the enforced legal limit.

The Use of Derivatives

Derivatives are financial instruments that allow controlling the risk associated with future price fluctuations. They are called derivatives because their value is derived from the value of other financial assets.

The financial results of the mining business are heavily affected by fluctuations of metal prices. For example, Anglo American's net earnings[33] increase or decrease by $107 million for each ±10% change in gold price and by $286 million for each ±10% change in the price of copper (data as of 2006). Earnings are so dependent on metal prices that some companies use derivatives to limit their fluctuations.

The LME and other organized market platforms provide financial instruments that allow companies to limit the risk of future variations in metal prices. They can, optionally, hedge metal prices (offset the variation between certain dates) or fix them (freeze the price level for a certain amount of their production). These instruments are similar to "price insurance"; therefore, they have a cost. Management determines whether the company will use these instruments in each case.

Each company must define the amount of risk that it wants to take. The decision on hedging or fixing the metal prices is a very strategic one. A mining company can decide to fix the price of its metal production during a given period of time. This will sharply reduce the volatility of its results, allowing a more accurate budgeting process and a more reliable financial forecast, facilitating financing negotiations with banks, and assuring acceptable (but not always maximal) return to shareholders. On the other hand, this option has drawbacks. Due to technical, social, or political circumstances (strikes, injunctions, accidents, etc.), the mine may fail to produce the amount of metal specified in the fixed-price contract. The balance would have to be purchased in the open market at the maturity of the operation at the prevailing price at that moment. If this price is higher than the one in the fixed-price contract, the difference would constitute a dry loss. If market price at maturity is higher than the fixing price, the competitors that decided not to fix will have a profit advantage, so the company would be negatively judged by analysts and financial experts with the risk of subsequent loss on share value. This is why the hedging policy should be approved, or at least explained, to the shareholders.

There are different arguments for and against hedging. Some mining companies claim that their shareholders buy their shares precisely because they seek exposure to the metal markets. Others prefer to adopt a policy of offsetting this risk, aiming to cover themselves against price slumps, thus giving up the benefits of price increases. The cost of the hedging operations, which can be significant, includes the broker's commissions plus the operating cost of the internal risk management cell.

From the operational point of view, the financial techniques used are not simple and need strict control. The danger of mismanagement really exists and should be integrated into the decision. Hedging operations introduce added accounting complexity and may reduce visibility into a company's financial activity.

Today, derivatives have an enormous weight on the regulated market platforms' operations and they are a significant factor in the price-setting mechanism. At the LME, the ratio of futures and options turnovers (expressed in metric tons, or t) to total world production has been 29 for aluminum, 34 for copper, 17 for lead, 22 for nickel, and 28 for zinc,[34] which means that for each physical ton actually produced, there are 17 to 34 t of futures and options traded.

Research and Technology

Although the common public perception is that mining and metals are low-technology sectors, this is false. Competition is fierce and mining companies are constantly seeking competitive advantages. Investing in research and development (R&D) is one of the best ways to achieve them.

Cost leadership is the main generic strategy in the minerals industry; therefore, streamlining the organization and investing in productivity and technology are vital for competitiveness. The environmental challenges the minerals industry faces drive research for cleaner technologies. Improvements in exploration techniques, rock mechanics, mining methods, and equipment are necessary to face competition.

Capital expenditures on minerals exploration are conceptually equivalent to research expenses. It has been observed that the amount of these expenditures closely correlates with the price of nonferrous metals.[35] The smelting sector is also facing new challenges from the high cost of energy and the increasing requirements for enhanced working conditions, lower environmental impact, and improved quality of products and processes.

To meet the challenge for technological improvements, companies may choose between in-house technology development and buying technology (if available) in the market. The in-house alternative is usually adopted when it may be justified by a high enough critical mass of research activity or when the technology is not available in the market or (if available) is more expensive. Internal research has the big advantage of exclusivity, because findings are not shared with competitors. Smaller companies can either pursue their research efforts through joint ventures with other companies (if competitive issues are not at play) or buy technology when available. Companies normally use a mix of these two options. The cost of the programs requiring exclusivity and the availability of technologies in the market are the main decision factors.

Patents and trademarks are intangible assets of mining and metals companies; therefore, they must be subject to adequate strategies and be managed to maximize the value of the company's portfolio of intellectual property. The business of selling technologies is rather incompatible with the mining activities (as observed in the past for some well-known companies). Selling competitive advantages to rivals is not the best way to make research investments profitable.

Before deciding whether to invest in product and process research, companies should identify the stage of maturity of the envisaged technologies given that investment effort levels are directly impacted,[36] as shown in the so-called "S" curve (Figure 4.12). Experience shows that

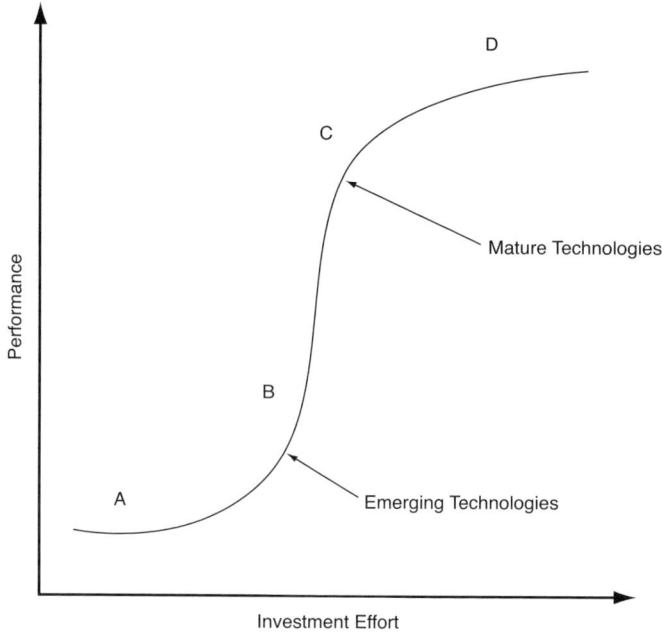

FIGURE 4.12   The S curve of cost performance in process and product research

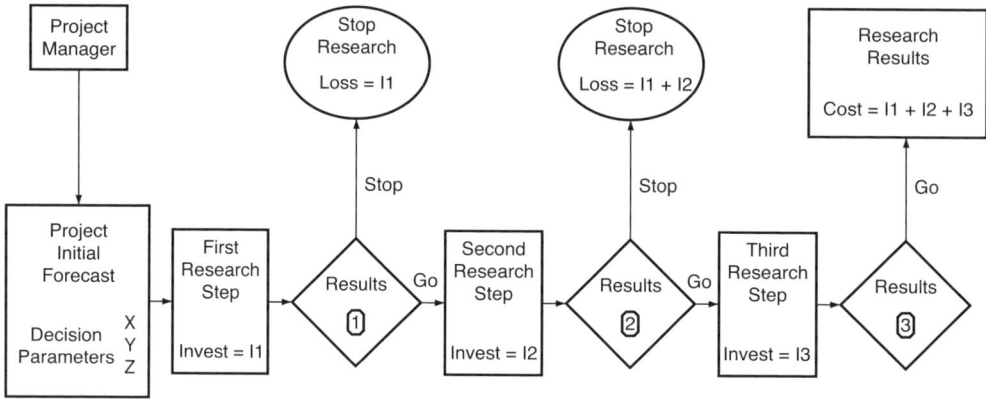

FIGURE 4.13   Managing research as a financial option

investment costs for developing emerging technologies are generally lower than those necessary to improve mature ones. According to this criteria, companies should favor the purchase of know-how (if available) in case of further improvements of existing process or product technology.

Because of its nature, the costs of research may be estimated, but the results are difficult to forecast. A way of optimizing the result/cost ratio is to use the methodology of financial options. Options are financial instruments commonly used in transactions. A buy ("call") option contract gives its owner the right (but not the obligation) to buy a good or service at a fixed price at any

R&D Investments Worldwide (2004)	
R&D Investment by Country in % of World Total	38.0% USA 31.1% EU 22.1% Japan 8.8% Others
Ratio: R&D Investment (% of Net Sales), by Region	4.5% USA 3.3% EU 4.0% Japan 3.4% Others
Ratio: R&D Investment (% of Net Sales), by Sector	13.5% Pharmaceutical, Biotech, Health 8.0% High Tech, Engineering, IT 4.5% Engineering, Chemicals 2.0% Others

NOTES:
The top 700 EU R&D investor companies spent $102 billion in 2004.
The top 700 non–EU R&D investor companies spent $213 billion in 2004.

FIGURE 4.14   How much to invest in R&D?

time on or before a given date. A sell ("put") option contract gives to its owner the right (but not the obligation) to sell a good or service at a given price at any time on or before a given date. Both operations have a cost, called the *premium*.

Decisions in research investments can be treated as options (Figure 4.13). With this concept, instead of deciding to go on with the entire research program from the beginning, the program may be planned in stages (each of a known cost), and at the end of each stage, the decision to go into the next stage is conditioned based on the results. If the decision is not to go on with the next stage, the company avoids the full cost of the program at the price of a premium, which is the cost of the first phase. In financial terms, one would say that the company "has not exercised the option."

The decision to set up a research program must be undertaken after estimating the profitability of a combined investment: research plus the capital cost corresponding to the implementation of the resulting measures. The cash flow is calculated by considering the cost reduction obtained.

How much a company should invest in R&D is a strategic decision. In fact, it very much depends on the strategic market and segment in which the company is operating.

The total investment of the mining sector in R&D is difficult to evaluate and is very variable. A close correlation has historically been observed between the R&D expenses and the price of metals,[37] where the expenditure on R&D and exploration is higher for periods of high commodities prices. Companies do not always explicitly publish their R&D investments, and there is no common criterion for the accounting of these expenses. If we include the exploration expenses in the R&D investments, we can estimate an average of 1% to 2% of their turnover. Figure 4.14 shows the results of a study published by the European Commission.[38] Data are related to the year 2004.

A significant share of the R&D expenditures in the mining sector are not made directly by the mining companies but by their suppliers and clients. The materials and techniques that the industry uses, like drilling, loading and hauling material, instrumentation (for bore sampling, rock mechanics control, grinding, flotation, etc.), automation and robotics, information technology, and so forth, are often developed and improved by service companies. The same situation occurs in the oil industry (geological and mineralogical logging, geochemistry, geophysics, mapping, etc.).

Organization and Leadership

Why is organization a strategic factor? Because it allows maximizing the efficiency of the company in reaching its strategic objectives.

Small corporations have less need for formal organization than larger ones. Lower density of information, multifunctional employees, narrow markets, and simplified formalities make complicated organizational schemes unnecessary. As size and complexity increase, corporations require ad hoc internal structures to be functional.

A company's organization can be compared to the mechanism of a machine that is engineered specifically to fulfill a given task. Its dimensions, shape, structure, strength, special position, and interrelation within parts allow the machine to perform the work for which it has been designed efficiently (e.g., an airplane to fly and a crane to lift weights). Everything is custom-made for its specific function.

Similarly, the strategy of a company calls for a custom-made organization, in which the weight and interrelation of the different functions is designed to provide overall optimal results. Pre-made, one-size-fits-all solutions are not efficient.

Organizations need leaders—individuals conducting the activities in each area of the company. The management style of these leaders must also be adapted to the company's culture and strategy. The two extremes of management style are fully autocratic and fully democratic. In the fully autocratic style, the information flows just from the top down; in a fully democratic style, it flows only from the bottom up. An intelligent mixture of both, with a good dose of transversal flows, will maximize the results.

The two most common structures in the minerals industry are the organization by business units and the matrix organization.

In the organization by business units (Figure 4.15), both functional and operational responsibilities are linked separately to the CEO. Although informal communication takes place directly between them, formal communication materializes through the CEO to the heads of the business units. Business unit heads are accountable for the results and balance sheet of their units. Central functions, which include human resources, legal, environment, strategic planning, control, sustainability, finance, and communication, report directly to the CEO. This organization has the advantage of simplicity and transparency. Each manager has a well-defined function and clearly identifiable responsibilities. It is very efficient in stable environments where transversal relationships are less important. Temporary task forces can be created to implement specific transversal programs like cost-reduction initiatives, new projects, and so forth. The designated manager has authority on the subjects related to the program.

The matrix organization (Figure 4.16) emphasizes the transversal flows of information. The name *matrix organization* comes from the fact that some of the functions have a double dependency, that is, to business unit managers and to country or geographical managers. Conflict arbitration corresponds to the CEO. Some tasks are distributed between business unit managers and country managers. Business unit managers run the business from strategy and production to marketing, and country managers take care of the institutional relations, compliance, environment, fiscal, and public relations. Both report to the CEO. This organization, considered to be more creative and adaptable than the previous business unit approach, asks for a more sophisticated and participative style. It generates some conflicts that need to be addressed, but it is still considered to be more efficient and competitive. Sometimes the function of the country managers is

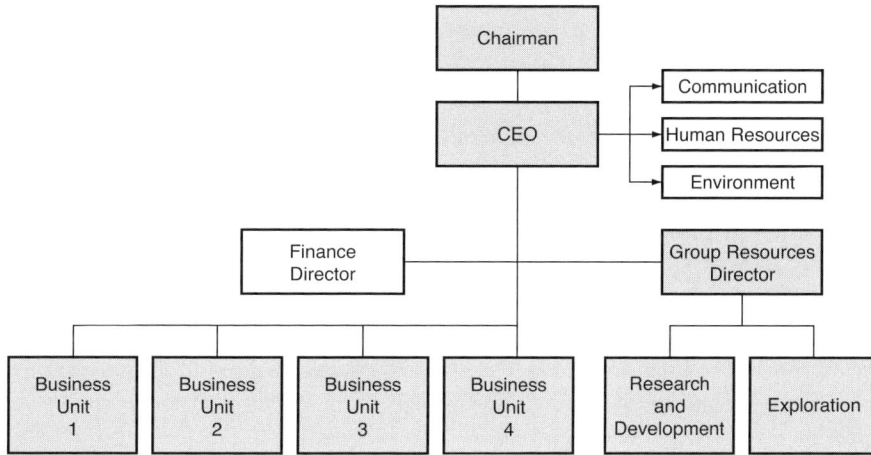

FIGURE 4.15 Organization by business units

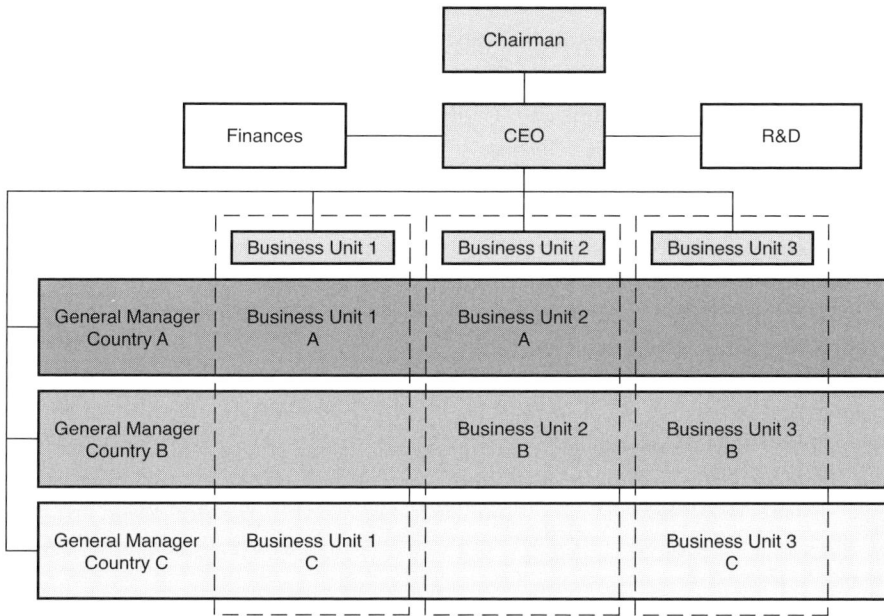

FIGURE 4.16 Matrix organization

defined as an advisory one without hierarchical power over the business units. The profile of the responsible manager is chosen accordingly.

The individuals in charge of running a company must have specific personal qualities that go far beyond technical and managerial skills. This particular personality allows the exercise of what has been called "inspirational leadership" that brings passion and direction to the group. Common qualities among business leaders are talent, initiative, charisma, team spirit, and optimism. Effective leadership in the industry requires explicit associated authority granted by the company's board. In a business such as mining, where cost reduction is the dominant strategy, the role of such leaders is essential.

Among the board's main tasks are to control the activity of the CEO and to enhance the team's functioning spirit. Of all the possible forms of corporate governance, having a supervisory board that controls the executive board offers the best formula for avoiding excessive individualism and providing support for the CEO's decisions.

The organizational structure of the minerals industry and the interfaces between company structure and sustainability strategy are addressed in Chapter 5.

Image and Public Perception

The flow of interactions between mining companies and their environment has been growing in recent years. Authorities, labor unions, citizens' organizations, NGOs, shareholders, and the public in general demand more detailed information every day about the activity of the mining companies.

The more mature the society where the activity is located, the more information is required. This flow of information should not be limited to the statutory requirements but extended to any facts that would help to build a trustworthy and favorable image of the company. Reputation is part of the intangible assets that contribute to obtaining the license to operate. It is hard to build and easy to destroy.

Many mining companies operate communications departments to deal with these issues in a professional manner (see the "Project Management and Stakeholders" section in Chapter 8). These departments control the information sent outside the company, verifying its coherence with past data and facts publicized, as well as with the image the company wants to convey. Everything from the annual reports to the analysts' meetings and special events is carefully prepared and presented.

Probably the most important collective communication initiative ever promoted by the minerals industry was the creation of the International Council on Mining and Metals (ICMM) in 2001, as a successor to the International Council on Mining and the Environment, created by a group of mining companies some years earlier. In an effort to improve the image of the industry, ICMM members adopted and publicized 10 basic sustainable development principles[39] in 2003, committing themselves to measure their corporate performance against these principles. These principles (referred to elsewhere in this book) establish the position of the members regarding ethical practices, corporate government, risk management strategies, health and safety principles, environmental management, biodiversity, sustainable product design and recycling, contribution to the development of neighbors' communities, and transparency in communications.

There may be big differences between the reality, the image that the company wants to convey, and the public's perception. Perception is what the recipient believes from the information received. Apparently small, unimportant details can create a biased perception of a company. Reputation plays a major role in mining activities. While governments rely on objective information to grant licenses to operate, other stakeholders could be more influenced by the image of the companies. The ability to build a favorable reputation creates an important competitive advantage in mining businesses.

A company needs to cover several main fields of activity in its communications strategy in order to build a favorable image:

- Financial situation: The information associated with the financial situation must be clear, objective, and rigorous. It should be established following internationally accepted accounting rules and presented consistently, using the same format every year to facilitate interpretation.

- Corporate strategy: This must always be communicated in a preferential place (often, the first pages of a company's annual report), be credible, and be explained consistently over time. Changes in strategy must be thoroughly explained and justified, not only by the ideas of a new CEO but by profound changes in the business environment.

- Markets and products: Information on markets and products contribute strongly to building the image of the company. A countervailing issue is the desire of companies to keep some information confidential for competitive reasons. The company must find the optimal equilibrium.

- Social and environmental issues: It has become increasingly common to publish specific documents covering these matters periodically, like "social and environmental reports," "sustainability reports," and so forth, which are separate from the company's annual report. These documents are legally enforced in some countries and their content is regulated. Objectivity and consistency must be the key qualities.

- Leaders' profiles and histories: The public in general and the stakeholders in particular find an increasing interest on the personal profiles of the corporate leaders. They want to know about each leader's biography, ideology, business approach, compensation, and so forth. Depending on the country the industry is located in, some of this information is either legally required or optional. In any case, it is always a good vehicle to convey a favorable image of the company.

Measures of Success

The most relevant measure of success of a strategy is the evolution of the market capitalization of the company (Figure 4.17). Every proceeding of the company has an effect on its share value. The share value reflects the future earnings of the company through its cumulative discounted cash flows. The market anticipates the future on the basis of the information and data available. An event that happens today but will have positive influence on future earnings will be acclaimed by the market by the corresponding increase on the share value, even if present results do not reflect this event.

By analyzing the evolution of the share price and comparing it with the evolution of the competitors' share prices and with the sector indexes, one can judge the success of a given strategy. This includes not only the intrinsic quality of the strategy but also the ability of the management to implement it.

A more qualitative instrument for evaluating a strategy is what is called "market consensus" or the common evaluative view of a financial market's analysts about the situation, perspectives, and value of the company.

Another, although more technical, instrument that can be used to measure the success of a mining company's strategies is the evolution of the cumulative cash cost curves with the time (Figure 4.18). These graphs represent the unit cost for a given product of all companies in the same market as a function of their cumulative production and were previously explained in this chapter, in the "Dynamics of the Strategy Formulations" section. They are produced by specialized firms that periodically survey the companies and build the curves after harmonizing data and correcting heterogeneities.

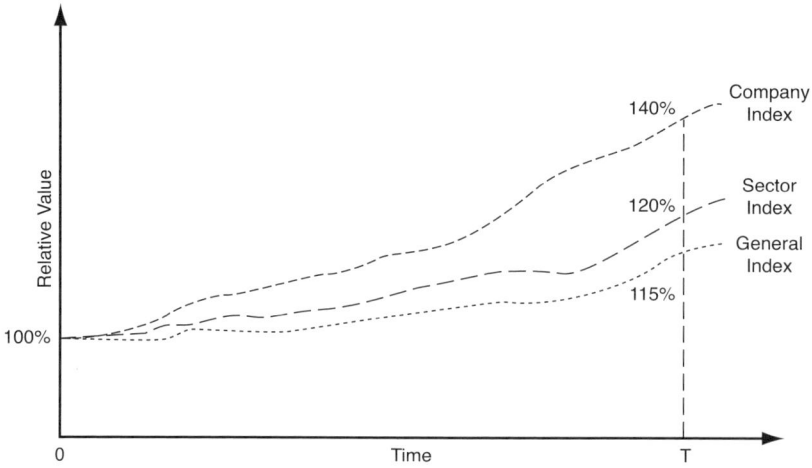

FIGURE 4.17   Measures of success: market value

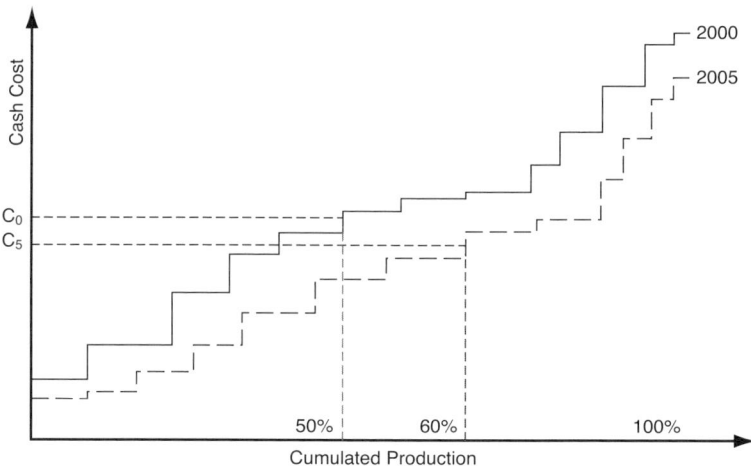

FIGURE 4.18   Measures of success: cash cost curve

Cash cost curves vary with time, reflecting the relative progress or deterioration of competitiveness of each company, as well as the evolution of unit cash costs of the different sectors in the market. The position of each company on these curves determines the margins it can obtain for a given product's price, as well as the general state of prosperity of the sector as a whole.

CONCLUSION—MINING STRATEGY AND SUSTAINABILITY

The leading idea of this chapter is that sustainability is becoming a fundamental pillar of a company's strategy. The profound reason that explains this is that today's society understands the consequences of the interaction between human activity and the ecosystem much better. The citizens are conscious of the importance of this issue for both present and future generations. As a consequence, the social license to operate mining ventures is becomimg more difficult to obtain, forcing companies to better integrate the concept of sustainability as well as the social and environmental factors into their strategy.

NOTES

1. *American Heritage Dictionary of the English Language*, 4th ed., s.v. "Strategy."

2. Michael E. Porter, "What is Strategy?" *Harvard Business Review* (February 1, 2000): 2–4.

3. Ibid.

4. Thomas H. Taylor, ed., *Corporate Strategy* (Amsterdam: North-Holland Publishing Company, 1982), v.

5. Gary Hamel and C. K. Prahalad, *Competing for the Future* (Boston: Harvard Business School Publishing, 1984).

6. George S. Day, *Strategic Market Planning: The Pursuit of Competitive Advantage* (St. Paul, MN: West Publishing Company, 1987), 129–131.

7. Porter, "What is Strategy?"

8. Michael E. Porter, *Competitive Strategy: Techniques for Analyzing Industries and Competitors* (New York: The Free Press, 1980), 35–41.

9. Commodities Research Unit Ltd. (London) and Brook Hunt (Addlestone, Surrey, UK).

10. Hamel and Prahalad, *Competing for the Future*, 195–217.

11. Jay Forrester, *Industrial Dynamics* (Waltham, MA: Pegasus Communications, 1961).

12. London Metal Exchange Web site: www.lme.co.uk.

13. Rio Tinto 2008 Chart Book, LME/WBMS (World Bureau of Metal Statistics).

14. Richard A. Brealey and Stewart C. Mayers, *Principles of Corporate Finance*, 5th ed. (New York: Irwin/McGraw-Hill, 1996).

15. Paul B. Downing, *Environmental Economics and Policy* (Boston: Little, Brown and Company, 1984), 121–126.

16. International Council on Mining and Metals, www.icmm.com.

17. European Association of Metals, www.eurometaux.org.

18. Independent Petroleum Association of America, www.ipaa.org.

19. American Petroleum Institute, www.api.org.

20. European Petroleum Industry Association, www.europia.com.

21. International Union for Conservation of Nature, www.iucn.org.

22. World Wide Fund for Nature, www.wwf.org.

23. Friends of the Earth International, www.foei.org.

24. Greenpeace International, www.greenpeace.org.

25. James Otto and others, *Mining Royalties: A Global Study of Their Impact on Investors, Government, and Civil Society* (Washington, DC: World Bank, 2006), 41–43.

26. Pierre-Noel Giraud, *Geopolitique des Resources Minieres* (Paris: Economica, 1983).

27. Paul Stevens, "Resource Impact: A Curse or a Blessing? A Literature Survey," *Journal of Energy Literature* 9, no. 1 (2003): 3–42.

28. John E. Tilton, Merton J. Peck, and Hans H. Landsberg, eds., *Competitiveness in Metals* (London: Mining Journal Books Limited, 1992), 9–10.

29. Scott Christofferson, Rob McNish, and Diane Sias, "Where Mergers Go Wrong," *The McKinsey Quarterly*, no. 2 (2004).

30. Rebecca Henderson, "Technology Strategy" Sloan School of Management, MIT: MBA Course, http://ocw.mit.edu/OcwWeb/Sloan-School-of-Management/15-912Spring-2005/CourseHome/.

31. Brealey and Mayers, *Principles of Corporate Finance*.

32. C. P. Robert and G. Casella, *Monte Carlo Statistical Methods* (New York: Springer-Verlag, 2004).

33. *Anglo American plc Fact Book 2006/7* (London: Anglo American plc, 2007): 8, www.angloamerican.co.uk/static/uploads/FInal%20PDF.pdf.

34. London Metal Exchange Data, 2003.

35. Vivek Tulpulé, *Outlook for Metals and Minerals: 2007 Full Year Results* (Rio Tinto February 1, 2007), 4–6, www.riotinto.com/documents/OUTLOOK_FOR_METALS_AND_MINERALS_2007.pdf.

36. European Commission, *R&D Investment Scoreboard 2004* (Seville: Directorate General Joint Research Centre, 2004).

37. Tulpulé, *Outlook for Metals and Minerals*.

38. European Commission, *R&D Investment Scoreboard 2004*.

39. *ICMM Sustainable Development Framework: ICMM Principles*, Document Ref: C 020/290503 (International Council on Mining & Metals, May 29, 2003), www.goodpracticemining.org/documents/jon/ICMM_SD_Principles.pdf.

CHAPTER FIVE

Integrating Sustainability into the Organization

J. A. Botin

INTRODUCTION

In an industrial context, management deals with optimizing the use of human, material, and financial resources in *operationalizing* strategy goals. Today, sustainability is a fundamental pillar of a mining company's strategy; therefore, sustainable mine management may be conceived of as the process of operationalizing the sustainability principles of a mining company. Most mining companies state their commitment to the values of sustainability in their vision declarations and policies. However, not many achieve an efficient integration of those values down into the operational levels of the organization.

Corporate commitment is an essential condition for integrating sustainability, but it is not sufficient. Another key condition is a business culture in which sustainability is a high professional and business value and objectives are implemented through commitment rather than compliance. Furthermore, the integration process requires of an organizational structure specific roles and integration mechanisms[1] and adequate information management systems.

Therefore, the sustainability challenge is (at least in a mining context) on the side of management. This is evidenced by the strong management focus of the sustainable development frameworks proposed by the International Council on Mining and Metals (ICMM),[2] the United Nations (UN) Global Compact,[3] and other multiparticipant initiatives offering policy guidance on sustainable development to the minerals industry, such as these examples:

- The ICMM's "10 Principles of Sustainable Development," a framework for the integration of sustainability into the company structure, focuses on concepts such as culture, structural integration, strategy implementation, continual improvement, and so forth, all of which are challenging management tasks.

- The UN Global Compact, a globally recognized framework for corporate responsibility and ethics, states: "To achieve sustainable development, environmental protection shall constitute an integral part of the business process and cannot be considered in isolation from it." Again, this implies complex management tasks and management systems.

- Equally, the "Towards Sustainable Development Initiative"[4] emphasizes business culture as a requirement for integration when it states: "Our actions must reflect a broad spectrum of values that we share with our employees and communities of interest, including honesty, transparency and integrity."

This chapter focuses on the integration of sustainability down to the operational levels of mining companies, the organizational structures, and the management roles and systems required for integration. The chapter is composed of six sections.

"Organizational Structure for Sustainable Management" deals with the organizational structure of mining companies, focusing on the positions, roles, and other integration mechanisms and systems that are required for the integration of sustainability values at operational levels.

"Integration of Sustainable Development into Mining Operations" provides the authors' perspectives on sustainable development and its incorporation into corporate strategy. It also attempts to make the critical link between sustainable development at a corporate level and how it is effectively translated and implemented on the ground in actual mining operations. Several useful views from mine managers on the challenges and nuances of making sustainable development work in a mining context in a developing country are presented.

"Managing for Stakeholders' Expectations: The Seven Themes of Sustainability" provides an approach to sustainable mine management that relies on combining the sustainability expectations expressed by local stakeholders and those identified in corporate plans and guidelines to sustainable development.

"Environmental Management System: A Sustainable Management Tool" focuses on environmental risks and describes the process of implementation of a certified environmental management system (EMS) in a mining operation and the advantages of its implementation during the preproduction stage. It also analyzes the advantages of operating under an EMS in terms of operational efficiency, public image, and involvement of stakeholders.

"Case Study: Partnership for Sustainable Development in Ghana" describes the various components of the strategy employed by Gold Fields Ghana to maximize the effectiveness of its sustainable development programs and to create a leading example of best practice in the international mining industry for community engagement and sustainable development.

"Case Study: Industrias Peñoles—A Large Organization Committed to Sustainability" describes the organizational structure of Peñoles and how the strategic objectives of sustainable development are integrated across the organization by means of the MASS System (Environment, Safety, and Health System).

ORGANIZATIONAL STRUCTURE FOR SUSTAINABLE MANAGEMENT

As a fundamental area of corporate strategy, sustainability must be integrated into the organization at all management levels. The integration process, often difficult and challenging, requires an organizational structure with integration mechanisms, management roles, and systems ensuring proper communication, coordination, and control.[5]

Conceptually, the design of the organizational structure of any company results from the following three business processes:

- A segmentation process, whereby decision authority and responsibility is divided into several operating units;

- An integration process, focusing on the implementation of certain mechanisms (integration mechanisms), ensuring that the operating units resulting from segmentation operate in alignment with the corporate strategy; and

- An empowerment process, through which the decision rights—executive and controlling—and responsibilities of each operational unit and each management level are defined.

The main characteristic of the organizational structure of a company is the criteria used for segmentation at the top management level (i.e., senior officers and vice presidents). These basic criteria identify the different types of organizations. Typically, mining companies may be segmented by functions (e.g., production, engineering, marketing), by divisions or business units (e.g., energy coal, base metals, gold), or through some combination of both descriptors. In some cases, segmentation is, to a certain degree, two dimensional (i.e., matrix organizations), where some positions have a double dependency (i.e., to the business unit managers and to the country managers) as the means to ensure that key strategic decisions are made after considering the local constraints.

Any organization needs *integration mechanisms* so that segmentation does not result in strategic misalignment among the segmented parts. Hierarchy is the main integration mechanism, but organizations require other means of integration, such as business culture, corporate plans and policies, staff roles, management information systems, and integration structures such as committees, task forces, and so forth.

The empowerment process aims to optimize the degree of autonomy or decentralization of decision making in the organization. The degree of decentralization and autonomy is mainly related to the size of the company, the geographical dispersion, and the nature and strategic diversity of its business but also on business culture and the information technology (IT) management systems available. Conceptually, maximum decentralization is always desirable, provided that decisions are coherent with company objectives and the decision-maker has the information, the professional capacity, and the motivation required for deciding. In each case, these constraints establish the limits to decentralization.

The structure of a company should focus on a correct balance of segregation versus integration, strategic coherence versus flexibility, and agile decision making versus information requirements.

Organizational Structure of Mining Companies

The criteria for the structural segmentation of a mining company follow. (Figure 5.1).

Small Mining Companies

Functional segmentation is a standard for small mining companies operating one or two mines in a single geographical area. In this structure, functional managers report to the top executive (i.e., chief executive officer [CEO], chief operating officer [COO], or executive vice president). For example, the company in the organization flowchart in Figure 5.1 is segmented into five functional areas (e.g., mine production, human resources, environmental health and safety, technical services, and finance and administration).

Also, the functional organization is an industry standard for unit mining operations within medium-sized and large mining corporations. For example (Figure 5.2), the Podolsky mine (FNX Mining Company) in Sudbury, Ontario, is organized in six first-level positions:

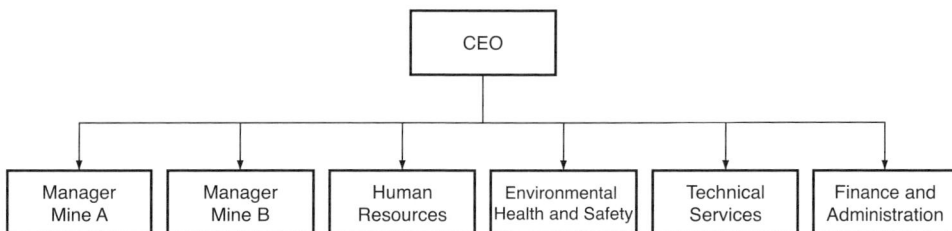

FIGURE 5.1   Typical flowchart of functional organizations

- The chief engineer, accountable for mine planning and engineering;
- The mine superintendent, accountable for production;
- The chief geologist, accountable for ore reserves management and grade control;
- The health and safety (H&S) coordinator, accountable for H&S coordination and control;
- The environmental coordinator, accountable for the coordination of environmental plans; and
- The cost accountant, accountable for the cost control system and cost reporting.

Also, as shown in Figure 5.2, unit mine operations use functional segmentation to organize the second and third organizational levels. In some cases, sustainability is considered a function and becomes the responsibility of a functional manager. In the organization outlined in Figure 5.1, this is the manager of environmental health and safety, who is accountable to the top executive.

Functional organizations are ideal for a single mine or a small mining company operating two or three mines within a small region. The functional managers reporting to the top executive (i.e., the general manager or the CEO) are specialists in their fields, thereby allowing for a simple and cost-efficient structure. However, being specialized professionals, functional managers often lack overall business vision, so the important decisions tend to get centralized at the top level. As the company grows larger, this dynamic can compound to make decision making slow and inefficient.

Because of this constraint, as mining companies grow larger and become the operators of several mines and projects and/or expand their activity into two or more geographical regions or countries, the organizational structure must change from functional to divisional, as described in the following paragraphs.

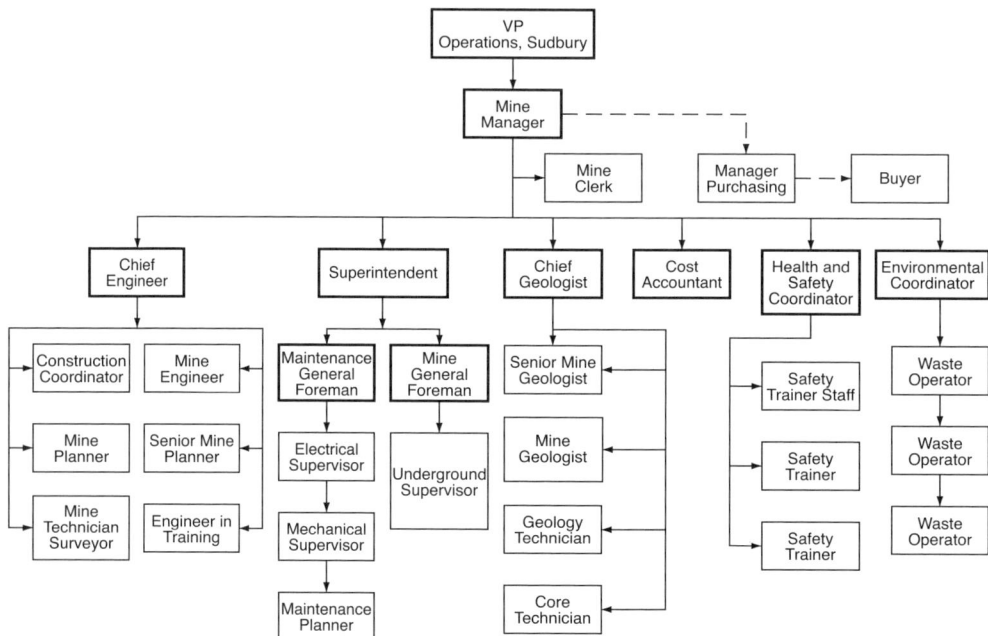

Courtesy of FNX Mining Company

FIGURE 5.2   Organizational flowchart of Podolsky mine

Medium-Sized Mining Companies

The organizational standard for most medium-sized and large mining companies is divisional segmentation, where the segmentation criteria may be the product (e.g., coal, copper), the region, the business unit, or a combination of these criteria.

As an example, Inmet Mining, a Canadian copper mining company that produced some 80,000 t of copper in 2007, is expecting to grow to a copper production of approximately 260,000 t in 2011.[6] The company has become global with operations in Canada, Turkey, Finland, and Spain and minority interests in Papua New Guinea and Panama. The company is organized in four operating divisions, as shown in Figure 5.3.

A typical divisional organization decentralizes operational responsibly down to the division level but remains centralized at the corporate (top) level for some strategically important functions in order to ensure alignment with corporate strategies and to prevent excessive redundancy of services. Typically, centralized functions are financial strategy (vice president of finance), corporate development; health, safety, environment and community (HSEC)/sustainability strategy; human resources (HR) policies; and others.

Furthermore, this increased decentralization of decision making down to the divisional levels requires the implementation of significant IT systems and the acceptance of some level of functional redundancy in the organization. In any case, an adequate balance between division autonomy, corporate strategy, and cost-efficiency must be the leading consideration.

Global Mining Giants

The organizational standards for the global mining giants (e.g., BHP Billiton, Xstrata plc, Rio Tinto, Anglo American, CVRD, and others), although formally divisional, goes one step further by allowing for some degree of strategic diversity. Although in the conventional divisional organization (Figure 5.3) the strategy is unique and centralized at corporate level, global mining giants become *business conglomerates* operating in two or more business sectors within the minerals industry (e.g., metals, energy coal), each with a specific business strategy, legal structure, and financial services. Here, the corporate (top) level retains the functions that are necessary to maintain a common business mission.

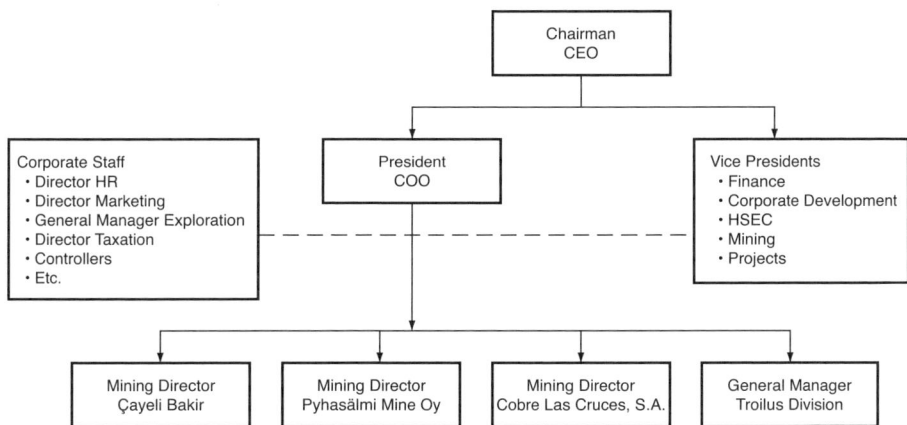

Source: Data from Inmet's Web site

FIGURE 5.3   Divisional organization flowchart of Inmet Mining Company

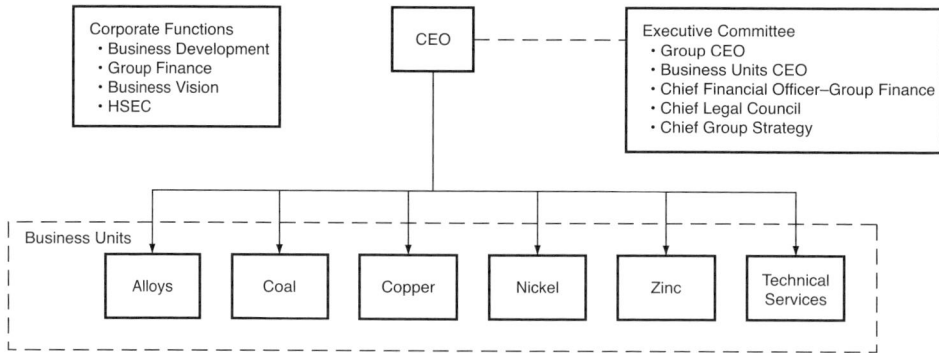

Source: Data from Xstrata's Web site

FIGURE 5.4   Organization flowchart of Xstrata plc

For example, Figure 5.4 shows the organization of Xstrata plc, a mining group with a market capitalization of more than US$50 billion. In 2007, Xstrata was organized into six business units. The group's business vision is summarized as follows:[7]

> We will grow and manage a diversified portfolio of metals and mining businesses with the single aim of delivering industry-leading returns for our shareholders. We can achieve this only through genuine partnerships with employees, customers, shareholders, local communities, and other stakeholders, which are based on integrity, co-operation, transparency, and mutual value-creation.

The six business units (alloys, coal, copper, nickel, zinc, and technical services) operate as independent businesses under their own direction, where the business unit strategy is designed to compete in a specific industry sector and to be aligned with the group's business mission. At the corporate level, Xstrata retains executive and controlling functions of group strategy and finance, business development, and sustainability (HSEC).

Another example (Figure 5.5) shows the organization of BHP Billiton, a mining company with a market capitalization of more than US$100 billion. The group is organized into 10 business units, which are named "customer sector groups" to emphasize customer focus. The group's business vision is summarized as follows:[8]

> To create long-term value through the discovery, development, and conversion of natural resources, and the provision of innovative customer and market-focused solutions. To prosper and achieve real growth, we must (i) actively manage and build our portfolio of high quality assets and services, (ii) continue the drive towards a high performance organization in which every individual accepts responsibility and is rewarded for results, and (iii) earn the trust of employees, customers, suppliers, communities and shareholders by being forthright in our communications and consistently delivering on commitments.

In this case, the segmentation criteria is the customer sector groups, where each sector group (e.g., aluminum, base metals, diamonds and specialty products, energy coal, iron ore, manganese, metallurgical coal, petroleum, stainless-steel materials, and global technology) operates as an independent company with its own direction, strategy, and financial and legal services.

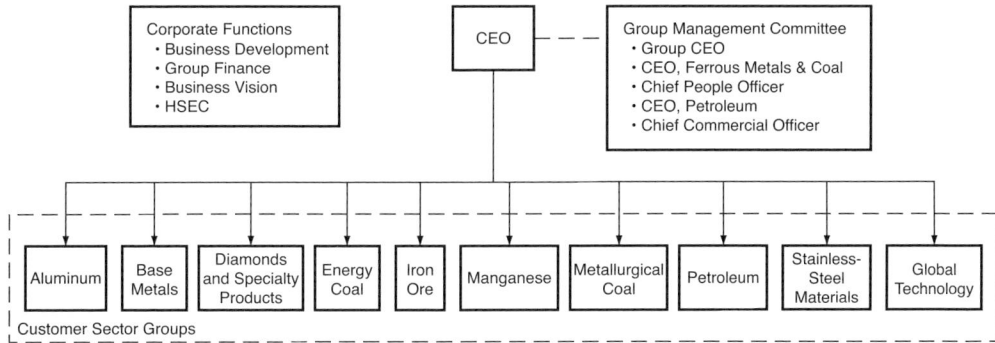

Source: Data from BHP Billiton's Web site

FIGURE 5.5 Organization flowchart of BHP Billiton

At the corporate level (Figure 5.5), the Group Management Committee retains the executive and controlling functions on group strategy, group finance, commercial, human resources and ethics, and sustainability (HSEC).

Most other mining giants are also organized as business holdings with some differences in the scope of the group business mission statement and corporate functions.

Sustainable Management Versus Company Structure and Culture

As stated previously, the meaning of sustainability in the minerals industry relates to the responsible and efficient management of mineral assets, human resources, health and safety, environmental, and community issues. To achieve this goal, the values of sustainability must be integrated across the organization down to the operating levels. Efficient integration of sustainability relates to structural and cultural factors, mainly the following:

- Business strategy (vision and mission) on sustainability
- Business culture and leadership
- Structure with specific integration mechanisms (roles, plans, and systems)
- Public reporting standards
- Independent assurance

In the following paragraphs, these factors are analyzed in more detail.

Business Strategy on Sustainability

The vision statement (issued by the CEO) is the directional guideline on the values and purpose of the organization. Here, sustainability must be properly addressed. Today, most mining companies' vision declarations refer satisfactorily to sustainability, as can be seen by the following examples:

- Alcoa:[9] Sustainability is "to deliver net long-term benefits to our shareowners, employees, customers, suppliers, and the communities in which we operate."
- Anglo American:[10] Sustainability is a management focus "on adding value for shareholders, customers, employees, and the communities in which the group operates."

- Newmont[11] values: "(i) Act with integrity, trust and respect; (ii) Reward creativity, a determination to excel and a commitment to action; (iii) Demonstrate leadership in safety, stewardship of the environment, and social responsibility; (iv) Develop our people in pursuit of excellence; (v) Insist on and demonstrate teamwork, as well as honest and transparent communication, and (vi) Promote positive change by encouraging innovation and applying agreed upon practices."

- Newcrest:[12] "To maintain its position as a leading producer of gold and copper, creating shareholder wealth in a manner that also benefits our employees and the communities and environment in which we operate."

- FNX Mining Company:[13] "FNX is committed to providing a safe and healthy workplace; environmental protection for the natural environment; and safety and well-being to our host communities. We will use proven health and safety and natural resource management tools and practices to minimize risk in project exploration, evaluation, planning, design, operation and closure."

Mission statements formulate strategy by targeting the position of the company in the market and the level of excellence required to achieve it. Typically, the business mission addresses sustainability by establishing a "sustainability framework," where a set of corporate policies on sustainability is defined. For example, sustainability frameworks put out policies on the following issues:

- Business excellence for economic sustainability
- Health and safety, human resources, and employee development
- Conservation of the environment
- Social license to operate (local communities, governments, and other stakeholders)
- Code of ethics, public reporting, and independent assurance
- Main sustainability challenges

Business Culture and Leadership

The integration of sustainability in a mining company may imply major changes in the company strategy, and often the company's ability to change the direction of its strategy can be seriously restricted by an inadequate business culture.

Sustainable management is best achieved when business culture is characterized by the following values:

- Sustainability is a self-motivated professional value.
- Sustainability objectives may be implemented through personal commitment of employees rather than compliance with policies and regulations.
- As a result of the first two values, decision making regarding sustainability may be decentralized to the lowest possible management levels.

Regarding leadership, sustainable management finds important synergies with the concept of "transformational leadership." Bass[14] defines transformational leadership as a type of leadership that "... starts with the development of a vision, a view of the future that will excite and convert potential followers. This vision may be developed by the CEO or may emerge from a broad series of discussions .... The most critical aspect is complete buy-in from the leader and/or team which has articulated and developed said vision."

TABLE 5.1 Mechanisms for sustainable development integration

Sustainable Development Integration Roles	Sustainable Development Integration Plans and Systems
Chief executive officer	Corporate charter (vision and mission)
Sustainability committee	Sustainability framework and policies
Vice president, sustainability	Sustainability standards
Division managers, sustainability	Sustainability management system
Site managers, sustainability	Management procedures
	Public reporting on sustainability
	Independent assurance (auditing)

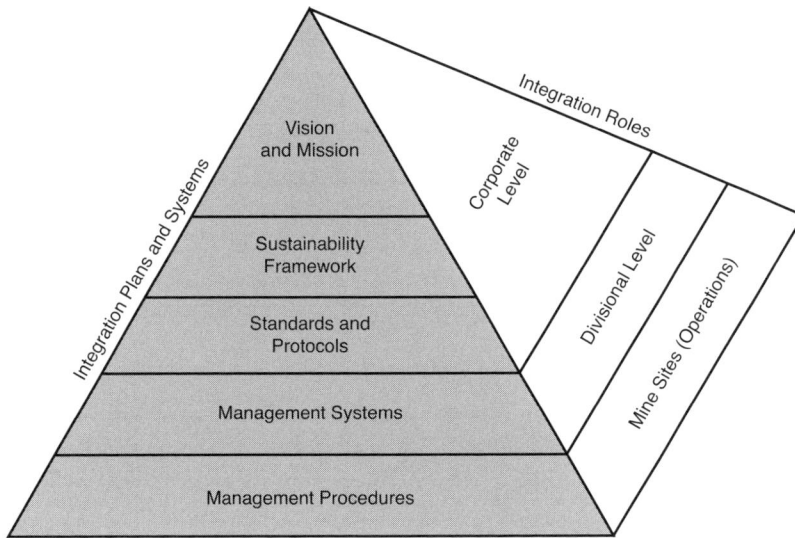

FIGURE 5.6 Hierarchy of sustainability management system

Integration Roles[15]

Methodological aspects of, and exemplar models for, the integration of sustainable development (SD) into the structure of mining companies will be addressed later in this chapter. The following paragraphs focus on the management positions and roles required to support the sustainable management process. These are listed in Table 5.1. The integration mechanisms in Table 5.1 refer to two types of mechanisms:

- Integration roles, i.e., the individual positions or ad hoc committees with accountability for the integration of sustainable development values and objectives, and
- Integration plans and systems, i.e., the policies, standards, and other management tools required to manage for sustainability at operating levels.

Both integration mechanisms—roles and plans/systems—are hierarchical (Figure 5.6), where integration roles act on the line of hierarchy and integration plans and systems integrate the planning, implementation, and control of the sustainability objectives.

As an example, Figure 5.7 shows a structural model for sustainable management of a large mining company (a "global mining giant"). The company in this example has a divisional structure; however, in the case of a small company operation with a functional organization (Figure 5.1),

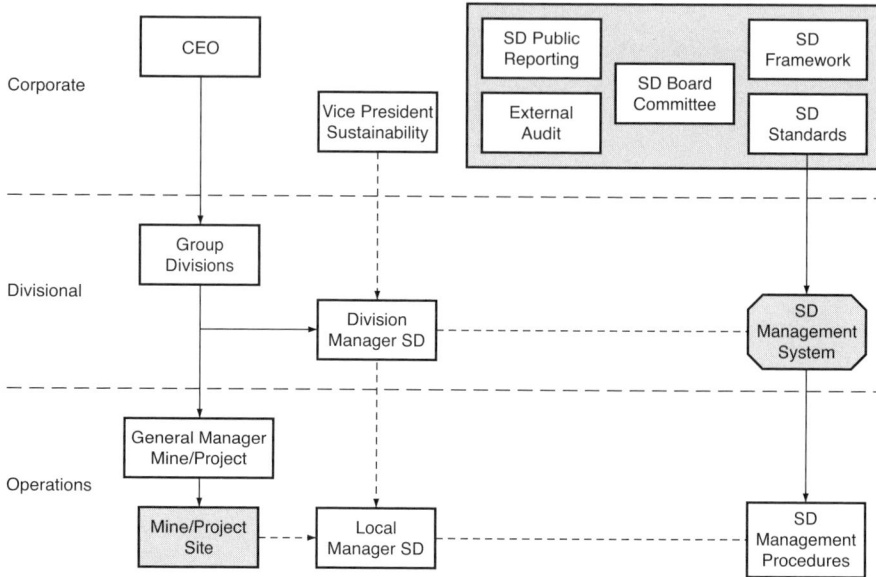

FIGURE 5.7   Structural mode for sustainable management of a large mining company

the model would be conceptually similar, but the divisional level would not exist. The model (Figure 5.7) includes integration mechanisms that are described in the following paragraphs.

The chief executive officer.   The CEO formulates a vision, oversees the implementation of strategic plans, and ensures that the company presents a strong, positive image to stakeholders. Regarding sustainability, the CEO presents and reinforces a motivating vision on sustainability and ethics, and formulates challenging goals.

The board committee on sustainable development.   Typically, the Sustainable Development Committee is composed of a group of three to five directors appointed by the board. Its main role is to assist the board in overseeing the sustainable development plans and performance, specifically the following:

- Sustainability risks (health, safety, environment, communities)
- Compliance with company's sustainability framework
- Evaluation of results in relation to sustainability standards
- Performance and leadership of the management and staff
- Public reporting and external audits on sustainability

The vice president, sustainability.   In most divisionally organized mining companies, the vice president, sustainability, is a staff position reporting to the CEO or COO and is accountable for setting strategy, establishing goals, and integrating sustainability down and across the organization.

This role is critical for the integration of sustainability at operational levels. It has the following functions:

- Leadership, cooperation, and assistance to the divisional heads on sustainability matters
- Leadership and cooperation with HR for the integration of sustainability values in the process of hiring, training, and motivating managers and staff at all levels of the organization
- Leadership of a team of divisional and local sustainability managers and staff

- Accountability for the development and implementation of the Sustainable Development Information Management System, external reporting, and independent assurance on sustainability matters
- Development and maintenance of an external and internal network for information and professional contacts on sustainable development matters

Divisional managers, sustainability.   Typically, the divisional manager is a staff member of the head of the division or business unit but maintains a functional dependence on the vice president, sustainability. He or she has overall accountability for sustainable development matters within a division and has a leadership and assistance role for the implementation and utilization of the SD Information Management System, sustainability audits, and SD reporting. In small, functionally organized companies, this position does not exist.

Site managers, sustainability.   The site manager is a staff position reporting to the general mine manager or project manager and has overall accountability for the management of sustainable development at the operational level, completing the integration process. In this position, leadership, motivation, and the ability to negotiate with site managers and supervisors are important.

Integration Plans and Systems

The implementation of sustainability plans at operational level requires the use of adequate IT-supported information management systems. Other important integration mechanisms are public reporting and independent assurance (Figure 5.7). Standards, management systems, and procedures are described in the following paragraphs.

Sustainable development standards.   The SD standards are a set of management standards issued to interpret and support the sustainable development framework and policy on sustainability at all management levels in the company. Also, the SD standards act to formalize the conceptual base for the design and implementation of specific SD management systems driving the continual improvement objectives of the company.

The SD standards refer to all stages of management (planning, leadership, and implementation and control). For example, Xstrata plc[16] has an SD system based on the following 17 standards.

- Policy and planning standards:
  - Leadership, strategy, and accountability
  - Planning and resources
  - Risk and change management
  - Legal compliance and document control
- Leadership and implementation standards:
  - Behavior, awareness, and competency
  - Communication and engagement
  - Catastrophic hazards
  - Operational integrity
  - Health and occupational hygiene
  - Environment, biodiversity, and landscape functions
  - Contractors, suppliers, and partners
  - Social and community engagement
  - Life-cycle management—projects and operations
  - Product stewardship

   – Incident management

   – Emergencies, crises, and business continuity

·   Control and corrective action standards:

   – Monitoring and review

In most cases, mining companies' SD frameworks and standards are designed to be consistent with international environmental H&S standards (ISO 14001,[17] OHSAS 18001,[18] etc.) and international SD frameworks (ICMM, UN Global Compact, etc.).

An SD standard should include the "intent" or management area that addresses the objectives or expectations and also refer to the management procedures for the control of results. For example, Table 5.2 shows an SD standard on health and occupational hygiene from Xstrata plc.[19]

Management system.   SD management systems are important tools supporting the integration of the SD management processes (planning, implementation, control, and reporting) at operational levels. An SD management system has several main objectives:

·   To support the planning, control, and reporting processes required for the implementation of the objectives and expectations derived from the SD standards for each individual mining operation;

·   To establish—for each SD objective—the measurement units and methodology, baseline values, performance targets, and management responsibility;

·   Using the "balanced scorecard" system, or other evaluation framework, to integrate system input data values into overall sustainability performance; and

·   To produce a hierarchical reporting system to report on performance to management at all levels and to allow for analysis and decision making.

Public Reporting Standards

Publicly listed mining companies are required by stock exchange commissions to issue an independently assured annual report on sustainability, but many nonlisted companies also consider independently assured public reporting essential for maintaining a positive public reputation.[20]

The annual sustainability reports of many mining companies are prepared in compliance with global reporting standards such as the Global Reporting Initiative (GRI),[21] the Mining Association of Canada (Towards Sustainable Mining), and others.

The use of a recognized reporting standard facilitates the inclusion of the company in sustainability indices (Dow Jones Sustainability World Index, FTSE4Good Index, etc.)—indices of companies meeting globally recognized corporate responsibility standards.

Among the public reporting frameworks on sustainability, the GRI is probably the world's most widely used. GRI is a public reporting framework providing guidance on how organizations can disclose their sustainability performance.

The GRI Framework (G3) is accessible online and is structured in five reporting sections:

How to Report

1.   Principle and guidelines

2.   Protocols (detailed reporting guidance)

What to Report

3.   Standard disclosures

4.   Sector supplements

5.   National annexes (unique country-level information)

TABLE 5.2   Xstrata plc sustainability management standard for health and occupational hygiene

Standard 9: Health and Occupational Hygiene
Intent: Systems, plans, and programs are established and implemented to identify, analyze, and evaluate, so far as reasonably practicable, and enhance the health and well-being of workers, contractors, and visitors, and provide a workplace that is free from significant occupational health and hygiene hazards. Public health risks affecting the local communities associated with our operations (including HIV, AIDS, and malaria) and initiatives are implemented to mitigate these in partnership with appropriate stakeholders.

	Requirements and Expectations

1	Occupational health assessment and surveillance systems and plans are established and implemented that include
	  i. Pre-employment health assessments that establish a baseline position and assess fitness for work;
	  ii. Regular health surveillance appropriate to the level of exposure;
	  iii. Communication of the results of health assessments and surveillance with due regard for confidentiality.

2	Occupational health and hygiene systems, plans, programs, and controls are established and implemented to
	  i. Identify occupational health and hygiene hazards including those associated with all work environments;
	  ii. Assess employee and contractor exposure to hazards with reference to internationally recognized monitoring;
	  iii. Eliminate, as far as reasonably practicable, or otherwise minimize exposure to hazards;
	  iv. Provide personal protective equipment where other controls do not effectively reduce the risks;
	  v. Drive continuous improvements in occupational health and hygiene.

3	An effective illness and injury management system is implemented that
	  i. Considers the location and nature of the operation, site, or project's ability to provide effective medical and first aid;
	  ii. Considers the physiological, psychological, and sociological elements of injury or illness;
	  iii. Ensures that health care is administered under the guidance of properly qualified professionals;
	  iv. Ensures that rehabilitation systems and procedures promote early intervention to assist optimum recovery from work injuries or illnesses and aids return to work;
	  v. Takes all reasonably practicable steps to assist or provide rehabilitation and suitable duties to employees who are injured at work;
	  vi. Maintains the injured person's job position for as long as is reasonably practicable or as specified under relevant laws.

4	Systems exist to identify significant public health risks and to assess the potential or actual impact on Xstrata's contractors and local communities associated with the operation, such as HIV and AIDS, tuberculosis, and malaria. Where high or rapidly growing prevalence of HIV and AIDS is identified as a risk, confidential, voluntary HIV testing and treatment programs are implemented in the workplace, complemented by education and awareness initiatives to provide community access to testing and treatment for HIV and AIDS to be undertaken where relevant, and with appropriate stakeholders.
5	The health and well-being of the workforce shall be promoted through access to health information and programs.

Source: Data from Xstrata plc

GRI has published a Mining and Metals Sector Supplement that was developed in cooperation with ICMM. This supplement is structured into five sections:

1. Context for Sustainability Reporting in the Mining and Metals Sector

2. Aspects to Be Reported Through Narrative Descriptions

3. New Economic Indicators for the Mining and Metals Sector

4. New Environmental Indicators for the Mining and Metals Sector

5. New Social Indicators for the Mining and Metals Sector

Independent Assurance

To accomplish its purpose, a sustainability report must be credible and generally accepted by all stakeholders and potential investors. Without independent assurance (audit), credibility is unlikely to be achieved.

Audits are normally performed on an annual basis by the internal audit team under the direction of an independent auditor. The objective is to provide independent assurance that the management practice and performance is in compliance with the SD standards and public reporting of the company.

Assurance practices of mining companies vary widely, but the use of standards such as AA1000 is increasing.[22]

Conclusions

Management is the key challenge with respect to making sustainability work—this is evidenced by the sustainability frameworks proposed by ICMM and other public initiatives relating to business culture, structural integration, continual improvement, and other complex and challenging management processes.

Today, most mining companies' charters reflect a total commitment to the values of sustainable development and business ethics. This is an essential condition for efficient integration of sustainability across the company, but it is not sufficient. Equally important—and much more difficult to achieve—is creating a business culture in which sustainability is a high professional value, so objectives may be implemented through personal commitment rather than compliance. A third condition is an organizational structure provided with specific roles, integration mechanisms, and adequate management systems.

Acknowledgments

Some of the material and ideas presented in this section are derived from information provided by and discussions with Paul Jones, general manager of sustainability for Xstrata; Anthony Macuch, COO of FNX Mining Company; and Ian Wood, vice president of sustainable development for BHP Billiton. The chapter author gratefully acknowledges these individuals' contributions to this discussion.

INTEGRATION OF SUSTAINABLE DEVELOPMENT INTO MINING OPERATIONS*

Any consideration of the integration of sustainable development strategy into corporate structure and the proximal cascading of this down to operational levels must, out of necessity, start by considering the definition of sustainable development. Definitions and deliberations related to quantifying and qualifying sustainable development are provided in other chapters of this book and are not repeated in depth here. If it is assumed that the overwhelming majority of mining and mineral processing companies worldwide are at least broadly aware of sustainable development, or are familiar with the concept, attempts can be made to tie down a corporate definition of it.

Sustainable development models adopted and practiced by mining companies worldwide are typically based on triple bottom line accounting or the capitals models, or variations thereof. Again, these models and the foundations for them are discussed elsewhere in this book. In these authors' experience, *sustainable development* as a term, a concept, has often evoked negativity and mild derision among some senior corporate players when presented as a fundamental component of their organization's long-term operational vision. It is seen as merely a public relations exercise to appease specific stakeholders without any real benefit. In addition, its origin within the ecological movement has tainted it with overly green or *ecocentric* hues, with many executives believing it to be merely another manifestation of environmental pressure impacting profits and economic growth.

---

*   This section was written by B. Johnson and D. Limpitlaw..

TABLE 5.3   Stages on the pathway toward sustainable development

Stage	Action	Description
1	Prepare	Prepare minimal sustainable development efforts, while assessing the issue, what other companies are doing, and potential opportunities.
2	Commit	Commit to moving forward in addressing sustainable development and choose a strategic direction for their sustainable development actions.
3	Implement	Launch programs consistent with their sustainable development approach.
4	Integrate	Make sustainable development part of everyday business processes.
5	Champion	Act as a leader and champion for others within industry on sustainable development.

Source: Global Environmental Management Initiative 2004

But, in reality and in its proper form, sustainable development transcends this. It is business, just done differently (i.e., not business as usual). And it is this doing business differently that is the essence of the sustainable management philosophy. Mining operations today have little choice but to operate differently—their long-term survival depends on it. Those mining companies that have made the business case link between sustainable development and short-, medium-, and long-term corporate strategy are the ones that will continue to remain financially sustainable and viable and will ensure the renewal of their respective social licenses to operate.

The current response from mining operations ranges from opposition to the idea of sustainable development to transformation of operations that respond effectively to the need for, and opportunities presented by, sustainable development.[23] The latter response typically occurs in stages, as indicated in Table 5.3.[24]

Challenge of Integrating Sustainability Principles into Business Operations

This challenge is considered in two ways: (1) integration into the mining corporate and operational management cycle, and (2) integration across the product value chain.

Integration Into the Management Cycle

In the past 10 years, the primary response from business to the challenge of sustainable development has been on three levels:

- To develop visions, policies, and mission statements reflecting commitments to sustainable development;

- To implement certifiable environmental management systems (e.g., ISO 14001); and

- To report on their financial, environmental, and social performance in safety, health, and environment or sustainability reports (i.e., triple bottom line reporting).[25]

These responses are important. However, in many instances, they are undertaken in isolation from an overarching sustainability strategy and/or framework for translating commitments into practice. Their effectiveness is therefore limited by poor integration into the full business management cycle. Going further, operational implementation is even harder if not held together by an overarching strategy.

Integration Across the Product Value Chain

Mining operations need to ensure that sustainability principles and actions are integrated across the full spectrum of the business and the product (or service) value chain.[26] The management of environmental and social impacts cannot be fully addressed by the actions of the Safety, Health, and Environment department alone. In fact, one of the key constraints of sustainable development that managers have raised (see "Operational Examples of On-the-Ground Sustainable Development Implementation" later in this section) is that these departments often operate as silos, which hinders the effective implementation of sustainable development.

Challenge of Incorporating Biophysical, Social, and Economic Elements into Strategy and Operations

Responding to the sustainable development challenge increasingly requires that business look beyond the short-term financial profit motive and contribute instead to meeting the following "triple bottom line" objectives:[27]

- Economic prosperity and continuity for the business and its shareholders and stakeholders;
- Social well-being and equity for both employees and affected communities; and
- Environmental (biophysical) protection and resource conservation, both locally and globally.

Challenge of Addressing External Political, Environmental, and Socioeconomic Drivers, Risks, and Uncertainties

Looking beyond the elements of sustainable development that lie within the realm of direct control and management of a mining operation, a number of external factors may influence how an operation is able to contribute to sustainable development. Some of these facilitate the contribution by business to sustainable development, whereas others may be limiting. It is important for companies to identify and recognize both these positive and negative drivers when developing a framework and strategy for integrating sustainability principles into their operations.[28]

There are a variety of examples of external drivers:

- International and national conventions and policies supporting sustainable development
- Initiatives to build a free and inclusive international market that is not distorted by subsidies, tariffs, or nontariff barriers
- Global trends and risks, for example, the effects of climate change, declining oil supplies, or the increase in uranium demand due to the increasing profile of nuclear power

Challenge of Creating Innovative Solutions

The World Business Council for Sustainable Development (WBCSD)[29] has developed and promoted eco-efficiency as a management philosophy that encourages business to search for environmental improvements that yield parallel economic benefits. The WBCSD's guidebook, *Eco-Efficiency: Creating More Value with Less Impact*, presents the argument that it makes good business sense to be more eco-efficient, as companies can thereby provide more value from lower inputs of materials and energy and with reduced emissions. The challenge to mine operations is to improve their eco-efficiency by, for example, reducing material consumption, reducing energy intensity, reducing and managing impacts, enhancing recycling, and maximizing use of renewable resources.[30]

These challenges apply both to large multinational companies and to small and medium-sized enterprises, operating in emerging economies and industrialized nations alike. It must, however, be borne in mind that eco-efficiency is not sufficient by itself because it integrates only two of sustainable development's three dimensions (i.e., economics and ecology). The third element, social equity and enhancement of quality of life, must also be addressed simultaneously. The challenge lies in using innovation as a way to positively influence all three elements of sustainable development, rather than as a driver for promoting increased consumption.

Challenge of Deciding on the Appropriate Level of Commitment to Sustainable Development

Mining operations need to determine the organization's level of commitment and management response to sustainable development appropriate for its situation and reevaluate them frequently. This decision is influenced by several factors, such as the values held by the company's leadership (see the Sasol example later in this chapter), owners, and investors; the risk appetite of the company; the financial strength of the company; and the degree of internal and external pressure the company experiences regarding its sustainability implications. The challenge becomes apparent when trying to balance and respond to multiple, often conflicting, expectations of the company's role and purpose in society.

Making Sustainable Development Operational: The Main Integration Problems

One of the primary objectives of this section is to provide a "coal face" perspective on some of the challenges of trying to operationalize sustainable development in the South African environment and then to consider its relevance within the global mining theater. Research and assessment[31] called upon for the purposes of this book (focused on selected southern African mining operations and those responsible for sustainable development on the ground) revealed interesting perspectives on operational sustainable development. Key to understanding how to realize sustainable development operationally is first to understand the constraints mine managers face when implementing sustainable development on a day-to-day basis.

Research undertaken for this book focused on discussions undertaken with a number of high-level managers responsible for sustainable development, either on their mine specifically or across operations (corporate SD managers). Responses to the question: "What are the biggest challenges faced when linking corporate sustainable development strategy and on-the-ground implementation?" elicited an interesting picture:

- Different personnel vary in their (operational) SD approaches. A mine in Namibia is currently undertaking three key upgrade projects involving, among other things, the construction of a new ore sorter and acid production plant. Each of the project managers responsible for these projects approaches sustainable development very differently (e.g., the generation of dust from the new ore sorter plant and the management thereof). This is often problematic because a consistent approach to sustainable development within and across the company at a corporate and organizational level is a cornerstone of its strategy.

- Differential application of sustainable development by mine personnel is largely dependent on the age of the staff member and the associated experience of sustainable development. Interestingly, there is often an inverse relationship between age (experience) and the application of sustainable development principles operationally. Typically, a full spectrum of approaches and attitudes toward sustainable development will be encountered in most mining companies. Again, the example of the mine with three new capital projects is illustrative. Out of the three capital appropriation requests that were completed for the three projects, only one incorporated SD principles into the decision-making criteria underpinning the capital request rationale (decision-making matrix). In this example, this project was led by the youngest of the staff, who ensured best practice technology was secured to optimize sustainable development. Anecdotal discussions have indicated that the application of sustainable development principles on the ground by older mine staff often does not come naturally because there is less familiarity with the concept and how to translate it operationally.

- Some companies may become overly comfortable with sustainable development. Linked to this is a trend in some mining companies for disbanding sustainable development awareness training programs and corporate sustainable development setting tools among mine staff. This generally occurs because sustainable development is deemed to have been fully incorporated operationally. The feeling among sustainable development operational staff is that this action may be premature and there is a continuing need for sustainable development awareness and training programs on the ground.

- Commitment to sustainable development both upstream and downstream is critical. Sustainable development buy-in is essential across the organizational supply chain in order for sustainable development to work operationally. For instance, if components of the (new project) supply chain upstream (e.g., contractors) are not well versed in the implementation of sustainable development in the planning and construction of, for instance, a new acid production plant, sustainable development will not be successfully translated on the ground. Similarly, downstream players (the mine's clients) can equally affect the realization of sustainable development. If, for example, International Organization for Standardization (ISO) criteria are applied by these buyers of a mine's products and upstream supply chain components are not aligned, the mine will not be operating within sustainable development principles.

- The right human capital is crucial to a sustainable development program. Corporate HR departments often display a lack of involvement, participation, and understanding of sustainable development in their support of mining operations. Having "the right people with the right attitudes" is absolutely pivotal to the implementation of sustainable development on the ground. This needs to be taken into account in HR training programs and in recruitment (see Chapter 6) There are further examples of HR's role in sustainable development. As a case study, an ongoing mine expansion program involves the expansion and securing of new accommodations for an increased workforce. Traditionally, HR would probably not get involved in operational aspects like this, but from a sustainable development perspective it is critical because the provision (building) of new facilities can have profound effects on the mine's surrounding social environment and its human capital. If the message received from other (nonoperational) organizational divisions is that sustainable development is not important, then it certainly will not work on the ground.

- Mining and corporate social investment. Corporate social investment and reporting are often the most challenging parts of sustainable development. They are too often seen as public relations exercises, especially if the business case for sustainable development has not been firmly established. For this particular aspect of sustainable development to work, senior management must be committed to it and it must have been developed from deep listening to communities on the ground.

Integrating SD into Strategy and Making It Happen Operationally: Getting Started

The development of a corporate-level vision to include a sustainable development ethos is generally the first step in ensuring that corporate-level sustainable development strategy is established and ultimately translated operationally. While articulating a vision may be regarded as often wooly and slightly nebulous (especially in the context of a book focused on operational sustainable development), it is nonetheless an important starting point. It reflects the values of the

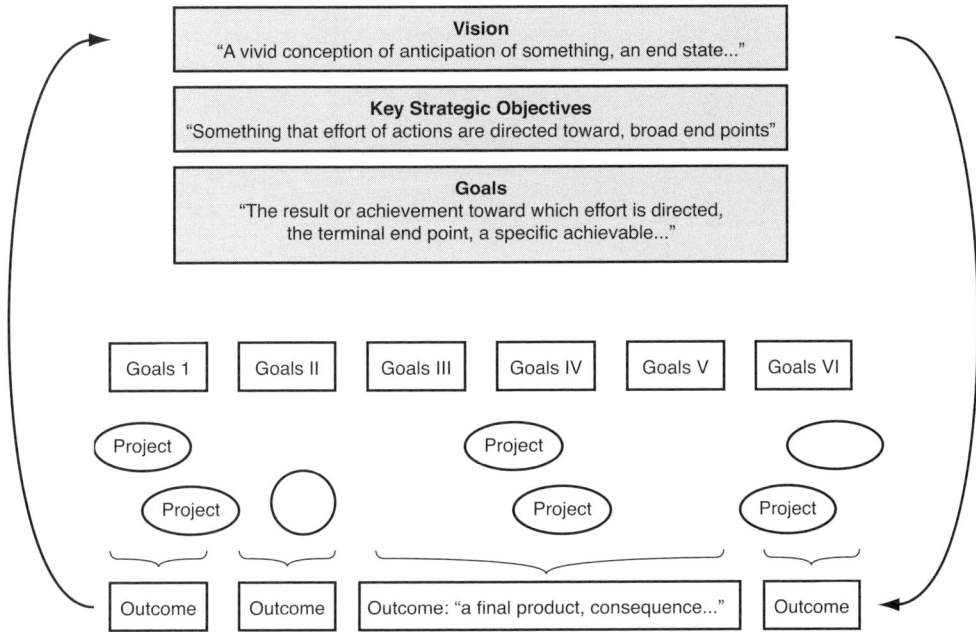

Courtesy of the Council for Scientific & Industrial Research

FIGURE 5.8   Vision-objectives-goals-projects approach used to establish strategy broadly

organization and sets the direction in which the company will move its priorities and values. Figure 5.8[32] illustrates how a vision sets the stage for the development of objectives/goals and hypothetically how this cascades to the operational (project) level.

Mechanics of Incorporating Sustainability into Strategy—A Potential Approach

One specific methodology with the potential for application to mining operations is the Integrating Sustainability Into Strategy (ISIS) process,[33] based on the strategic business planning tool developed by Chantell Ilbury and Clem Sunter.[34,35]

The underlying premise of both the Illbury and Sunter process and the ISIS process is that in order for companies to make informed strategic business decisions, it is necessary first to gain a common understanding of the objectives being pursued and then unpack and understand what influences the achievement of these objectives. The difference between the two processes is that Illbury and Sunter focus on supporting business to operate more effectively in the traditional pursuit of profit maximization, whereas the ISIS process aims to support business in integrating sustainability principles into its operations.

Ideally, the ISIS process is undertaken after a company or sector has defined its high-level commitments to sustainable development. The process can then be run as a scoping exercise to inform the development of a sustainability framework or strategy that outlines how the company or sector will translate these high-level commitments into "on-the-ground" actions. These different stages form part of the initial stages of a management system that promotes continuous improvement of management actions through regular monitoring, reporting, and review. Figure 5.9[36] indicates where the ISIS process fits into the business management cycle.

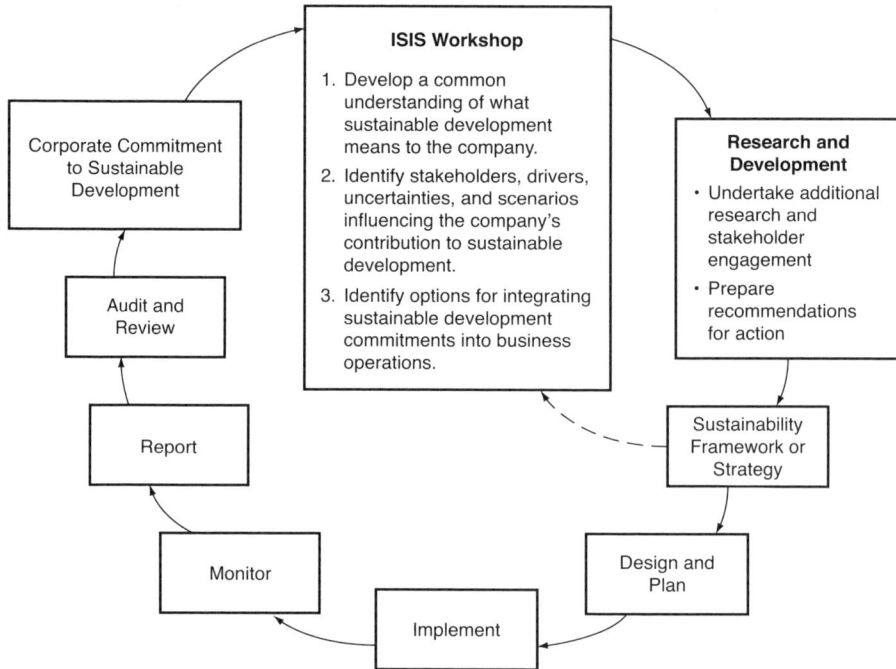

Courtesy of the Council for Scientific & Industrial Research

FIGURE 5.9   Locating the ISIS process in the business management cycle

The ISIS process is implemented as a facilitated, multistakeholder workshop that aims to assist an organization or sector to think holistically about the concept of sustainable development, enhance the generation of ideas regarding implementation, and achieve buy-in delivery on broad-based sustainability objectives. The workshop typically brings together individuals from across different operational units within a company (e.g. human resources, finances, procurement, and environmental management) and/or different stakeholders influencing sustainable development within a particular sector (e.g., businesses, government, nongovernmental organizations).

In the case of external contractors and project managers working on a specific project (e.g., the mine expansion in Namibia described previously), representatives could also be drawn from these groups. This would be useful given the long-term working relationships with mining contractors and mining operations. The four steps in the ISIS process are shown in Figure 5.10[37] and are described in the following paragraphs.

Step 1: Define the Scope of Sustainability for the Company

The development of any sustainability framework, strategy, or report needs to start with the mining operation having a clear understanding of what sustainable development means to the organization and its operations, who the stakeholders are that influence its response to the sustainable development challenge, and the drivers that support or constrain sustainability.

Understanding the meaning of "sustainable development" for the business.   The purpose of this step is to stimulate awareness and understanding among workshop participants of what the generic concept of sustainable development means to the company specifically (i.e., understanding the scope of economic, social, and environmental issues that a company needs to be

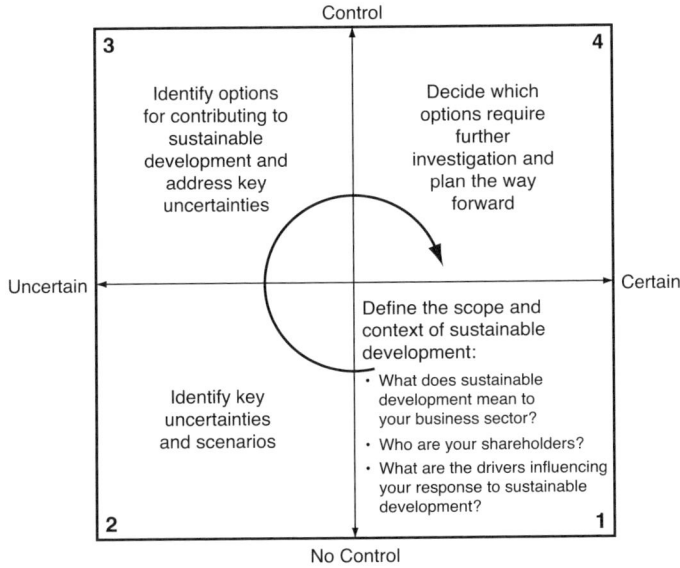

Courtesy of the Council for Scientific & Industrial Research

FIGURE 5.10 The ISIS process, showing the four main steps

responsive to across the product value chain). This would obviously differ for different mining organizations in the specifics but would be very similar at broad generic levels.

Identifying the stakeholders influencing the sustainable development strategy.   The purpose of this step is to identify the range of stakeholders who could influence the company's response to sustainable development. Stakeholders—both internal and external to the company or sector—representing economic, social, and environmental issues should be identified. These stakeholders need to be considered in developing and implementing a sustainability framework, strategy, or report.

Identifying the drivers influencing the business's response to sustainable development.   Drivers are the factors that motivate, support, or constrain the integration of sustainability principles into business strategy. They may be internal to the company (e.g., existing company policies and operating practices) or they could be external to the company and could apply at a range of scales from an international to a local level. Drivers could include the following: legislation, policies, guidelines and standards, initiatives, trends, or changing societal values.

The purpose of this step is to identify proactively the specific external and internal drivers that influence business's response to the sustainable development challenge and that need to be factored into the development of a sustainability framework or strategy.

Step 2: Identify Key Uncertainties and Scenarios
In order to ensure that the sustainability framework or strategy is robust, it is important for a company to identify key uncertainties and future operating scenarios that may influence the implementation and effectiveness of the strategy.

Identifying the key uncertainties that influence the sustainability strategy.   The purpose of this step is to identify the two key uncertainties that have the greatest potential influence on a company's contribution to sustainable development in order to use these as the basis for defining future scenarios.

Defining plausible future scenarios. Scenarios provide an image of possible, plausible futures, reflecting different outcomes to current uncertainties. They serve as a useful tool for considering how current actions may influence the realization of one or another scenario, as well as considering and deciding on what actions to take today, taking into account different possible futures.

The purpose of this step is to define four plausible future scenarios in order to take these different possible futures into account in developing a sustainability framework or strategy. Lochner provides further illustrative examples of scenarios that have been encountered by organizations using the ISIS model.[38]

Step 3: Identify Options

The crux of the sustainability framework and strategy is the identification of a range of possible projects or actions that can be implemented by the business in order to translate its commitments to sustainable development into practical actions, taking into account the outcomes of the preceding steps. The ISIS process provides an opportunity for the company to brainstorm ideas, based on its own experience, as to what can be done to support sustainable development. These are the options that require further investigation to determine their practicality and feasibility.

The purpose of this step is to identify a range of options for supporting sustainable development. This is not to say that all these options should or will be implemented. However, it provides a starting point for further investigation in developing the sustainability framework or strategy. Importantly, options should stretch beyond the boundaries of current actions and support continuous improvement in a company's contribution to sustainable development. A number of key questions would be asked at this stage:

- Where would the company locate itself on the sustainability spectrum?
- What is the range of options for translating sustainable development principles and commitments into actions?

Step 4: Make Decisions and Plan the Way Forward

Typically, this stage of the process would involve reaching agreement among workshop participants on how the outcomes of the workshop should be taken forward and who would be responsible for different actions. Time frames and resources required to implement the decisions taken should also be agreed upon.

The purpose of this step is to identify and agree to the way forward, in particular in terms of follow-up actions and investigations that may be required to inform the sustainability framework, strategy, or reporting process. Several more key questions would be asked at this stage:

- Which of the identified issues, drivers, uncertainties, and options require further investigation?
- How will the outcomes of the workshop be taken forward?

When conducting the ISIS workshop, it is important to identify links through the process—from the issues and drivers to the options identified and the decisions taken.

Operational Examples of On-the-Ground Sustainable Development Implementation

Sustainable development requires integrating environmental and social considerations into technical and economic decision making. Sustainable development issues cannot be addressed using the old paradigm of handouts; difficult and unpopular choices often need to be made.

Practical Examples of Changes in Mindset Required for Sustainable Development

The example of an isolated mining operation located in an arid, rural African setting can be used to illustrate the types of choices involved in sustainable development in mining operations. Mines are substantial consumers of power and water and operators make significant investments in infrastructure to ensure an uninterrupted supply of both to the mine site. Underdevelopment in the surrounding countryside often means that local women have to walk long distances to collect water; in addition, no electrical power is available. There is usually great pressure on the mine operators to install a reticulated water system to supply water to the local community. The first consequence of such water provision may be an immediate increase in the carrying capacity of the area, with a subsequent dramatic increase in population. This increases the level of dependency on the mine; on mine closure, the large settlement that has developed is completely unsustainable and may require involuntary resettlement.

In mature mining areas, new mines may become viable with increases in commodity prices. Newly arrived companies are sometimes tempted (or pressured) into repeating the mistakes of the past. In many places, unsustainable mine villages have been created. Over the years, with closure and/or downscaling of the older mines, these villages have tended to collapse and are dependent on continuous support from mining companies. New mines should ensure that their activities do not create new dependencies that only defer the settlement's collapse to the date of closure of the last mine to open and increase the magnitude of the problem.

Any decision to support local communities must be made while taking broader ecological, social, and economic factors into account and must consider the impact of the removal of mine support. Settlements and infrastructure investments must be made in the light of broader development plans for the region so that these can be handed over to the state or a third party for management upon closure. Where such management is not available, infrastructure should be removed and the site rehabilitated.

Other Responses

Although the ISIS model is one approach to establishing and putting sustainable development principles into operation, variations to this model have been used to good effect locally. South Africa provides a prime environment from which to draw examples of the successful translation of corporate-level sustainable development strategy operationally. As a country with highly developed and technically evolved mining operations in a deprived developmental context (with problems such as unemployment, HIV/AIDS, and water scarcity), the challenges of translating lofty corporate visions and ideals onto the ground are arguably more difficult than elsewhere in the world.

Moving from the hypothetical model described previously, a brief consideration of one of South Africa's leading companies, Sasol,[39] illustrates how the link has been made from integrating sustainable development into the organization's vision and values to the day-to-day operational risks it faces (and the measures it takes to manage them). Sasol is an integrated oil and gas company with substantial chemical interests. It mines coal in South Africa and produces gas in Mozambique, converting these into synthetic fuels and chemicals.

As an example, it is illustrative to look at the organization's sustainable development commitment (vision), which is articulated in this way:

> Sasol commits to sustainable development as defined by the World Commission on Environment and Development. We aim to meet the needs of today's generation without compromising the ability of future generations to do the same. Our operations are conducted with sensitivity towards the economic, social and environmental needs of

stakeholders. The business is a healthy mix of financial prosperity, balanced with environmental stewardship and social responsibility.[40]

Figure 5.11 illustrates the corporate organizational hierarchy in Sasol that sets and executes sustainable development policy on the ground (modified[41]). Sasol's sustainable development management framework covers its international construction, exploration, production, and marketing operations in all countries in which it operates. Globally, the international Sasol group's commitment to sustainable development is managed at corporate level and implemented at business level, with ultimate responsibility residing with the board of directors (see Figure 5.11). Flowing from the executive committees' function are strategic focus areas and long-term (2015) goals. Figure 5.11 provides a perspective on the placement of these components and how they cascade down to on-the-ground actions. To realize their vision, Sasol's managment has set specific goals to be achieved for each of the six priority focus areas (Table 5.4).[42] Progress toward the attainment of these is measured by, among other things, indicator reporting according to GRI guidelines (discussed earlier in this chapter).

Resolving the first focus area and goal, it becomes clearer as to how exactly the organization is translating sustainable development from this corporate level to on-the-ground actions. A key priority for Sasol in terms of sustainable development and health (and indeed South Africa as a whole) is the question of HIV/AIDS. To this end and in order to meet its first strategic focus area listed in Table 5.4, Sasol developed its integrated Sasol HIV/AIDS Response Programme (SHARP). Launched in September 2002, the program focuses on reducing the impact of HIV and AIDS at an operational level. A key objective is to extend the quality of life of infected employees by providing managed health care. Businesses, trade unions, community representatives, and independent experts all contributed to the design of SHARP.

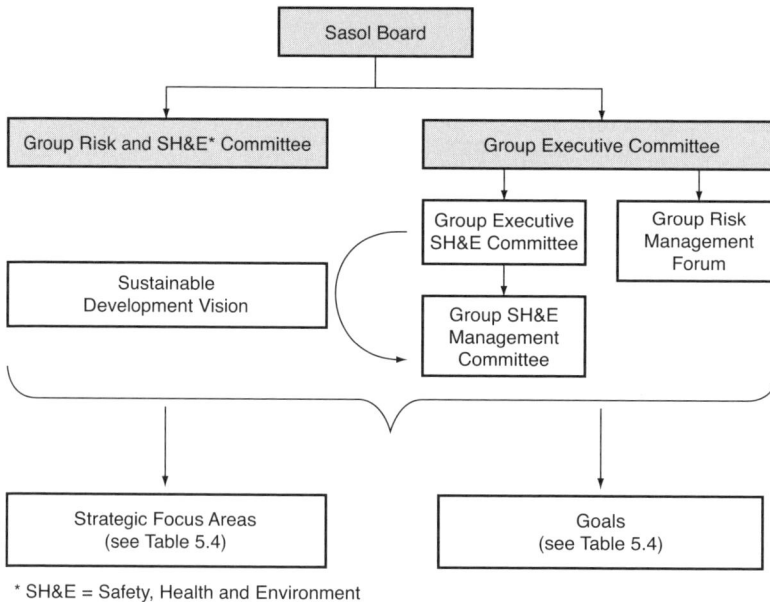

* SH&E = Safety, Health and Environment

Courtesy of Sasol

FIGURE 5.11 Sasol's organizational structure and its links to developing operational SD

TABLE 5.4   Sasol's strategic focus areas and goals

Strategic Focus Areas	Goals for 2015
Safety and health	Zero fatalities and zero injuries for employees and contractors.
Performance and technical	Meet all group and business unit Safety, Health and Environment (SH&E) targets.
Climate change and greenhouse gases	Reduce greenhouse gas emissions by 10% per ton of product and implement carbon dioxide capture and storage initiatives according to plan.
Proactive legal compliance	Ensure full and continuous legal compliance on a continuous basis and meet the SH&E minimum requirements globally.
Governance and assurance	Full SH&E and sustainable development assurance provided to the board.
Stakeholder relations	Stakeholders are satisfied with Sasol's SH&E and sustainable development performance, and communities value Sasol's presence.

Source: Sasol

Specifically, SHARP is coordinated at group level by a steering committee and chaired by a group executive committee member. At appropriate operational levels, this initiative:

- Implements measures to eliminate discrimination on the basis of a person's HIV/AIDS status;
- Encourages a behavioral change through HIV/AIDS education and awareness programs;
- Provides access to free and confidential voluntary counseling and testing (VCT);
- Provides treatment of opportunistic illnesses such as tuberculosis, as well as treatment of sexually transmitted infections;
- Provides managed health care, including anti-retroviral treatment for employees; and
- Reduces and manages the total cost to Sasol of the business impact and response to HIV/AIDS.

In its most recent (2007) sustainable development report, Sasol reports a number of measurable outcomes from this program, which can be viewed as SD being effectively realized on the ground:

- During the 2005–2006 year, 83% of employees underwent education and VCT. The HIV incidence rate was 7%.
- A 3-hour training course has been provided for about 5,000 service-station employees, and each service station has been provided with information on their nearest available public health resources for testing, counseling, and treatment.
- Through its corporate social investment department, Sasol has partnered with numerous community-based organizations to increase awareness and improve access to care in the communities in which the organization operates. SHARP was recognized as one of the top five workplace programs in the country at the Khomanani Excellence Awards[43] in 2005.

Conclusions

This section has provided a short, high-level perspective of sustainable development and its incorporation into corporate strategy as it relates to mining operations. Critically, it has also attempted to make the link between sustainable development at a corporate level being effectively translated and implemented on the ground in actual mining operations. There are many examples of this realization and this section is certainly not exhaustive—any longer discourse would have been beyond the scope of this book. Similarly, a detailed discussion on sustainable development strategy and its link to corporate and operational strategy was not possible, and the authors have relied on subjective perspectives based on professional experience. Despite this limitation, the section presents a useful view on some of the challenges and nuances of making sustainable development work in the context of mining in a developing country.

Measures Required at Operational Level

It is critical that sustainable development policy formulated at corporate level and decisions made during the design phase of the project are properly implemented. To ensure this, a detailed environmental and social baseline that informs an environmental and social management plan is required. This management plan must be implemented under the supervision of a resident sustainable development manager, and social and community affairs should be managed by an experienced staff member in a dedicated senior post. The senior site manager must be directly accountable for the implementation, and it should be one of his or her key performance areas.

Staying on Track—Monitoring and Auditing

The mining company, especially the people responsible for the sustainable development policy and design decisions, should receive an annual audit report compiled by the sustainable development manager and this should be verified every 2 years by an external audit. The audit should measure progress against a range of indicators, including rehabilitation targets, effluent and emissions objectives, and the success of community development projects. Detailed monitoring programs are required to enable such auditing.

Staying in Touch with Stakeholders—Reporting

Sustainable development is largely about consensus. Development that is sustainable in the long term can only be achieved if it is based on extensive consultation and takes views from divergent stakeholders into account. To ensure that these views influence the project, extensive public participation is required. Such participation must be informed by a comprehensive stakeholder consultation plan that includes information dissemination, grievance mechanisms, and participative monitoring and reporting.

Acknowledgments

The development of material and ideas presented in this section benefited from discussions with R. Schneeweiss (Rossing Uranium), Paul Warner and Conri Moolman (BHP Billiton), Professor May Hermanus (University of the Witwatersrand), and Elize Swart (Department of Minerals and Energy). The contributions of Sasol and specifically Stiaan Waandrag are also gratefully acknowledged. Colleagues at the Council for Scientific and Industrial Research in South Africa, including Paul Lochner and Mike Burns, are duly acknowledged and appreciated.

MANAGING FOR STAKEHOLDERS' EXPECTATIONS: THE SEVEN THEMES OF SUSTAINABILITY*

Many modern mining companies have clear visions and goals with respect to the application of sustainability principles throughout their corporate structures. The implementation of these visions and goals is more difficult to accomplish at all levels of a corporation, especially at the level of operations and maintenance management. Daily pressures associated with operations and maintenance management can blur the strategic corporate vision and goals. Consistent

---

* This section was written by D. Van Zyl.

application of sustainability concepts must be an expectation of all operations personnel. One approach to achieving this is to provide simple tools to the operators that can be used for making and evaluating all project decisions. This section presents such a potential tool.

Any approach to sustainable mine management relies on the identification of sustainability expectations expressed by local stakeholders. Community relations personnel at the corporate and mine level must identify these expectations in their engagement with stakeholders. Local sustainability expectations must be combined with those identified in corporate guidelines to operations. This total set of sustainability expectations must be communicated to the whole workforce at a mine and integrated into the daily management of the mine. The seven themes described in this section provide a powerful approach to integrating sustainability into all designs and decisions.

Seven Questions to Sustainability

As part of the North American regional process of the internationally funded Mining, Minerals and Sustainable Development (MMSD) organization, a practical approach was developed to assess how a mining/mineral project or operation contributes to sustainability. Specifically, the following objectives were established for the activity:[44]

- Develop a set of practical principles, criteria, and/or indicators that could be used to guide or test mining projects throughout their life cycle in terms of their compatibility with concepts of sustainability; and

- Suggest approaches or strategies for effectively implementing such a test/guideline.

A multidisciplinary team participated in this MMSD activity and the outcome reflects this broad input, producing the seven questions key to sustainability:[45]

1. Engagement: Are processes of engagement in place and working effectively?

2. People: Will the project/operation lead directly or indirectly to maintaining or improving people's well-being?

3. Environment: Will the project/operation lead directly or indirectly to maintaining or improving the biophysical systems?

4. Economy: Will the project/operation contribute to the long-term viability of the local and regional economy?

5. Traditional and nonmarket activities: Will the project/operation contribute to the long-term viability of traditional and nonmarket activities in the implicated community and region?

6. Institutional arrangements and governance: Are rules, incentives, programs, and capacities in place to address project or operational consequences?

7. Synthesis and continuous learning: Does a full synthesis show that the net result will be positive or negative in the long term, and will there be periodic reassessments?

The original study recommends that a formal application of the seven questions approach include project-specific objectives, indicators, and metrics under each of the questions. It also suggests including the potential uses of this assessment template:

- Early project appraisal related to assessment for acquisition of a potential mining property and its feasibility for development;

- Overall project planning and identification of who should be involved and when;

- Thorough identification of risk as a basis for financing and insuring;

- Guidance for processes of licensing and approval;
- Internal corporate reviews of performance during operation;
- Guidance for corporate reporting to shareholders and others; and
- Guidance for external review of projects.

Hodge and Van Zyl[46] describe a number of case studies where the proposed template has been applied.

Seven Themes of Sustainability

Van Zyl et al[47] presents the development of an evaluation process based on the seven themes that form the basis for the seven questions. These seven themes are very useful reminders (a checklist) of the issues that must be considered in all decisions and activities. The ultimate goal is for the mine to contribute to sustainability at a local and regional level.

The seven themes include the four major items that are commonly referred to in all sustainable development considerations: community, environment, economics, and governance. However, it also highlights other topics that are very significant in a mining context, namely traditional and nonmarket activities, as well as synthesis and continuous learning.

The following paragraphs describe some of the aspects associated with each of the themes. It is essential that local issues drive this process rather than a set of generic statements developed by corporate personnel. The discourse during the development of local issues associated with each theme is part of the overall process of considering the contributions of a specific project toward sustainability.

Engagement

Engagement with the stakeholders of a project is one of the most important activities that must be undertaken by all mines. This process, which includes the identification of stakeholders and active participation in discussions, can lead to better implementation of local expectations in the project design, development, operations, and closure. Engagement is not a "one-way street" public relations exercise; rather, it is the willingness to listen and respond with actions based on the expectations. It can result in some loss of the power of independent decision making, but the outcomes will be better appreciated and accepted. It is important to realize that stakeholders will change as the project develops and that the expectations for these stakeholders will also change. Engagement is never a static process. Ongoing communications, reporting, verification, and other activities are necessary.

People

This includes all the people coming into contact with the project, whether directly or indirectly. Workers' health and safety are regulated in most jurisdictions; however, it is important to go beyond just regulatory compliance. It is always better to be proactive than reactive to all issues affecting nearby communities, such as wind-blown dust from a tailings impoundment or potential impacts on the availability or the quality of drinking water. The Millennium Ecosystem Assessment[48] listed the following components of human well-being: security, basic material for good life, health, good social relations, and freedom of choice and actions. It further identifies the role of ecosystem services as supporting (e.g., nutrient cycling), provisioning (e.g., food), regulating (e.g., climate), and cultural (e.g., recreation). These aspects are part of all human interaction with the earth, including mining activities.

Environment

All managers of mining operations are aware of the importance of environmental protection and stewardship. Extensive regulatory regimes have been developed to protect the environment affected by mining operations. Ongoing monitoring and the consistent reporting of the results to regulatory agencies is an important activity at all mines. Going beyond the regulatory requirements by implementing concurrent reclamation and other activities meant to enhance the well-being of the environment should be on the radar of mining operations.[49] Sometimes this includes a better understanding of the local ecosystem by funding research.

Economy

All mining projects must make money and this is essential in the positive contribution of the mine to local and regional sustainability. The economic successes of a mine are shared through employment, local taxes, and so forth. Ongoing operational reviews to improve efficiencies at a mine are a big part of this theme. Many mining companies are also regular contributors to local activities such as schools, libraries, and universities. Contributions to sustainable causes must be a consideration; multi-generational thinking is an essential part of sustainable development.

Traditional and Nonmarket Activities

Understanding the traditions and customs of local indigenous people must take high priority at all mines. Having access to certain plants or places at a mine site may be essential to local indigenous people. There are also many other nonmarket issues associated with mines, such as volunteerism associated with local fire departments, emergency medical activities, churches, or other social activities, such as search and rescue, and so forth. Mining operations are usually very aware of the need to participate in such activities.

Institutional Arrangements and Governance

Mining companies operate in well-regulated environments and regulations for land ownership, corporate financial management and reporting, taxes, labor relations, safety, and so forth are typically well developed in most jurisdictions. Regulatory compliance is the first order of business for all companies. However, many companies now have corporate policies and requirements for sustainable development, environment, health and safety, and many other aspects. Many of these policies go well beyond the governmental and regulatory frameworks.

Synthesis and Continuous Learning

Ongoing review and continuous improvement is part of the culture of most mining operations. Regular synthesis of all aspects of the sustainability policies and applications at a mine site should be done to review whether the contribution to people and the environment will be net positive in the long term. This theme also includes the ongoing review of project and activity alternatives to improve operational efficiencies and improve overall project performance with respect to the six other themes.

The seven themes should be implemented in the management of mines as a tool to ensure that an integrated approach is followed with respect to the sustainable management of projects and activities. Using the seven themes as a project or decision evaluation template can provide a useful checklist for daily decision making. Broad management considerations can be identified for each of the themes, such as those listed in Table 5.5; however, it is essential that project-specific management considerations be developed for each activity, making sure that local sustainability expectations are addressed.

TABLE 5.5   Management considerations associated with the seven themes of sustainability

Theme	Management Considerations
Engagement	Involving stakeholders in the design, operations, and closure of mine waste management facilities; acknowledging the importance of a license to operate
People	Worker and surrounding population health, training of personnel, implementation of modern management approaches that value employees
Environment	Site-specific ecological information, evaluation of environmental stress as a result of the mine activities
Economy	Project and life-cycle costs, contributions to the local and regional economies
Traditional and nonmarket activities	Knowledge of traditional (indigenous) activities and customs, protection and access to significant traditional resources
Institutional arrangements and governance	Compliance with regulatory requirements, active participation in ongoing regulatory development, compliance with corporate policies, participation in voluntary regulations (e.g., the Cyanide Code)
Synthesis and continuous learning	Consideration of project/facility alternatives, a commitment to ongoing reviews and improvement

ENVIRONMENTAL MANAGEMENT SYSTEM: A SUSTAINABLE MANAGEMENT TOOL*

An environmental management system (EMS) is a standard methodology for the integration of sustainable environmental management throughout the corporate structure. EMS includes the organizational structure, planning activities, resources, and procedures for the implementation of the environmental policy as an integral part of the management process. It is a useful tool to implement in order to comply with legislation, address stakeholder pressure, improve corporate image, and raise awareness of environmental issues.

EMS is a problem-identification and problem-solving tool, based on the concept of continual improvement. It can be implemented in an organization in many different ways, depending on the sector of activity and the needs perceived by management. In particular, standards for EMS have been developed by the ISO and by the European Commission Eco-Management and Audit Scheme (EMAS).

This section describes the complex management process and practices that lead to the implementation of a certified EMS, indicating the advantages of implementation during the pre-production stage. It also shows what the implementation process of a certified EMS process consists of and its benefits in terms of improvement in operational quality and efficiency, public image, involvement of employees, administration, customers, suppliers, the media, neighbors, and other interested parties.

The final goal is to enable the information provided in this section to assist in the decision to implement an EMS in the first stages of a mining project.

Why Implement an EMS?

There are two main reasons a company decides to implement an EMS:

- As a management tool aiming to achieve sustainable management of environmental issues throughout the organization; and

- An interest in possessing an external certification of the EMS under an international standard (for example, ISO 14001), either for commercial or other reasons.

---

* This section was written by P. Cosmen.

TABLE 5.6   Advantages of environmental management systems

Improved company image:
- Improved company reputation and public image
- Competitive advantages of companies having EMS over those that do not
- Greater awareness and involvement of workers in achieving environmental goals
- Supportive management in decision making where changes are required
- Increased confidence of legislators, investors, and insurers

Improved environmental management and control:
- Reduced environmental risk and facilitation of preparedness
- Greater control of raw material and energy use
- Reduced risk of accidents
- Reduced energy consumption, water use, emissions, and operating costs
- Proper management of waste generated by the company
- Improved environmental action and reduction of impacts
- Identification of weak points in the system

Improved management of permitting and legal issues:
- Obtaining permits, authorizations, and subsidies made easier by improved level of compliance with legal environmental requirements
- Greater legal security: familiarity with applicable environmental law and upcoming environmental regulations
- Reduced risks, thus making insurance coverage easier and more economical

The guidelines provided here are mainly focused on mining companies that really want to implement an EMS because they wish to improve their sustainability performance. In addition, implementing an EMS may often be justified on the basis of improved operational efficiency or financial and legal considerations. The actual benefits that may be obtained are widely variable and dependent on specific circumstances. Table 5.6 shows the main potential advantages of an EMS.

However, an EMS needs dedicated staff and material resources for its implementation and operation and requires operating staff to comply with standard procedures. When management culture lacks motivation toward sustainability, EMS may be considered as distracting and time-consuming.

Costs and Savings

To estimate the costs and savings derived from the implementation of a certified EMS, the following must be taken into account:

- Costs of external consulting to support the system's implementation;
- Cost of external certification (e.g., ISO 14001); and
- Costs associated with company staff required to manage the EMS.

On the other hand, an EMS is expected to help improve operational efficiency, mainly from the following perspectives:

- It is more effective to prevent pollution than manage it after it has been generated.
- It is better to do something correctly the first time than fix it later.
- It is cheaper to prevent a spill in the first place than clean it afterward.

An EMS may also generate direct cost savings through these considerations:

- Optimization of water and energy use, sale of by-products, and so forth;
- Public aid for environmental actions;
- Reduced industrial risk and, as a result, low cost of insurance;
- Advantages in credit applications;
- Prevention/reduction of legal actions, fines, legal costs, and third-party liability; and
- Government subsidies that can be obtained for investment in an EMS.

When to Implement the EMS

A company may be quite certain of the need for an EMS from the preliminary feasibility stages of the mining project life cycle. In this case, EMS planning and design can and should be undertaken as part of the feasibility and construction stages.

Planning an EMS during the preproduction stages may be advantageous because it will help integrate many environmental aspects into the project engineering. By doing so, preventive measures can be adopted in many cases, which will be more efficient and less costly than subsequent corrective measures. Moreover, management will establish a commitment to environmental behavior, define environmental policy, and begin to define the procedures to be integrated into the EMS.

Abstaining from implementing an EMS during the construction phase may incur several disadvantages associated with the highly changeable character of any construction stage, where the short-term participation of numerous contractors and subcontractors will make them unwilling to accept full environmental commitment.

How to Implement an EMS

Once the decision has been taken to implement an EMS in accordance with an international standard, it is important to determine the baseline conditions for the organization because this will greatly influence the process:

- The stage of the project (feasibility study, permitting/engineering stage, construction, or already in operation);
- The company structure and, specifically, whether the company has an environmental management department with adequate resources and expertise; and
- The previous experience with an EMS that may be internally available within the company, the business unit, or the corporate structure.

General Starting Conditions

The way in which to implement an EMS will be different depending on previous circumstances. For example, in cases in which a management system of any kind exists, it will only be necessary to adapt the existing system to comply with the standard under which the EMS is to be implemented. This is generally much easier than starting from zero.

When an EMS is to be implemented during the preproduction stages, it will normally be necessary to start from zero unless internal expertise is available within the corporation.

Key Aspects to Be Defined for Implementation of an EMS

The following aspects will have to be defined for planning how the implementation process is to be executed:

- Implementation phases and calendar;
- Necessary resources (internal and/or external);
- System scope (the entire company, a production center or an area);
- Standard to be employed (international standard such as ISO 14001, European standard such as EMAS);
- System documentation model;
- Level of environmental behavior to be implemented; and
- The definition of the organizational structure and task distribution.

Planning	Establish the objectives and procedures required to achieve results in accordance with the organization's environmental policy.
Doing	Perform process implementation.
Acting	Take action to continually improve EMS performance.
Checking	Carry out tracking and measurement of the processes with respect to environmental policy, the objectives and goals, legal requirements and other requirements, and report on the results.

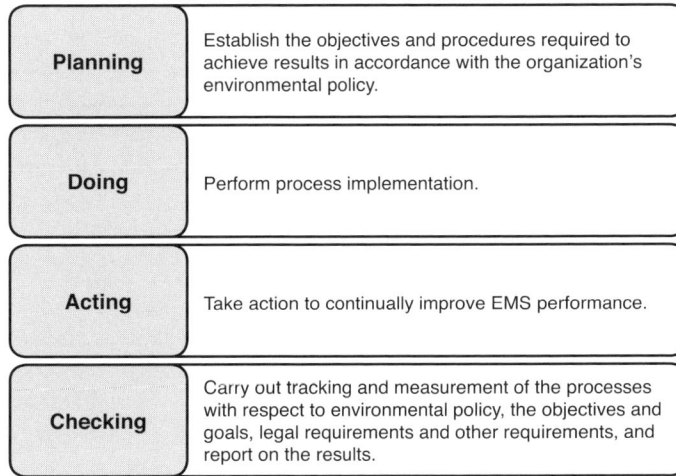

Source: After ISO 14001:2004

FIGURE 5.12   EMS methodology

One of the most important aspects is the decision to implement the system with company personnel or whether support from an outside specialist consultancy will be obtained. It is clear that outside consultancy support will enrich the system because of its experience in implementing that system in other companies. On the other hand, it must also be taken into account that even with the support of a consultancy, company personnel must actively participate in the implementation process, because a system developed exclusively by consultants, without the involvement of in-house personnel, will have virtually no chance of success.

Key Aspects in EMS Implementation in Accordance with ISO 14001

The international ISO 14001 standard establishes how an EMS should be implemented and is based on the methodology known as planning–doing–checking–acting.

Methodology.   Figure 5.12 is a summary of the planning–doing–checking–acting methodology, which is usually similar in most EMSs.

Requirements for a Certified EMS

The ISO 14001 standard establishes that "the organization must establish, document, implement, maintain and continually improve the EMS and must also determine how to comply with these requirements."[50]

In order to accomplish this, a series of steps must be followed until system certification is finally achieved (Figure 5.12). The implementation of an effective EMS (Figure 5.13) should take into account the guidelines set out in this section.[51]

Phases.   The usual recommended implementation phases for an EMS, after the management commitment is available, are shown in Table 5.7.

The time required for system implementation will largely depend on the company's specific circumstances. However, a general guideline schedule can be given, as in Figure 5.14, with a few months' variation from case to case.

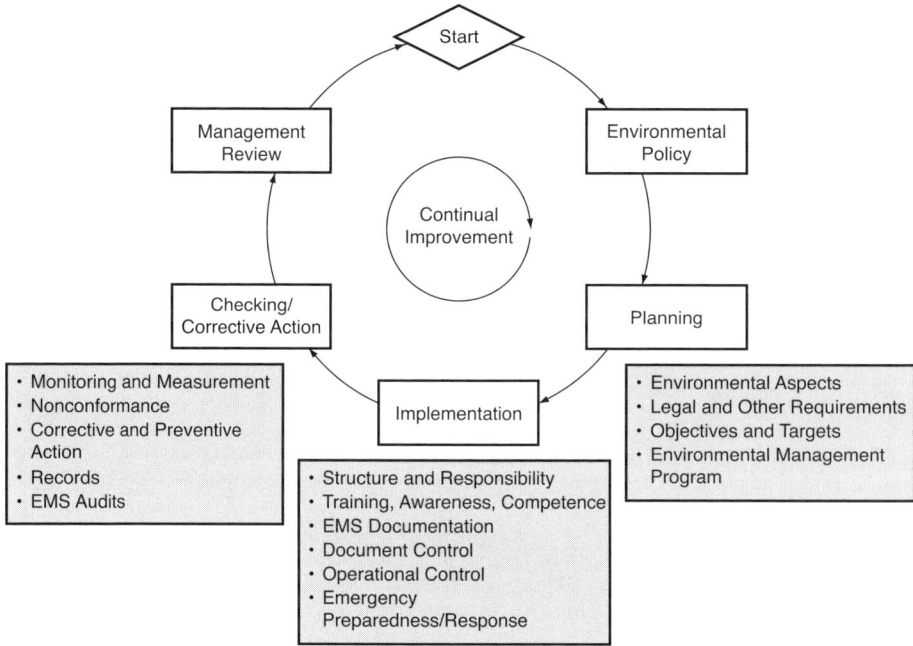

Source: After ISO 14001:2004

FIGURE 5.13   EMS requirements

TABLE 5.7   Main implementation phases for an EMS

Phase I	Review of the baseline environmental situation and existing management system Identification of requirements demanded by the standard Organization of the existing management, whether or not documented procedures exist Identification and evaluation of environmental aspects
Phase II	Definition of the EMS model and documental structure Definition of the organization and general responsibilities
Phase III	Preparation of the environmental management manual Preparation of general procedures Definition of environmental objectives and goals
Phase IV	Preparation of detailed documentation: - Specific procedures - Technical instructions
Phase V	Documentation review and verification
Phase VI	System implementation and personnel training
Phase VII	Internal audits and corrective measures
Phase VIII	External audits, corrective measures, and obtaining the certification or register

Source: Adapted from ISO 14001:2004

Principal Issues

Three issues are relevant in the conceptual definition and the implementation of an EMS and are described in the following paragraphs.

Initial environmental review.   The purpose of this review is to define the environmental situation at the beginning of the EMS implementation, to detect any existing environmental problems, and to determine what degree of environmental management exists within the company

Phases/Months	1	2	3	4	5	6	7	8	9	10	11	12	13	14	15	16	17	18	19	20
Phase I	▓	▓	▓	▓	▓															
Phase II				▓																
Phase III					▓															
Phase IV					▓	▓	▓	▓	▓											
Phase V										▓	▓									
Phase VI								▓	▓	▓	▓	▓	▓							
Phase VII														▓	▓					
Phase VIII																	▓	▓	▓	

FIGURE 5.14   Guideline schedule for EMS implementation

(whether documented procedures exist or not). For example, it may be found that not all the government permits required for the mining activity have been obtained; in this case, a full review of the permitting process must be carried out. (See Chapter 8.)

Environmental aspect evaluation.   ISO 14001 defines an environmental aspect as an element of the organizational activities that could interact with the environment and cause an environmental impact. An environmental impact is a change in the environment directly or indirectly caused by the activities performed by an organization. This change may be beneficial or harmful.

The organization must identify the environmental aspects and determine the most significant ones. After the significant ones are identified, these must be treated as priorities within the EMS.

When an EMS is to be implemented during the construction phase, the identification of environmental aspects presents a series of problems. On the one hand, the environmental aspects relating to the construction phase have to be identified, but environmental aspects of operations must also begin to be identified. On the other, their quantification is usually difficult because, on many occasions, quantitative data is not available to make the evaluation.

There are many approximations for the identification of environmental aspects. Table 5.8 shows what is considered most suitable for mining operations, assuming that operations encompasses the mine (underground or open pit) and the plant where the mineral is separated from the tailings and, in some cases, mineral processing.

Given that the evaluation is going to identify a large number of environmental aspects, a method and criteria must be established to identify the most significant environmental aspects. There are many methods for identifying the most significant environmental aspects and all may be considered valid if the obtained results can be considered coherent.

EMS documentation in mining operations   The EMS will have a manual,[52] which will include the description of the company organization, together with the organization of technical procedures and instructions that are going to make up the system. It will also include a series of technical procedures and instructions, together with support documentation, as shown in Figure 5.15.

Table 5.9 shows the main documents that are usually found in an EMS, including the section of ISO 14001 with which they comply listed in the left-hand column.

The process is initiated after the certificating organization receives the application and comprises five phases (Figure 5.16):

TABLE 5.8 Main environmental aspects in a mining project

Aspect	Description	Related Activity
Consumption	Water	Mine and plant
	Electricity	Mine and plant
	Diesel oil	Mine and plant
	Other consumptions in the mine (grease, solvents, etc.)	Mine
	Raw material consumption at the plant (acid and lime, etc.)	Plant
	Lesser consumptions (paper and secondary consumption, etc.)	Mine and plant
Discharge to rivers	Discharge from the plant	Water treatment plant
	Discharge from workshops	Mine workshops
	Discharge of treated town wastewater	Mine and plant
Emissions to air	Emission of particles and metals	Material extraction and transport
	Emission of particles	Grinding and milling
	Emission of particles and gases from heavy equipment	Mine
	Emission of gases	Plant
Hazardous waste	Used oil from heavy equipment maintenance	Mine and plant
	Tailings	Plant
	Oil filters	Mine
	Solvents	Mine and plant
	Batteries, accumulators, fluorescent lamps, and packaging, etc.	Mine and plant
	Packaging with dangerous wastes	Mining extraction
	Absorbent material impregnated with dangerous waste	Mining extraction
	Other waste (depending on mineral treatment)	Plant
	Electrical and electronic equipment	Mining extraction
Non-hazardous waste	Cardboard	Mine and plant
	Scrap	Mine and plant
	Nonseparated solid town waste	Mine and plant
	Wood pallets	Mine and plant
	Used tires	Mine
	Packaging from nondangerous waste	Mine and plant
	Purification sludge from town wastewater	Mine and plant
	Electrical and electronic material	Mine and plant
Mine waste	Inert material	Mine
	Noninert material	Mine
Noise	Mine machinery	Mine
	Operation of various types of plant equipment	Plant
Vibration	Emission of vibration to the exterior due to blasting	Mine

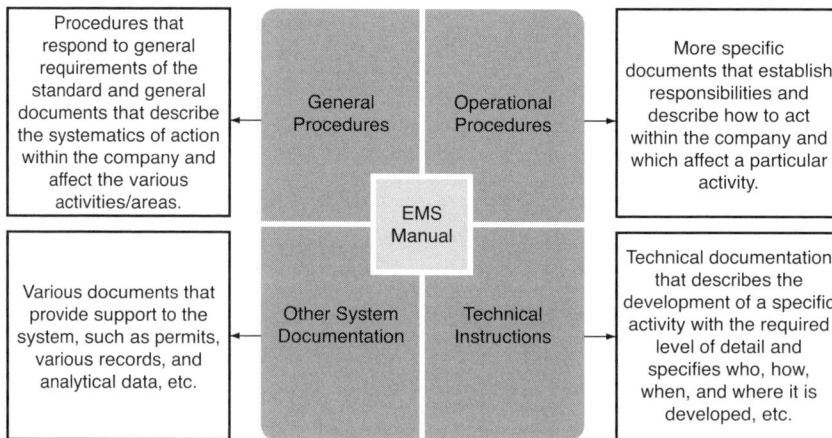

Source: After ISO 14001:2004

FIGURE 5.15 Main documentation forming part of the EMS

TABLE 5.9   Usual documentation in a mining operation EMS

ISO 14001	Document
	System Manual
4.1	Environmental policy, objectives, and goals
	System scope
	Resources, functions, responsibilities, and authorities
	Company organization
	Description of the main system elements
	Documents and records required by the standard
	Documents and records required by the organization
	General Procedures
4.3.1	Identification, evaluation, and recording of environmental aspects
4.3.2. and 4.5.2	Identification, evaluation, and recording of legal and other requirements (permits)
4.3.3	Establishment of objectives and goals, and management program preparation
4.4.2	Training and awareness
4.4.3	Control of internal and external communications
4.4.5	Documentation control, distribution, and preparation
4.4.6	Operational control
4.4.6	Tracking and measurement device control
4.4.7	Preparation and response to emergencies
4.5.1	Tracking and measurement
4.5.3	Management of nonconformities, together with corrective and preventive actions
4.5.4	Recording control (included in other procedures)
4.5.5	Executing internal audits
4.6	System review by management
	Operational Procedures
	Application of the environmental monitoring program
	Mine tailings management and control
	Management and control of nonmining waste (dangerous, soild town waste, and others, etc.)
	Management and control of surface and underground water
	Water management control
	Technical Instructions
	Control of fuel and chemical product storage in the mine
	Control of emissions into the atmosphere
	Discharge control
	Noise control
	Environmental management in offices
	Response in the case of unauthorized dumping or spills
	Response in the case of exceeding authorized limits (emissions, discharges, and noise, etc.)
	Maintenance control (mining and plant equipment, environmental control equipment, etc.)
	Raw material and resource consumption control
	Mine tailings management control
	Restoration work control
	Various samplings and measurements (make technical instructions for each type)
	Other System Documents
	Documents on legal requirements
	Audit plan
	Operational control programs
	Training plan
	Environmental aspect evaluation
	Environmental monitoring program
	Obtained permits
	Internal emergency plan

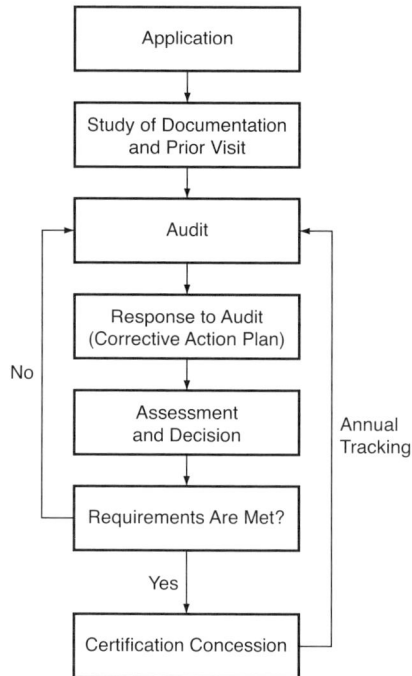

Source: After ISO 14001:2004

FIGURE 5.16   Scheme of the certification process in accordance with ISO 14001

1. Documentation analysis. The audit team studies the system documentation in order to assess its coherence and suitability to the requirements of the standard, with any detected observations included in a report.

2. Prior visit. The auditors visit the company to assess the action carried out by the company in response to the observations contained in the analysis report, together with an evaluation of the degree of implementation and adaptation of the company's EMS.

3. Initial audit. The audit team evaluates the EMS in accordance with the requirements of the standard. All nonconformities encountered and comments are included in a report that is given to the company at the final audit meeting.

4. Corrective action plan. In a situation in which the audit detects nonconformities in the system, the company is allowed an established length of time to present a corrective action plan intended to correct any nonconformities detected during the audit.

5. Concession. The audit report and corrective action plan are evaluated, followed, as applicable, by the concession of the certification.

Conclusions

It may be concluded that although certified EMS implementation is a complex process that requires significant dedication and cost, it also provides a series of important benefits:

- Improves the company's image and appeal to employees, government, customers, suppliers, media, social partners, and neighbors;
- Increases confidence of legislators, investors, and insurers;

- Reduces energy consumption and raw material costs, optimizes water use, and reduces emissions and costs of waste disposal; and
- Improves compliance with legal environmental requirements.

In summary, the benefits to a mining company of operating within an EMS to internationally recognized standards are sufficient to outweigh the complexity and costs entailed in implementing and running the system. Correct environmental management systems by mining companies are increasingly viewed as absolutely essential steps to be implemented as early as possible during project development.

CASE STUDY: PARTNERSHIP FOR SUSTAINABLE DEVELOPMENT IN GHANA—DEVELOPMENT PROGRAMS AT THE TARKWA AND DAMANG GOLD MINES*

The Tarkwa and Damang gold mines, operated by Gold Fields Ghana Ltd. and Abosso Goldfields Limited (Gold Fields), are located in the Wassa West District of the western region of Ghana. The Wassa West District has been the site of gold mining activity since the early 1900s and also hosts two other major mines (AngloGold Ashanti's Iduapriem mine and Golden Star's Bogosu mine). Gold Fields acquired the Tarkwa gold mine from Ghana's State Gold Mining Company in 1993, at which time there was an underground mine on the 200-km^2 mineral concession producing approximately 60,000 ounces of gold per year. During the next 5 years, a large-scale exploration program was carried out while the underground mine continued in operation. In 1997, based on significant additional reserves defined during exploration, construction of open-pit heap leach facilities began, and surface mining operations commenced in 1998. Underground operations became uneconomic in 1999 and were suspended. A series of phased expansions, including acquisition of a portion of the adjacent Teberebie gold mine from Teberebie Goldfields Ltd. in 2000 and construction of a carbon-in-leach mill in 2005 (Figure 5.17) increased production from the Tarkwa gold mine from 200,000 ounces per year to 750,000 ounces per year.

The Damang gold mine was originally built and operated by Ranger Minerals Ltd. in the 1990s. Gold Fields acquired the Damang mine and its 50-km^2 mineral concession from Ranger in 2002 and has maintained production of approximately 250,000 ounces per year since that time, using open-pit mining and carbon-in-leach milling. The Tarkwa and Damang mineral concessions are contiguous and offer significant potential for definition of additional reserves. The Government of Ghana owns 10% of both operations, IAMGOLD Corporation is an 18.9% shareholder, and Gold Fields, the operator, holds the remaining 71.9%. Total production from the Tarkwa and Damang mines exceeds 1,000,000 ounces per year in total, making Gold Fields the leading gold producer in Ghana. The Tarkwa mine has an estimated lifespan of 20 years and the Damang mine currently has more than 5 years of reserves. The two operations employ more than 4,000 people directly and indirectly through contractors in Ghana. Gold Fields implemented ISO 14001–compliant environmental management systems at all of its operations worldwide in 2002 and 2003 and became the first major gold mining company in the world to achieve this milestone.

Gold Fields' early social responsibility efforts in the vicinity of the two mine sites were primarily aimed at enhancing public infrastructure in these communities, with a focus on education, water and sanitation, and health. Although these efforts were in large part successful and appreciated by the communities, in some cases the sustainability of this infrastructure and/or its

---

* This section was written by T. Buchanan and T. Aubynn.

Courtesy of Gold Fields Ghana Ltd.

FIGURE 5.17   Tarkwa carbon-in-leach gold treatment plant

function was at risk due to insufficient capacity in beneficiary communities and local government to manage and maintain these facilities. This situation created a risk of permanent and unsustainable dependence on company resources. Additionally, early programs did not integrate formal monitoring or evaluation systems.

Gold Fields embarked on a focused effort to integrate international best practice and leading development expertise and experience into the design and implementation of its sustainable development programs. This effort was undertaken to realize a better return on the company's social investment in terms of improvement in the quality of beneficiaries' lives, strengthen relationships with stakeholders, leverage the company's resources through partnerships that provided access to development expertise and experience and complementary resources, and maximize synergies with other development programs. In the following case study, the various components of the strategy employed by Gold Fields Ghana to maximize the effectiveness of its sustainable development programs and to create a leading example of best practice in the international mining industry for community engagement and sustainable development are described.

Socioeconomic Setting

The Wassa West District, like much of Ghana, is poor in terms of average household incomes, and human development indicators reflect the full spectrum of development needs found in most of the world's developing economies. The Tarkwa and Damang mines each have eight primary stakeholder communities with a total population of 15,000 (30,000 in total for both mines). Primary stakeholder communities are defined by Gold Fields as communities that are located on or immediately adjacent to the Tarkwa and Damang mining concessions and have been affected to varying extents by Gold Fields' mining operations. The individual populations of these communities vary from less than 1,000 to nearly 6,000 residents; two are peri-urban and the other fourteen are rural in nature.

In the Wassa West District, subsistence agriculture engages 47% of the workforce, 50% of the population has only a primary school education, and only 6% have any schooling beyond grade nine. In rural areas, school infrastructure and supplies are inadequate to nonexistent, 70% of the teachers are untrained, and on average there are two teachers for six classrooms. Access to potable water is poor (52%), very few communities have any sort of toilet facilities, and the water sources used (rivers, streams, ponds) are generally polluted with fecal coliform bacteria at levels far above the World Health Organization standard. No refuse collection/disposal services exist. Agriculture in the area, while it employs the majority of the workforce, is characterized by poor productivity, poor market access, lack of access to credit, high input costs, and poor farm management skills. The artisanal mining sector is attracting labor from agriculture; however, in addition to its illegal status, this sector often uses child labor and suffers from a variety of safety, environmental, and health problems. District and regional governance evolved in the 1990s in Ghana from a central government management system, and local service delivery programs related to water, sanitation, health, and transportation are underfunded and ineffective, with rural areas often receiving little to no meaningful assistance. The average lifespan for males is 58 years and for females is 60 years, infant mortality is 54 per 1,000 live births, and waterborne diseases and infectious skin diseases are the leading causes of death. District government staff do not have adequate access to capacity-building programs, and members of local government and local communities do not have an adequate awareness of their respective functions, roles, and responsibilities as related to local governance.

Internal Management Systems and Structures for Sustainable Development

In conjunction with a major effort to strengthen its sustainable development programs in Ghana, Gold Fields established the Gold Fields Ghana Foundation in 2002 to support the socioeconomic development of the communities in the vicinity of its operations and in Ghana as a whole. The foundation is recognized as a nongovernmental organization (NGO) and registered charity by the Government of Ghana and enjoys tax-exempt status, and its activities are governed by the applicable laws of Ghana. Funding for socioeconomic development projects undertaken by the foundation is provided by contributions to the foundation by the Gold Fields operations in Ghana through a formula related to ounce production and net profit and contributions from other sources.

A board of six managing trustees was established to oversee the activities of the foundation. This board includes four senior Gold Fields management staff, two members of Parliament from the Wassa West District, and the chief executive officer of the Ghana Chamber of Mines. The input provided by the three external foundation trustees was considered essential in ensuring that a broad and locally relevant perspective was brought to bear on the activities of the foundation. The foundation's trustees meet an average of three times per year and review/approve proposals for project funding developed by Gold Fields staff and development partners, receive progress reports on projects in progress, and set priorities for funding in terms of project focus areas. Formal annual reports describing all foundation-funded programs are produced and distributed widely to stakeholders in Ghana.

Gold Fields established community consultative committees (Figure 5.18) for both the Tarkwa and Damang mines, which are led by the general managers of the two mines and include representatives of local communities (traditional leaders, assemblymen, farmers' groups, women's groups, youth groups, artisanal miners, etc.), local government, government ministries, regulatory agencies (Environmental Protection Agency, or EPA), and other relevant stakeholders. The environmental managers and community relations managers for Tarkwa and Damang also participate in the committee meetings. The two committees meet three to four times per year and were established to provide an open forum for two-way discussion of issues related to mining operations and sustainable development programs. Any issues of concern held by a community member can be aired in a committee meeting, and the company uses these meetings to inform

Courtesy of Gold Fields Ghana Ltd.

FIGURE 5.18   Community consultation

stakeholders of developments related to mine operations and to solicit their input where potential impacts are identified. In addition, Gold Fields maintains an open-door policy for local residents to ensure that a qualified company representative can be accessed at any time for the
purpose of discussing issues or incidents of concern.

Gold Fields also established an internal sustainable development committee comprised of
senior management staff from the corporate office in Accra and the Tarkwa and Damang mines,
including the vice president and managing director for Gold Fields Ghana, the corporate manager of public affairs and social development, the manager of sustainable development Ghana,
the AIDS coordinator, and the general managers, environmental managers, and community relations managers and superintendents from both mines. Other senior company personnel from the
mines are invited to committee meetings when relevant, including mining managers, mill
managers, security managers, engineering managers, and so on. The Sustainable Development
Committee meets every 2 to 3 months to receive project progress reports from relevant managers and development partners, develop frameworks for new initiatives, analyze performance of
on-going projects, evaluate the potential impact of mine operations activities in terms of local
communities and sustainable development, and to discuss any pressing social issues.

The committee was extremely effective in bringing focus to the myriad issues related to sustainable development in terms of prioritization and definition of roles and responsibilities; additionally, it provided an important means of driving corporate social responsibility concepts into
company culture and alignment of performance with these concepts at the operational level. The
manager of sustainable development Ghana and corporate manager of public affairs and social
development were responsible for general oversight of all related programs, including community and local economic development, and for establishing and maintaining the associated relationships and partnerships with development NGOs, donor agencies, and so on. Quarterly
status presentations on sustainable development programs in Ghana were delivered to senior corporate management from the Johannesburg corporate office during their routine visits (CEO,
executive vice presidents for international operations).

Biweekly sustainable development reports were issued by the SD managers to all SD committee members and other interested internal parties, including corporate management. These

reports contained a status update on all active or planned SD initiatives/programs and other information relevant to the company's sustainable development activities in Ghana, including government activities, NGO/donor agency activities, international developments, trends and best practice, company activities in other regions of the world, and so on.

Drivers for Partnership in Sustainable Development

Gold Fields identified the following drivers for seeking partnership in implementation of sustainable development programs in Ghana:

- Core competencies did not exist internally for design and management of large, holistic community development programs or specialized local economic development projects.

- There was a requirement for maximum return on social investment in terms of improvements in beneficiaries' lives and community relations.

- Gold Fields did not want to relearn the same lessons learned by the development field over the last 50 years.

- Poorly designed and implemented development programs can degrade beneficiaries' lives and undermine community relations.

- Little development work was being done in the Wassa West District by government, donors, or NGOs; Gold Fields was seen as the "development agency" in the primary stakeholder communities. In addition to being unsustainable, this situation was creating unrealistic expectations of the company.

- There was a need to address "soft" and "brick and mortar" issues related to health, education, and livelihoods in a holistic fashion to ensure maximum potential for sustainable community development—training, capacity building, and mentoring required in support of all initiatives.

- Formal monitoring and evaluation systems needed to be integrated to measure impacts of social investment and define success.

- There was a desire to maximize the benefits of social investment by leveraging company resources.

- Gold Fields realized that the same level of technical and management expertise and effort employed in the core business areas should be brought to bear on social responsibility programs.

Partnership for Sustainable Development

Gold Fields conceived, led, and provided financial support for the establishment of the Ghana Chamber of Mines Sustainable Development Forum to encourage a broad public dialogue in Ghana on sustainable development issues related to mining and to facilitate identification of potential development partners for mining areas. Participation in the forum included other mining companies, government ministries (to the minister and vice president level), district and regional government leaders, traditional leaders (chiefs), development NGOs, other NGOs, donor agencies (e.g., the United States Agency for International Development [USAID], United Kingdom Department for International Development [DFID], European Union), UN agencies, and other interested parties including media (attendance was not restricted). Half-day forum public events were held approximately every quarter, with a specific focus topic selected and speakers/presenters invited; a typical audience included 100 attendees/participants.

The forum events were successful in broadening a dialogue that had generally been one-sided in the media ("mining is destructive to communities and the environment"), calling attention to the activities of more progressive mining companies in terms of community/sustainable development programs and their willingness to find solutions through collaboration and consultation, and in publicly advocating for more equitable distribution of mining revenues by the Government of Ghana to mining areas for development and greater transparency in the flow of those revenues. As a result of forum events, a number of development organizations began to express interest in learning more about opportunities to work with mining companies where common goals existed.

Senior Gold Fields staff visited dozens of development NGO and donor agency offices in Ghana to discuss interest in and potential areas of collaboration. Following these visits, those organizations who appeared to be the most significant in terms of common goals, ability/desire to work in the vicinity of the Tarkwa and Damang mines, organizational capacity and experience, and willingness to work with the mining industry were targeted for field visits to project sites where they were involved in development programs. Third-party evaluation reports were obtained where available to access expert opinion on the effectiveness of these organizations' development programs.

The partners and programs selected are described in the following paragraphs.

Sustainable Community Empowerment and Economic Development (SEED) Program

Gold Fields partnered with Opportunities Industrialization Centers International (OICI) to design and implement a results-focused, sustainable, and integrated community development program (SEED) (Figure 5.19) that focuses on economic growth, wealth creation (Figure 5.20), quality of life improvement (Figure 5.21), and empowerment through education, capacity building, and infrastructure development. OICI is a large international development organization that works in 19 African countries, has worked in Ghana for more than 30 years, and has been a major USAID contractor for many years. To ensure sustainability and community ownership of the program, the design of SEED incorporated active engagement between Gold Fields, OICI, the District Assembly, relevant government ministries, other development NGOs, and community representatives in order to secure and maintain the required support (human, financial, and material) and participation in the implementation process.

After completion of updated socioeconomic assessments and surveys in all 16 communities, Gold Fields personnel worked closely with OICI staff in workshop settings to analyze root causes of identified developmental problems and strategic intervention areas using a logic modeling approach, define strategic objectives and related intermediate objectives and program activities, develop performance indicators and a monitoring and evaluation plan, and elaborate 5-year community development implementation plans. The community development plans were reviewed with community and government representatives to ensure that they reflected community-felt needs and, where applicable, other stakeholders' views. SEED implementation, which began in 2005, is designed to provide for socioeconomic development in the 16 primary stakeholder communities around the Tarkwa and Damang mines. The SEED program has a variety of components, with a value of more than US$6 million for the period 2005–2010:

- Agricultural livelihoods: Oil palm, cocoa, vegetables, and livestock, including provision of inputs, technical training, group capacity building, and establishment of micro-credit groups;

Courtesy of Gold Fields Ghana Ltd.

FIGURE 5.19   Community registration SEED livelihood program

Courtesy of Gold Fields Ghana Ltd.

FIGURE 5.20   SEED cabbage farmers, Damang

- Processing/value-adding livelihoods: Palm fruit processing, cassava processing, aqua-culture, soap-making, and beekeeping, including provision of inputs, technical training, group capacity building, and establishment of micro-credit groups;
- Water and sanitation: Wells and toilets in rural communities, formation, technical training, and capacity building for water and sanitation (WATSAN) committees;

Courtesy of Gold Fields Ghana Ltd.

FIGURE 5.21   Potable water well at Tarkwa

- Health: Community education and capacity building for all pertinent health issues, health clinics in rural communities, maternal and child health programs, AIDS programs, training and support of community health workers, health infrastructure in district capital; and

- Education: Formation and strengthening of school management committees; school buildings and supplies in rural communities; scholarships for secondary school, vocational school, and university; annual teacher skills training; teachers' quarters in rural communities (Figure 5.22).

A dedicated project team of nine development professionals, all Ghanaian, were hired by OICI for the SEED project (or transferred) and relocated to Tarkwa, where a SEED office was established. Senior Gold Fields staff participated in the selection process for these individuals. The OICI SEED program manager attends Gold Fields' Sustainable Development Committee meetings and the OICI country director attends foundation trustee meetings. Weekly meetings were held between OICI field staff and Tarkwa and Damang community relations staff to ensure that day-to-day program management and logistics were optimized and that any problems identified were resolved efficiently. OICI and Gold Fields staff commonly visit and work in communities together, and OICI personnel wear shirts and hats to brand the program clearly as a Gold Fields/OICI initiative. All vehicles used by OICI are branded with the Gold Fields/OICI SEED logo.

Ghana Responsible Mining Alliance

Gold Fields Ghana acted as a primary driver in formalizing a US$9.5 million USAID Global Development Alliance (GDA) partnership in 2006 with USAID/Ghana and Newmont Ghana Gold, which facilitates collaboration on development activities of mutual interest. The alliance leverages the financial resources and field experience of the mining companies with the technical capacity and development expertise of USAID/Ghana in a program to create measurable socio-economic improvements in Ghanaian mining communities.

Courtesy of Gold Fields Ghana Ltd.

FIGURE 5.22   School block built by Gold Fields

The GDA agreement commits the alliance members to collaborate over a 4-year period in implementing development activities with the overall aim of promoting beneficial and interactive relationships between the mining industry, host communities, civil society, and local governments. In addition, the alliance members will share lessons learned and best practices in environmental stewardship and social development with communities in Ghana and other West African nations facing similar challenges. There are three major alliance activities:

- Local capacity-building of governance structures. The objective of this activity is to increase the capacity of both formal and informal governing bodies within the mining areas to plan and implement community development. This objective will be accomplished in the Tarkwa/Damang area through implementation of USAID's Government Accountability Improves Trust (GAIT II) program, designed to increase participation of civil society in local governance and clarify roles and responsibilities of various actors in local government and communities. In conjunction with the GAIT II program, the Wassa West District Assembly receives technical capacity training to build skills in development planning, local revenue mobilization, project management, and implementation. Additionally, a certain amount of capacity-building resources are targeted at the local chiefs and the more "traditional" local governance structures.

- Promoting economic growth through a vibrant, local private sector. The objective of this activity is to foster a vibrant private sector of small- to medium-scale enterprises in the mining regions by implementing culturally appropriate and economically significant livelihood programs for impacted communities. This objective is linked to activities planned within the SEED program with OICI and to programs being implemented by USAID's Trade and Investment Programme for Competitive Export Economy.

- Participatory multistakeholder process—"best practices for mining in Ghana." This alliance activity is designed to establish a clear guide to social and environmental "best practice" for mining operations in the Ghanaian context that, although voluntary, would

Courtesy of Gold Fields Ghana Ltd.

FIGURE 5.23   Artisan miners at Damang

become a road map to responsible mining for mining companies operating in Ghana, their stakeholder communities, and government. Through the existing Ghana Chamber of Mines Sustainable Development Forum, the alliance will convene mining companies, NGOs, development agencies, government officials, and community leaders to discuss and agree on a set of guidelines for best practice in issues such as resettlement, compensation, mine closure, and artisanal/small-scale mining. This multistakeholder process would adapt the content and guidance already available from international sources (e.g., International Finance Corporation [IFC], ICMM, WBCSD) for application in the unique sociopolitical setting of Ghana. Rather than seek to change existing mining or land tenure laws, the group would create practical, on-the-ground modalities for the process of negotiating and resolving critical issues facing mining companies and their stakeholder communities. The resulting identified best practices will clearly outline how a leading company would engage with communities and stakeholders in the Ghanaian context given the challenges presented by the national legal system.

Artisanal Mining

The artisanal and small-scale mining (ASM) sector in Ghana is dominated by gold production, and it is estimated that ASM employs more than a quarter million people in the country (45% women). The Wassa West District is a major and historic center of legal and illegal ASM in Ghana. The sector is troubled by environmental degradation, health and safety issues, AIDS, excessive child labor, gender inequality, generally poor relations with large-scale mining, and a lack of institutional capacity and financial and technical support. Gold Fields has developed and maintained cooperative working relationships with ASM sectors in the vicinity of their operations (Figure 5.23).

Three of the primary stakeholder communities near the Damang mine have a large contingent of artisanal miners as residents. In order to address the negative and unsustainable aspects of ASM in these communities, Gold Fields partnered with the U.K. DFID to implement a

Courtesy of Gold Fields Ghana Ltd.

FIGURE 5.24   Aquaculture pilot project at Tarkwa

US$400,000 pilot program to implement activities designed to create alternative livelihood opportunities; improve environmental and safety aspects of the ASM sector; educate ASM participants about health, safety, and environmental impacts; and address child labor conditions. The Ghana Mines Department and Ghana EPA are designed to be partners in this program.

Postmining Land Use

As a part of open-pit mining operations at Tarkwa and Damang, Gold Fields has or will create hundreds of hectares of water bodies, including open pits, water storage dams, and lined processing ponds. Water quality in these water bodies has been demonstrated to remain within surface water quality guidelines, and with proper management, these water bodies can be transformed into productive resources for aquaculture and fisheries as they are successively decommissioned (Figure 5.24). Gold Fields had provided support for small-scale aquaculture initiatives in conjunction with livelihoods programs at Tarkwa and Damang, and in the process contributed toward building the capacity of the Wassa West Fish Farmers Association. Marine fish stocks in Ghana are under severe strain because of unsustainable exploitation levels, and the Ministry of Fisheries is actively promoting aquaculture to reduce this strain and meet the country's increasing demand for fish products.

Building upon these pilot activities, Gold Fields approached the Food and Agriculture Organization of the United Nations, the Ghana Water Research Institute, the Ghana Ministry of Fisheries, and the World Fish Center to study further the potential for aquaculture in decommissioned mine facilities at Tarkwa. Several visits to the mine sites were organized, and a number of studies were undertaken to gain a better understanding of the dynamics of the current aquaculture industry, market opportunities for aquaculture products, and possible technological options. A technological feasibility study[53] was then undertaken by the partners with a goal of developing a design for postmining land and facility use that would serve as a catalyst for expanding the aquaculture sector in the Wassa West District. The feasibility study incorporated current best management practices that have been developed in the commercial aquaculture sector and addressed some of the impediments to economic optimization of such projects in the field, such as fingerling quality and quantity, genetics of fish seedlings for maximum yield, sexing of fish to maximize yield, and sources of fully balanced feed.

The results of the feasibility study indicated the potential at mine closure to develop profitable integrated aquaculture production facilities using decommissioned mine infrastructure, which would yield more than 700 t of food fish per year and employ more than 200 people at twice the national average wage. This plan was based on use of 48 ha of pit lakes, 28 ha of ponds, and 20 ha of dams and other water bodies for fingerling and food fish production, and included incorporation of an on-site hatchery and food production facility at former mine mill sites. The feasibility study was used to modify the "Costed Reclamation Plan and Decommissioning Study"[54] prepared for the EPA (and updated every 2 years) to include aquaculture as a postmining land use.

Enterprise Development

Gold Fields embarked on a program in 2004 to identify potential large-scale projects that would provide for sustained economic vitality in stakeholder communities in the vicinity of the Tarkwa and Damang gold mines and help ameliorate the economic contraction that would occur with mine closure. Subsequent to a broad review, agribusiness was chosen as the focus area for investigation of opportunities in Ghana to establish a market-led, commercially sustainable business of significant size in the area of the Tarkwa and Damang gold mines due to its demonstrated potential to generate meaningful employment for the rural poor, the significant cultural familiarity with agriculture, the demonstrated potential of this sector for commercial viability, and complementary government policies and programs designed to stimulate the sector.

Gold Fields contracted with TechnoServe, a leading development NGO with significant experience in large-scale agribusiness projects in Africa, including Ghana, and Latin America to assist with completing preliminary screening studies[55] to identify the agribusiness subsector that had the highest potential for establishment of a commercially viable and sustainable agribusiness operation in the Tarkwa/Damang area. A total of 34 subsectors were investigated, including tree crops, staple food crops, herbs and oils, and livestock. Numerous screening criteria were used to rank the sectors. During the study, consultation was undertaken with traditional leaders, local and national government officials including the Ministry of Agriculture, and management of established agribusiness operations.

The screening studies indicated that oil palm offered the most favorable opportunity for commercial agribusiness, and a more detailed evaluation of key focus areas for a potential model of an oil palm business was subsequently completed in 2005 by TechnoServe. The results of this study were used to gain board approval for a full feasibility study. Unfortunately, despite numerous prior consultations with local leaders and traditional landowners in the area planned for an oil palm plantation and mill development (Figure 5.25), the project was abandoned in 2006 when a land rush for speculative farming activities (in anticipation of compensation by Gold Fields) made development of the project uneconomic. Although this setback resulted in starting again from scratch, it did provide a very satisfactory experience in working with TechnoServe to identify opportunities for local economic development based on sound business principles and socioeconomic development goals. Gold Fields Ltd. has now formed an agricultural holdings company, Agrihold, in South Africa to pursue large-scale agribusiness developments in the vicinity of Gold Fields operations worldwide.

Courtesy of Gold Fields Ghana Ltd.

FIGURE 5.25   Oil palm SEED program

Conclusions and Lessons Learned

Although most of the sustainable development programs described in this case study are still in progress and their ultimate outcomes remain unknown, Gold Fields' experience indicates that partnership for sustainable development brings meaningful and valuable benefits in terms of effective program design and implementation. Development is a science, and accessing up-to-date development technical expertise and field experience through partnerships with proven development organizations allowed Gold Fields to improve greatly the quality of its social investment. Additionally, the implementation of holistic programs for community development had a measurable positive impact on community relations and relationships with other stakeholders in Ghana. Company representatives were often asked by members of the press and others how the company had been able to maintain such a high level of cooperation with its stakeholder communities, and the government of Ghana on several occasions publicly mentioned Gold Fields as an example of responsible mining practices. Other mining companies also approached Gold Fields for advice and insight into the company's approach, and Gold Fields gained a leading position in the eyes of a variety of stakeholders in Ghana in terms of its social and environmental practices.

Some important lessons were learned from Gold Fields' journey:

- Contracts/agreements with partners must clearly define exactly what activities or outputs are expected of each party and the resources that will be dedicated to them (financial, human, equipment, services, etc.). The potential for the need to modify the contract/agreement by mutual agreement of the partners should be incorporated in initial agreements.

- The roles and responsibilities of the company and partners in terms of project management must be clearly established and documented.

- Clear accountability for partnership/relationship management must be established by both the company and the partner in terms of defining respective internal roles and responsibilities that are communicated between the parties.

- Those assigned responsibility for program management must have adequate time and resources to fulfill this role properly.

- A frequent consultation/communication regime between partners is necessary to optimize partnership effectiveness and ensure prompt decisions on actions and resolution of issues of concern—this should include frequent field-level meetings with implementation personnel and regular management-level meetings to ensure that all parties are kept abreast of program progress and challenges, and that the company is involved in management of the program.

- Development programs must be designed for and retain some degree of flexibility in order to adapt to changing conditions on the ground and actual implementation experiences and beneficiary feedback.

- A longer contract/agreement period will provide a superior base for an NGO partner to plan, manage, and deliver a development program in an optimal fashion. For example, facing the prospect of a contract/agreement not being renewed every year can be an impediment in terms of staffing a project with the best personnel and other related resource allocation such as vehicles, and so forth.

- Do not rush into any development program—adequate background research, stakeholder consultation, and accessing appropriate expertise are absolute requirements for moving forward with an initiative.

- The company's interest is strongly served by encouraging, assisting, or, when necessary, challenging other companies operating in the same jurisdiction to improve their social and environmental performance and adopt international best practice.

- Performance and impact measurement systems are essential in optimizing development program effectiveness and in maintaining internal and external support for these programs.

- Company (or industry) leadership in sustainable development initiatives may often be required to achieve meaningful results. Although it may be desirable for government or the community to lead, if they do not provide meaningful and results-oriented leadership, the benefits of company (or industry) assumption of this role will often far outweigh the costs in terms of human and financial resources and, more importantly, perceptions of the company or the mining sector.

- Past environmental and social performance has a strong correlation to the willingness of donors and development NGOs to work with a company or industry—the better the record, the better the quality of partners from which to choose.

CASE STUDY: INDUSTRIAS PEÑOLES—A LARGE ORGANIZATION COMMITTED TO SUSTAINABILITY*

Industrias Peñoles S.A.B. de C.V. is a mining and metallurgical company founded in 1887 and part of Grupo BAL, a privately held diversified group of independent Mexican companies. Peñoles has integrated operations in the extraction, smelting, and refining of nonferrous metals and the production of inorganic chemicals. It is the world's largest producer of refined silver, metallic bismuth, and sodium sulfate, and is a leader in Latin America in the production of refined gold, lead, and zinc.

Its mining and milling operations are located across 12 states in central and northern Mexico, with more than 20 work centers comprised of mining units, plants, and offices. Although the corporate headquarters is located in Mexico City, its principal metallurgical complex, Met-Mex, has been in the city of Torreón since 1901. In addition to Mexico and the United States, the

---

* This section was written by L.E. Ortega and A.C. Zomosa-Signoret.

company has been successful in entering such varied markets as Europe, Japan, and China, as well as diversifying its operations. As of December 2007, Peñoles had a workforce of 7,818 people, as well as more than 4,600 contractors in such diverse roles as transport, raw materials, and services.

Challenges of a Large Organization

The magnitude of its operations is not the only challenge facing the company. As a producer of basic raw materials for other industries, it has a critical role in the supply chain, and moreover must produce high-quality products at low cost to distinguish itself from its competitors. The inherent characteristics of the mining and metallurgical sector are significant challenges in themselves: the social and environmental impact resulting from operations are easily identifiable and, therefore, the target of recurrent questioning by numerous groups. There is also a shortage of human resource talent in the earth sciences; the company's existence depends on the discovery of new deposits, as well as on ongoing technological improvements that facilitate the processes and make them more efficient. These factors imply considerable high-risk investments.

Corporate Structure for Sustainable Development

Peñoles celebrated its 120th year in 2007. A large company faces great challenges, and those described previously require the company to manage all different kinds on a daily basis: logistical, financial, operational, functional, communications related, technological, and of course, organizational. How has it overcome these?

The answer can be found in the commitment and responsibility of Peñoles's management team and in its capacity to adapt to new circumstances, reflected in a functional and flexible organizational structure. The company has had to adapt itself according to the size of its needs, as well as to adopt the best technology available to improve its production processes and address its environmental, safety, and health priorities.

In 1998, the company faced a delicate situation when it discovered that children in the populated areas around its Met-Mex metallurgical plant in the city of Torreón had levels of lead in their blood that were higher than the permissible limit (10 μg/dL). The company acknowledged its responsibility to the community and designed daily action plans, among which the most important were to reduce all types of emissions, a large-scale hygiene/diet education program for the neighboring population, a program to clean and remediate the historical accumulation of residues, and the acquisition of the highest risk zone, in which green areas were constructed. The results speak for themselves: from 1999 to 2007, the number of children with more than 25 μg/dL of lead in their blood was reduced from 2,765 to 28; in the same period, the percentage of children with less than 10 μg/dL rose from 16.2% to 80%.

As a result of this great lesson, Peñoles's commitment to environmental, safety, and health issues areas went well beyond specific programs: the company made important changes to its overall orientation and institutional structure. In addition to such aspects as economic profitability, a focus on customers, and the development of its own capabilities, Peñoles directed its overall strategy on the continuous improvement of its processes. In this regard, sustainable development was assimilated as a key factor in the business strategy. The tangible expression of this is outlined in the company's mission—"To add value to nonrenewable natural resources in a sustainable manner"—and in its vision—"To be the most recognized Mexican company worldwide in its sector for its global focus, the quality of its processes and the excellence of its people."[56]

As can be seen in the company's organizational chart (Figure 5.26), sustainable development is today one of the basic structural pillars, in both operating and functional areas. In addition, based on the internal concept of its "value chain," Peñoles designed a highly functional organizational

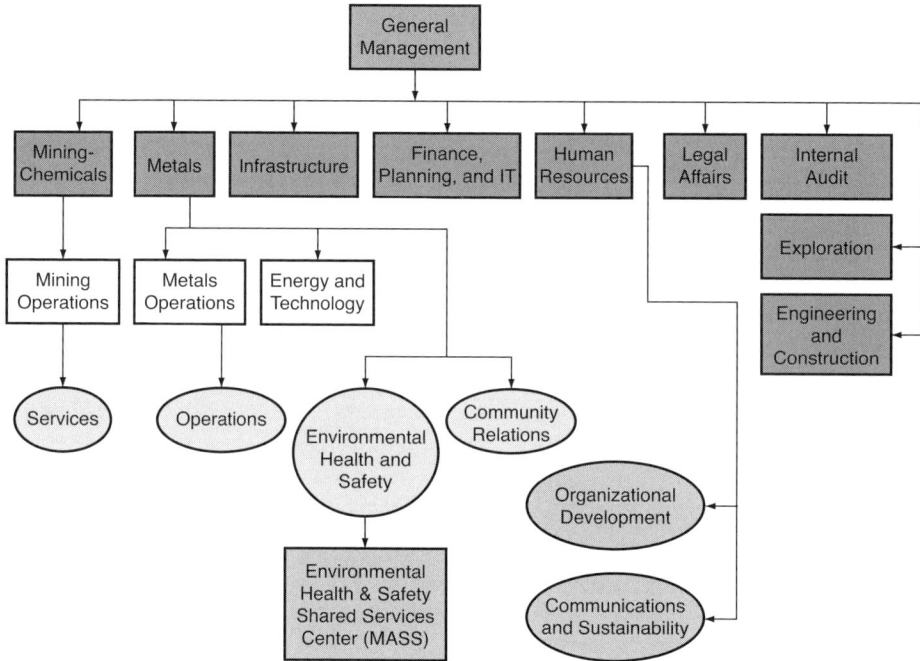

Courtesy of Industrias Peñoles

FIGURE 5.26 Organization of sustainability in Peñoles

structure tied to its strategic objectives (Figure 5.27). It has nine divisions: three operating divisions (i.e., Mining-Chemicals, Metals, and Infrastructure), two "growth-driving" divisions to drive growth processes (i.e., Exploration and Engineering & Construction), and four support divisions (i.e., Finance, Planning, & IT; Internal Audit; Legal Affairs; and Human Resources). All these ensure that the necessary functions exist for the proper management of the company and, within this framework, serve the needs and expectations of its stakeholders—shareholders, customers, personnel, suppliers, and communities—who have become the strategic axis.

Peñoles has found structural solutions to logistical problems related to its value chain through numerous corporate engineering initiatives. For example, the short-haul Coahuila-Durango rail line provides service to some of Peñoles's mines, and the Termimar maritime terminal is used to ship its export products. The Center for Technology Research and Development, which has been in operation since 1922, is responsible for research projects in areas directly related to the nature of the company's operations: mineral and chemical processes (which focuses on the Mining-Chemicals and Exploration divisions) and metallurgical processes (which focuses on the Metals division). In addition, by leveraging its experience, Peñoles has diversified its activities: an example is the companies created by the Infrastructure division, including Bal-Ondeo and its affiliates in the management of municipal drinking water, sewage, and wastewater treatment systems, and Termoeléctrica Peñoles for the generation and supply of electricity for the company's own operations.

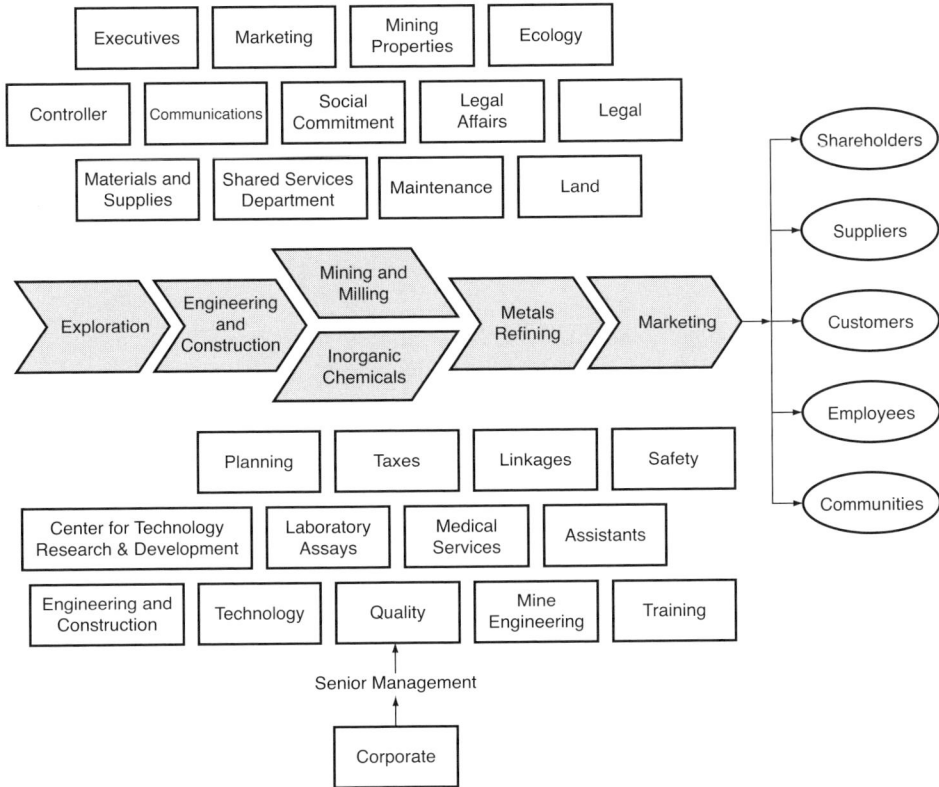

Courtesy of Industrias Peñoles

FIGURE 5.27   Peñoles value chain

MASS System: Environment, Safety, and Health

The historical trajectory of each of Peñoles's business units is a clear example of the concern for issues of environment, safety, and health. Each has developed important efforts to improve the efficiency of water and energy consumption, the responsible management of materials and residues, the prevention and control of emergency situations, and the safety of its employees and the communities where the company operates, among other initiatives.

Approximately three decades ago, these practices began to be systematized in specific areas: first, by centralizing them in the divisions in operation at the time—Mining, Chemicals, Metals, and Refractories—and later, through the design and implementation of management systems that evolved to become the Environment, Safety, and Health System in place today (*MASS System*, for its Spanish acronym).

The roots of the MASS System date back 80 years, when the Safety, Hygiene, and Environmental Control system was created for the mining units and the Ecology, Safety, and Occupational Health system was created for the chemical and metallurgical plants. Both were focused on regulatory issues, particularly those related to workplace safety. Environmental issues were gradually gaining ground, leading to the creation of the Corporate Environmental Management Group in 1991. This area reported to Legal Affairs and was focused on compliance with environmental regulations.

In 1995, work began in environmental management systems in each of the business units, through ISO 14001 certifications. Safety and environmental practices were gradually integrated into the existing systems. A department of Environment, Safety, and Health (Figure 5.28) was created due to the importance that care, compliance, and coordination of these functions already had, in order to have an organizational structure that had the corporate responsibility and necessary authority to make decisions for all of Peñoles's operations. The Environmental Policy was also designed, as well as for Safety and Health, Environmental Principles, and subsequently, the Principles of the Philosophy of Safety, that with their respective updates continue to govern the conduct of the company's employees. At that point, the areas that worked independently to provide services to the different divisions began to report to this new department through three corporate managers: in Ecology, Safety, and Medical Services, with the objective of aligning and standardizing the prevention and control efforts in the business units, where in turn, those with direct responsibility for those issues were appointed.

The *MASS Shared Services Center* (CSC-MASS for its Spanish acronym) was created in 1993 as a unit within the Environment, Safety, and Health Department. This established the main body responsible for supporting the efforts of the operating areas of Peñoles by providing advisory and coordination services to the work centers in all the divisions—Mining, Chemicals, and Metals—on subjects such as operating discipline, accident investigation, precontingency action plans, risk analysis, and environmental management and protection, as well as occupational health and safety. The MASS System soon led to the creation of a CSC-MASS team as a working program based on international standards (ISO 14000 and OHSAS 18000[57]) and best practices known as Operating Discipline, Zero Accident Tolerance, and Safety Training

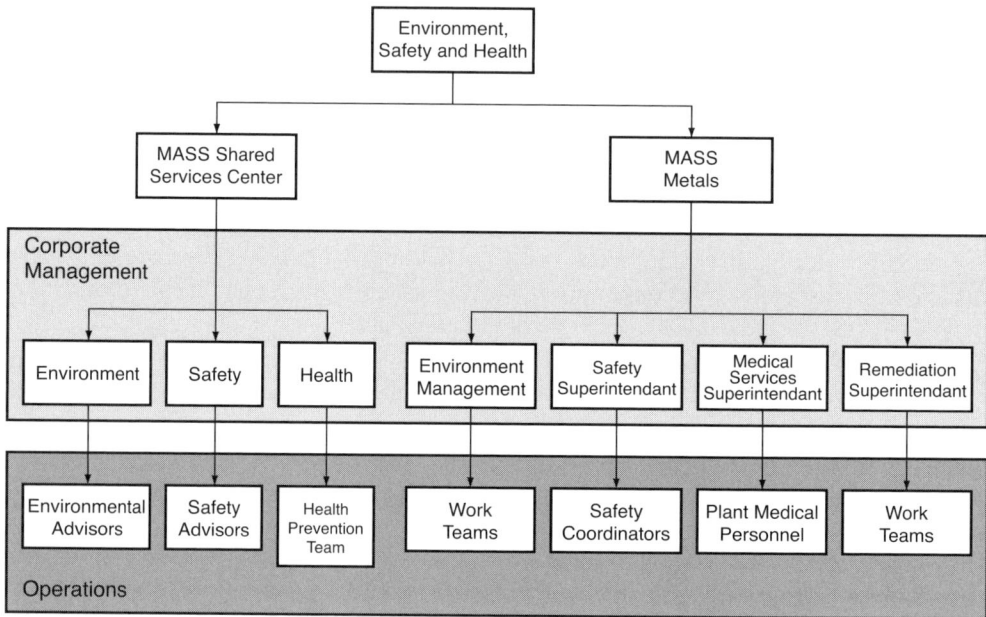

Courtesy of Industrias Peñoles

FIGURE 5.28 Organization of the Environment, Safety, and Health Department

Observation Program (STOP).[58] The MASS acronym was selected to impart the sense of joining efforts and working together.

Since then, environmental, safety, and health advisors work hand in hand with the coordinators in each business unit who, in turn, report to the operating heads in each plant, mining unit, or office. CSC-MASS has provided support in areas such as the design of the MASS System; identifying, updating and disseminating applicable regulations; issuing guidelines on environmental matters and workplace risks; training personnel in the business units in applying clean and safe operating procedures; supporting the search for technologies or mechanisms that minimize the environmental impact of operations and the risk to safety and health; conducting internal audits and handling administrative procedures with governmental authorities.

In addition, the MASS System is designed to establish training, awareness, and skill programs for employees related to significant environmental impact or who are exposed to health and safety risks. It also defines the mechanisms for internal communications among all employees and external communications with interest groups, and establishes procedures for recording and controlling documentation, records, and data in order to keep all the necessary information generated up to date (Figure 5.29).

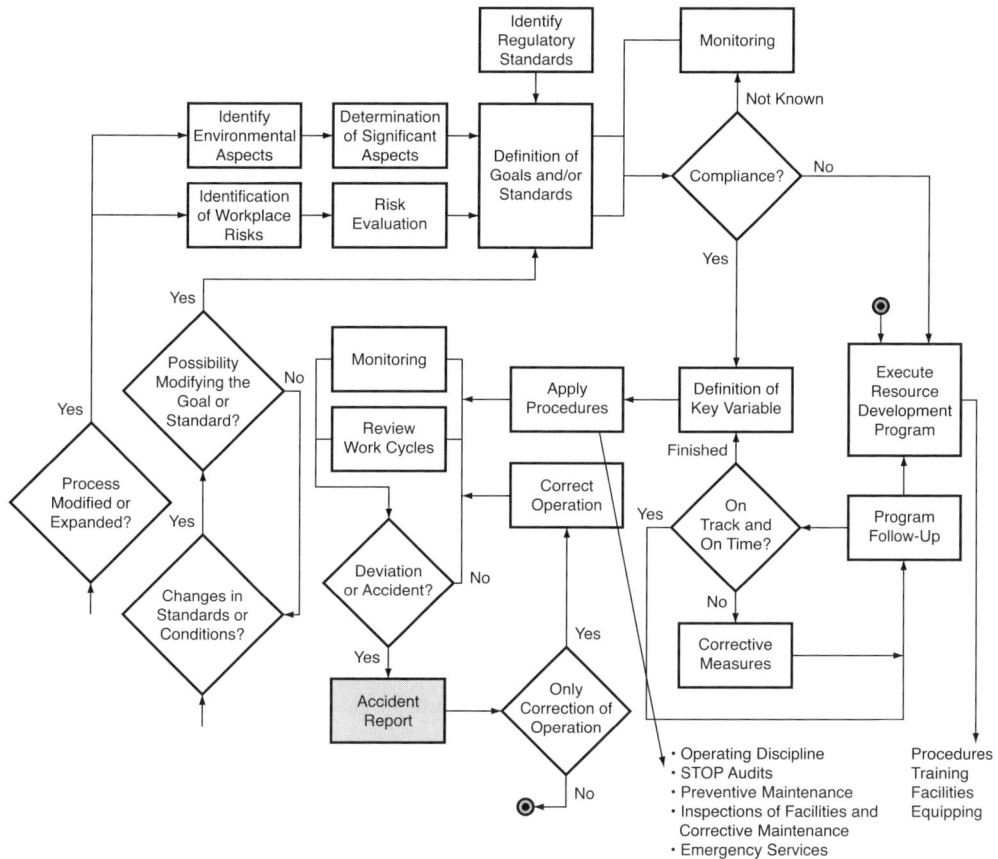

Courtesy of Industrias Peñoles

FIGURE 5.29   Diagram of the decision-making process in the MASS System

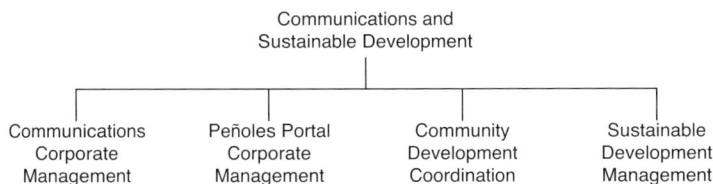

Courtesy of Industrias Peñoles

FIGURE 5.30   Organization of the Department of Communications and Sustainable Development

Throughout Peñoles, the implementation of the MASS System has strengthened the culture of preventing environmental impact, and has caught the attention of employees in reducing the occurrence of accidents and illnesses. Decisions are now made on the basis of systematic and predictable results derived from the implementation of the MASS System through a group of employees called directors or team leaders.

At the same time, attention to the various matters involved in the relationship between the company's operations and the communities in which they operate has taken on greater importance, becoming part of the strategic actions with specific plans and objectives. In this context, the communications programs for external audiences and internal personnel required greater alignment and specialization in order to manage messaging and communication channels, and have become a priority for Peñoles. In order to respond to these functional needs with greater organizational weight, the Department of Communications and Community Development was created in 2005 at the corporate level and recently evolved into the Department of Communications and Sustainable Development, whose initial focus on communities has been broadened to include comprehensive sustainability strategies.

This group, which reports to the company's Department of Human Resources and Sustainable Development, is in turn responsible for communications and relations with external communications media and is comprised of four main areas: Corporate Communications Management, which designs internal communication strategy and channels; Community Development Coordination, which coordinates the social development projects at the various work centers; and Sustainable Development Management, which explores global trends in the area of sustainability (Figure 5.30).

Among the communications media recently developed are electronic bulletin boards for employee notices, the integration of the company's various intranet pages into a Peñoles Portal with functionality and an institutional image, and electronic employee newsletters. Among the initiatives for outreach to the communities where it operates, the company periodically undertakes perception studies of its external image and reputation; conducts employee surveys in order to learn about satisfaction rates, and relies on a system for addressing complaints; executes cultural and health campaigns; and conducts workshops on the prevention of emergency situations.

Implementation of the MASS System at Bismark Mining Unit

The existing operations and new projects that Peñoles designs and implements are directly aligned with its strategic plan, the objectives and performance indicators of which reflect the company's priorities. Among the objectives are the "formal processes that ensure the sustainable development of the business" through strategic actions in the areas of "safety and occupational health," "clean operations and environmental care," and "outreach to communities and authorities."

The company conducts continuous monitoring and risk analysis that simulate adverse conditions that may occur, as well as their consequences on the environment, safety, and health. After these risks have been evaluated, modifications are made to the processes or projects with the commitment that if such risks cannot be eliminated or reduced to acceptable minimums, these processes will be completely redesigned or suspended.

A representative example of this can be seen in the strategic priority of "occupational health and safety" through programs of operating discipline and the reduction of risk levels. With the commitment to "maintain the MASS System in all mining units," within the operating plan of the Mining-Chemicals division, there is a commitment to work on self-managed safety projects in the mining units.

Minera Bismark S.A. de C.V. is one of Peñoles's mining units and is engaged in the mining and milling of zinc, copper, and lead minerals. Located in Chihuahua, it employs 587 workers and contractors. In 2000, it incorporated the self-managed program developed by the Secretary of Labor and Social Security (STPS for its Spanish acronym).

In this context, and in concert with Peñoles's MASS System, the Management System for Workplace Safety and Health was developed. This is based on continuous improvement in such programs as STOP, Take Two, Electrical Safety, Zero Accident Tolerance, and Operating Discipline, and has been successful in creating a culture of accident prevention characterized by the identification of workplace risks, the adoption of training programs, updating of applicable regulations, and a monitoring program on the part of line management.

In the initial diagnosis, Bismark achieved 85% compliance with existing regulations, and areas of opportunity were identified that were gradually taken into consideration. Today, advances in this system have enabled Bismark to create a culture of prevention, resulting in a reduction in accident indices from 3.29 in 2005 to 1.25 in 2006 and 0.87 in 2007, as well as a greater degree of participation and commitment from union and nonunion workers and from members of the Safety and Hygiene Commission of the mining unit.

Bismark has today become a model for preventing workplace risks in Mexico. The advances achieved from these programs have led to recognition from labor authorities as the only successful operation in the area of safety and hygiene within the underground mining industry that is functioning in the country. The mining unit remains in the self-managed program and will be reevaluated by STPS in April 2008.

Conclusions

With the historical heritage of its 120 years of existence, Peñoles has consolidated its leadership position in the mining and metallurgical sector in Mexico. In addition, its overall concern about the environment, health, and safety has become a real and daily consideration in each of the operating and functional activities of the company. This is reflected in every one of its processes and projects.

As the organizational structure of Peñoles indicates, structural relationships are critical to the implementation and control of the company's strategic objectives, in which sustainable development has become a key factor in the business strategy.

Peñoles continues to place emphasis on sustainability by conducting programs and campaigns, the application of clean production technologies, and the responsible management of minerals and metals with the objective of balancing its economic, social, and environmental performance. The MASS System has been a successful case in the search for synergies among the efforts that each Peñoles business unit has traditionally undertaken to prevent and control its environmental impact and workplace risks.

Beyond its corporate and operational structure, there are numerous tangible examples of the organization's efforts to guarantee its sustainability—among others, active participation in international initiatives such as the Global Compact to promote fundamental values like human labor and environmental rights, as well as national initiatives such as the Greenhouse Effect Gas Mexico Plan to address climate change.

NOTES

1. J. R. Galbraith, *Strategy Implementation: The Role of Structure and Process* (St. Paul: West Publishing, 1978).

2. International Council on Mining and Metals, *ICMM Sustainable Development Framework* (London: ICMM, 2003).

3. UN Global Compact, "The 10 principles," www.unglobalcompact.org (accessed February 2008).

4. Mining Association of Canada, *Towards Sustainable Mining (TSM)* (Ottawa, Canada: Mining Association of Canada, 2004), www.mining.ca.

5. Galbraith, *Strategy Implementation*.

6. Inmet Mining Web site, www.inmetmining.com (accessed February 2008).

7. Xstrata plc Web site, www.xstrata.com (accessed February 2008).

8. BHP Billiton Web site, www.bhpbilliton.com/bb/home.jsp (accessed February 2008).

9. Alcoa Web site, www.alcoa.com (accessed February 2008).

10. Anglo American Web site, www.angloamerican.co.uk (accessed February 2008).

11. Newmont Web site, www.newmont.com (accessed February 2008).

12. Ibid.

13. FNX Mining Company Web site, www.fnxmining.com (accessed February 2008).

14. B. M. Bass, "From Transactional to Transformational Leadership: Learning to Share the Vision," *Organizational Dynamics* (Winter 1990): 19–31.

15. Galbraith, *Strategy Implementation*.

16. Xstrata plc Web site, *HSEC Standards* (accessed February 2008).

17. International Organization for Standardization (ISO), ISO 14001, Geneva, Switzerland. www.iso.org/iso/iso_catalogue/management_standards/iso_9000_iso_14000.htm.

18. Occupational Health and Safety Standards (OSHAS), OSHAS 18001, www.ohsas-18001-occupational-health-and-safety.com/.

19. Ibid.

20. J. Mawson, "Art of communication," *Mining Journal* (April 2004): 18–20.

21. Global Reporting Initiative, www.globalreporting.org/Home.

22. AA1000 Assurance Standard Revision Process, 2008, www.accountabilityaa1000wiki.net/index.php/AA1000_AccountAbility_Commitment_and_Principles.

23. P. Lochner, K. Govender, S. Heather-Clark, and C. Will, *Sustainable Business Contributes to Sustainable Development—Discussion Document*. CSIR Report No. ENV-D-I 2004-005. CSIR Internal STEP Research, Stellenbosch.

24. Global Environmental Management Initiative (GEMI), *Clear Advantage: Building Shareholder Value*, 2004 report, Item No. EVI-101, www.gemi.org/resources/GEMI Clear Advantage.pdf.

25. Lochner, *Sustainable Business Contributes to Sustainable Development*.

26. Ibid.

27. J. Elkington, *Cannibals with Forks: The Triple Bottom Line of 21st Century Business* (Oxford: Capstone Publishing Limited, 1997), 69–97.

28. Lochner, *Sustainable Business Contributes to Sustainable Development*.

29. World Business Council for Sustainable Development (WBCSD), *Eco-Efficiency: Creating More Value with Less Impact* (Switzerland: WBCSD, 2000).

30. Lochner, *Sustainable Business Contributes to Sustainable Development*.

31. South African Department of Minerals & Energy (DME), Sustainable Development Through Mining Programme Web site, www.sdmining.co.za.

32. Paul Lochner et al., *Integrating Sustainability into Strategy (ISIS): A Process to Inform Sustainability Strategies, Frameworks and Reports* (Johannesburg: Annual National Conference of the International Association for Impact Assessment, 2006).

33. Lochner, et al., *Integrating Sustainability into Strategy*.

34. Chantell Ilbury and Clem Sunter, *The Mind of a Fox: Scenario Planning in Action* (Johannesburg: Human & Rousseau, 2001).

35. Clem Sunter, *Games Foxes Play: Planning for Extraordinary Times* (Johannesburg: Human & Rousseau, 2005).

36. Lochner, et al., *Integrating Sustainability into Strategy*.

37. Ibid.

38. Ibid.

39. Sasol, Annual Financial Statements 2006, "Sustainable Development," sasol.investoreports.com/sasol_ar_2006/review/html/sasol_ar_2006_82.php.

40. Ibid.

41. Sasol, *Sustainable Development Report* (Johannesburg: Sasol, 2007).

42. Ibid.

43. South Africa Department of Health, Khomanani Excellence Awards, 2005, www.doh.gov.za.

44. International Institute for Sustainable Development (IISD), Seven Questions to Sustainability: How to Assess the Contribution of Mining and Minerals Activities (Winnipeg, Manitoba: IISD, 2002).

45. R. Anthony Hodge, "Mining's Seven Questions to Sustainability: From Mitigating Impacts to Encouraging Contributions," *Episodes 27*, no. 3 (September 2004): 177–184, www.episodes.org/backissues/273/177-184.pdf.

46. Dirk J. A. Van Zyl, "Contributions of Mining Projects to Sustainable Development," in *Proceedings of 20th World Mining Congress*, November 7–11, 2005, Teheran, Iran.

47. D. Van Zyl, J. Lohry, and R. Reid, "Evaluation of Resource Management Plans in Nevada Using Seven Questions to Sustainability," in *Proceedings of the 3rd International Conference on Sustainable Development Indicators in the Mineral Industries*, ed. Z. Agioutantis (Milos Conference Center, 2007): 403–410.

48. Millenium Ecosystem Assessment, *Strengthening Capacity to Manage Ecosystem Sustainability for Human Well-Being*, 2005, www.sciforum.hu/file/mehta_goverdhan-2.ppt (accessed March 30, 2008).

49. Australian Government, Bes*t Practice Environmental Management in Mining—Checklists for Sustainable Minerals*, 2005, www.ret.gov.au/Documents/Checklists_for_Sustainable_Minerals 20051123122545.pdf (accessed March 28, 2008).

50. International Organization for Standardization. ISO 14001:2004, *Environmental Management Systems: Requirements with Guidelines on Usage*. Geneva: ISO.

51. International Organization for Standardization. ISO 14001:2004, *Environmental Management Systems: General Directives on Principles, Systems and Support Techniques*. Geneva: ISO.

52. Cobre Las Cruces, *Environmental Management Manual*, September 2008.

53. Gold Fields (Ghana), Food & Agriculture Organization, United Nations (FAO), the Ghana Water Research Institute, the Ghana Ministry of Fisheries, and the World Fish Center, *Technological Feasibility Study on Post-Mining Land Uses of the Tarkwa and Damang Operations*, unpublished report, 2003.

54. Environmental Protection Agency (EPA), Costed Reclamation Plan and Decommissioning Study of the Tarkwa and Damang Mines Closure, unpublished report, 2002.

55. TechnoServe, A Preliminary Screening Investigation of Opportunities for Sustainable Agricultural Business in the Area of the Tarkwa and Damang Gold Mines, unpublished report, 2004.

56. Peñoles, Mission and Vision, www.penoles.com.mx/penoles/ingles/about_penoles/profile/mission.php.

57. Occupational Health and Safety Standards (OSHAS), OSHAS 18000, www.ohsas-18001 -occupational-health-and-safety.com/.

58. Dupont de Nemours, *The STOP System*, www2.dupont.com/Safety_Consulting_Services/ es_MX/ (accessed April 2008).

CHAPTER SIX

Human Resources Management

L. W. Freeman and H. B. Miller

INTRODUCTION

The effective management of human resources, including employee recruitment and development, is as critical to the long-term success and economic viability of a mining company as is the development of new orebodies and exploration targets. Whether they are professional/administrative staff, highly skilled production employees, or general mine labor, these individuals comprise the backbone of any mining and resource company, where their skills, effort, and personalities greatly influence the ability of these organizations to succeed and create wealth. While the impact of labor productivity and cost on the economics of mining operations have long been a major focus of management, the traditional philosophies engrained in the mining industry toward human resources and workforce issues have recently undergone substantial change. What precipitated this change is debatable; however, many believe the cause is symptomatic of societal changes in employee attitudes and the daunting challenges facing mining companies as a consequence of a shrinking talent pool, government regulation, negative public image, and the rapidly increasing skill competencies required of today's miners. One thing is clearly evident: the expectations of employees toward employment and company management are very different than they were just two decades ago. This condition often transcends cultural and political boundaries, where the perceived responsibilities of mine management toward labor continue to expand and evolve as a function of globalization, government mandates, and societal expectations. The role of mine managers and supervisors is no longer oriented solely toward maximizing the potential utility of employees in terms of performance and work quality. In addition, they must foster a work environment that is conducive to attracting and retaining high-quality employees and to providing the necessary training to meet the needs of these highly mechanized and technology-dependent operations. As such, the ability to manage talent and understand the psychological and physical needs of a given workforce are now skills that must be fundamental to all levels of mine management.

Throughout most of the world, one of the most prominent risks now facing mining and resource companies involves their ability to develop and maintain a social right to operate. Although the legal authority to mine a specific property is embodied in the regulatory consent granted by governmental agencies, this "right to mine" is only as valid as the explicit or implicit social license granted by the potentially affected communities and stakeholders. In exchange for this social license, mining companies make a commitment to local communities to provide tangible benefits and improve the quality of life of residents during and after mining. Although often overlooked, employees play a critical role in facilitating this social license. Employees serve as a conduit for disseminating information about the company and its practices to the general public. The nature of this exchange extends beyond issues involving mine operations and environmental stewardship to how the company values its employees. Everything associated with employee talent, ranging from recruitment to skills development and safety, impacts a company's reputation and how it is perceived by local communities, either directly or indirectly. This perception and the relationship between a company and its employees are often the primary drivers on whether communities allow individual operations to continue to operate successfully. In addition, employees also convey important information back to the company about social and community issues. This feedback is critical for assessing the outcomes of specific social programs and identifying key factors that might indicate community needs or potential conflicts.

The primary theme of this chapter is to present topics that are fundamental to the creation of a corporate culture that promotes the productive utilization of human resources and the development of sustainable business practices that produce tangible benefits for the company, employees, and the communities in which they operate. This balance of this chapter is divided into four sections.

"Values-Based Principles" reviews corporate values of mining companies, focusing on achieving a business culture that fosters ethics, social license, and safety, and clearly represents a commitment to upholding standards and codes of conduct.

"Employees as Portals" analyzes the role of employees as "portals" through which each party to the social license can communicate and the importance of employees as interpreters and communicators of the needs of the community.

"The Quiet Revolution" discusses the challenges facing the minerals industry as a consequence of shortages of skilled labor and managers with the types of leadership skills and experience necessary to address the growing social responsibilities and demands they now must face.

"Talent Toolbox" discusses the management principles and tools in relation to developing a corporate culture that nurtures employees to succeed, maximize their potential, and play pivotal roles as "portals" to the community.

VALUES-BASED PRINCIPLES

Corporate values form the cornerstone of any business culture that promotes and fosters the productive utilization of human resources. These corporate values comprise the core beliefs, operating philosophy, and guiding principles that govern how a company interacts with other businesses, its employees, and society. Although these values are prominently featured in the mission statements and annual reports of most companies, they should also be readily apparent to outside observers. Within most mining companies, corporate values typically include considerations regarding ethics, social license, and safety, and clearly represent a commitment to upholding standards and codes of conduct relative to company employees, local communities, the environment, and stockholders/investors. It is essential that these values be understood by every employee within a company, regardless of position or classification, and be discussed regularly. Employees should be recruited and developed with respect to these values, and held accountable for ensuring that they are followed. The most valuable leaders are those employees, regardless of title, who manifest these values. Compromises should be unacceptable, where small breaches are as grievous as large ones, in as much as they reflect quality of character. In the vernacular of safety, all incidents and potential hazards are indicators of accidents, regardless of outcome. Such is the case for all values-based principles. Employees have the inherent responsibility to adhere to the corporate values and ensure that their peers do as well, where transgressions should be dealt with in a direct and transparent manner. All aspects of human resources management start and end with these corporate values.

Ethics

A company's ethics are inherently defined by the behavior of its employees. Accordingly, ethics play a critical role in how a company recruits, develops, and evaluates employees. Ethics is a surprisingly complex subject that goes well beyond the scope of this section. However, a few comments serve to introduce this extremely important topic. Although there are often well-defined boundaries that establish the limits of ethical conduct based on employee morals, regulatory policies, industry standards, and corporate values, a company's ethics are usually defined by the actions and behavior of its employees with respect to "gray areas." A gray area is where there is no absolute boundary separating ethical and unethical behavior or when, in the absence of prescriptive policy or standards, employees define ethics based on their own interpretation of the situation. For example, breaches of ethics such as theft and lying are easy to define. However, what about exaggeration for personal gain? Most people believe that exaggeration skirts the edge of lying to the degree that personal gain is directly involved. How directly? Imagine the case of a geologist who is considering an opportunity to acquire a mining property adjacent to an operating mine. Is it ethical to ask for a tour of the mine for the sole purpose of deciding whether the acquisition is warranted? Most industry professionals would likely believe this is an unethical act, particularly if the intent to acquire the adjacent property was not disclosed in advance of the mine tour. How about corporate intelligence? Is it ethical to spy on a competitor's drilling activities? Most people in industry would probably indicate that this type of behavior borders unethical conduct but might be situational. Conversely, what about requesting technical and economic information from a contractor regarding the activities of a competitor? In this case, most would believe it is ethical to ask, but unethical for the contractor to divulge information that would disadvantage his or her client or reveal anything deemed confidential. Is it ethical to withhold information from a competitor with regard to systems that might improve safety? Most would agree that all values-based aspects of business are to be shared and should not be held as a corporate advantage. Suffice it to say, ethics is a complex subject that often deals with issues that are not black and white. It is the gray areas that define a company's culture. As such, the ethics of any corporation or business are the product of its employees and are greatly influenced by the character of the employees the company recruits and develops. In addition, codes of conduct and company values need to be discussed regularly, particularly with regard to these ethical gray areas, to ensure every employee in a given company understands the expectations and responsibilities that accompany employment.

Social License

Social license is a covenant drawn up between the company and the affected communities and epitomizes a dynamic partnership crafted on the basis of trust and shared vision. Normally formulated to mitigate risk, the nature of this partnership is unique for each mine and community; it is dynamic in as much as both the circumstances of the mine and the community are constantly changing. As is often the case, this sophisticated covenant attempts to balance the concerns of individual stakeholders who are participating in this partnership. From the perspective of most companies, financial risk is normally a paramount issue. From the perspective of the community, any potential risk that may result in the degradation of quality of life will likely be a major concern.

How can such a covenant be maintained in the face of constant change and between two disparate parties? Employees are the key to maintaining a social license. They are the common element linking the mine/company to the community. Employees serve as the company's portal to the community, as well as the community's portal to the company. The actions and treatment of employees are realities with respect to the value systems of the parties engaged in the partnership, where employees attest to day-to-day realities of the company. Accordingly, recruitment, development, and retention of employees are critical components for the process that defines corporate culture and social license.

Although the specifics in each social license will vary, the principle of shared vision with rights and responsibilities upon which the license is developed is constant.

Each social license has two phases:

1. *The initial phase under which the mine is granted a right to begin operations.* The initial social license is granted on the basis of anticipated development needs of the mine and the changes in the community associated with this development. Lacking real-life experiences from which to draw, it is granted on the basis of promises and trust. The terms of the social license are established as part of the formal mine permitting process. The relationship between the mine and the community starts with first contact during exploration and continues through the evaluation process. The "realities" upon which trust is established are manifested through these early periods of activity and grow as individuals from local communities are hired as employees. The initial phase ends with a new reality associated with the surge of employment at the beginning of mine construction. Trust, a commitment to work together, and the ability to develop a shared vision are the key ingredients to success in the initial phase.

2. *The operating phase under which the mine functions as a business entity.* This relationship follows the initial phase and ends with successful mine closure and reclamation. It typically lasts from half a dozen years to decades, during which time there are an unlimited number of events that define the actual relationship between the mine and the community. Maintaining trust, working together, and sharing a constantly evolving vision are required for a successful operating phase.

The conceptual model for the social license is the triple bottom line (TBL). In this context, TBL refers to the process by which a company gauges success. Traditionally, success has been measured relative to financial (economic) performance and compliance with stockholder expectations. Although financial performance is still a central tenet of corporate success, TBL implies companies should also be held accountable with regard to their actions and contributions toward social and environmental factors. As such, measuring the success of a company should encapsulate all three of these criteria: economic, environmental, and social (Figure 6.1). From a community's perspective, environmental and social factors are quality-of-life issues that extend to employment, social stability, standard of living, and the health and welfare of people and their environment. For many companies, the concept of the TBL is the vision for sustainable mining and social license, and is the basis for their corporate culture.

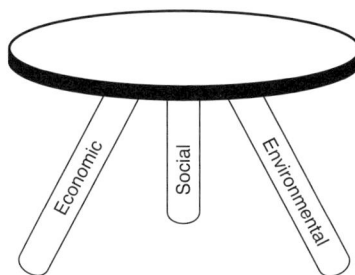

FIGURE 6.1   Triple bottom line

In this scenario, employees play a complex but integral role. They deliver the capacity for a company to succeed economically through the application of their skills and talent. Through their actions as representatives of the company, employees also directly impact the social and environmental components of the TBL. As discussed, one of the more interesting roles employees fill is that they confirm to the community that the mine is operating under the terms of the social license. Conversely, as members of the community, employees also serve as a conduit from the community back to the company.

Safety Management: Building a Culture of Prevention

It is widely acknowledged that the most efficient mines are also the safest ones. In most major operations, very systematic and comprehensive planning, monitoring, and control processes are implemented to provide continuous improvements in the health and safety of miners, both on and off the job. These processes also have direct implications toward incremental improvements in labor productivity and cost. Today, a total loss control perspective pervades mine management philosophies, and the general definition of an accident as anything that occurs that was not planned is widely held.[1]

Most major multinational mining companies have committed to the principles of sustainability and have embraced the importance of health, safety, environment, and community as part of their social responsibility and license. This commitment is often emphasized in the mission statements of these companies and in their annual reports to stockholders. In many cases, these companies have also established performance targets as a way to gauge the success of their programs and serve as corporate objectives. For example, BHP Billiton has targeted zero fatalities, zero fines, and zero prosecutions.[2] The BHP Billiton management process is systematized corporation-wide, as summarized by the following excerpts from a 2004 report:

> All sites [are] to undertake annual self-assessments against the BHP Billiton HSEC [Health, Safety, Environment and Community] Management Standards and have plans to achieve conformance with the Standards (p. 8).

> To help us better understand and manage HSEC risks that are critical to our business, risk registers are in place and being maintained at all sites and at Customer Sector Group and Corporate levels of the Company, in line with our HSEC target. Work was also undertaken to better align HSEC risk assessment processes with our Enterprise-Wide Risk Management processes to improve the efficiency of assessments.

Although program details and specifics may vary, this type of corporate philosophy and commitment has been adopted by most of the major mining corporations and information is readily available on their Web sites (e.g., www.angloamerican.co.uk, www.newmont.com, www.fcx.com, and www.riotinto.com).

Following a decade of record accomplishments in mine safety, the U.S. underground coal industry was rocked by a series of multiple-fatality disasters in 2001, 2006, and 2007. The 10-year recent history of the 3-year rolling average number of fatalities and of the 3-year rolling average fatal incidence rate in the mining industry are shown in Figure 6.2 and Figure 6.3 respectively. Among several studies scrutinizing mine safety, the National Mining Association (NMA) established the Mine Safety Technology and Training Commission, which studied a 25-year history of U.S. mine disasters and focused on requirements for an approach to prevent lost-time injuries and fatalities systematically. With the consensus of a diverse membership, the commission's 2006 report[3] recommended that "a comprehensive approach, founded on the establishment of a culture of prevention, be used to focus employees on the prevention of all accidents and injuries."

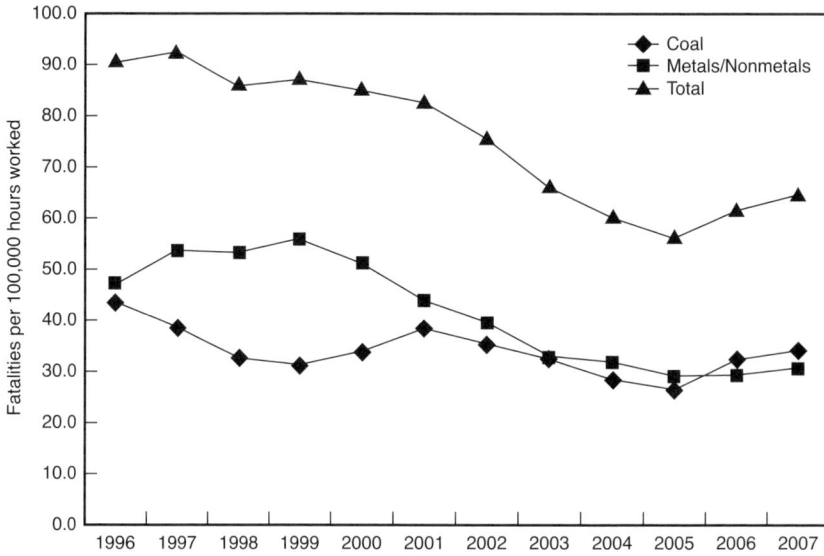

Source: Mine Safety and Health Administration database

FIGURE 6.2   Three-year rolling average number of fatalities in mining, 1996–2007

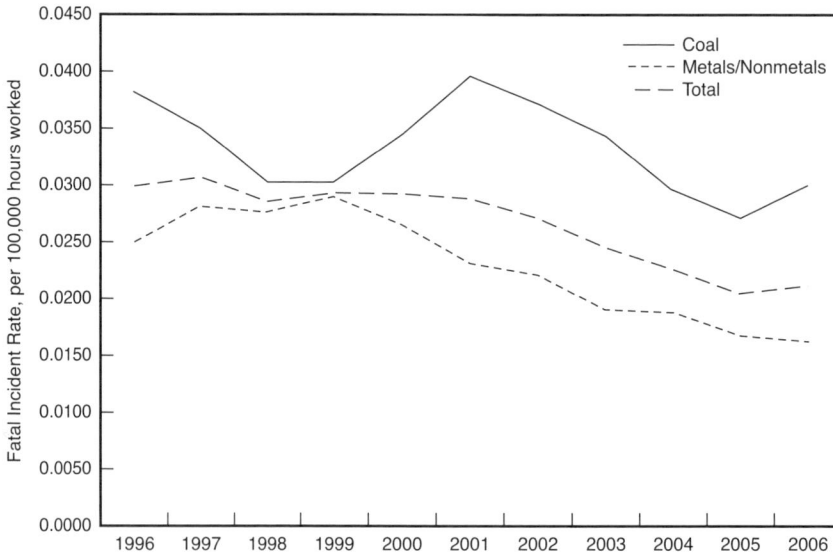

FIGURE 6.3   Three-year rolling average Fatal Incidence Rate in mining, 1996–2006

Noting the Australian adoption of risk management and the industry's significant improvement of mine safety performances, the report further recommended "every mine should employ a sound risk-analysis process, should conduct a risk analysis, and should develop a management plan to address the significant hazards identified by the analysis."

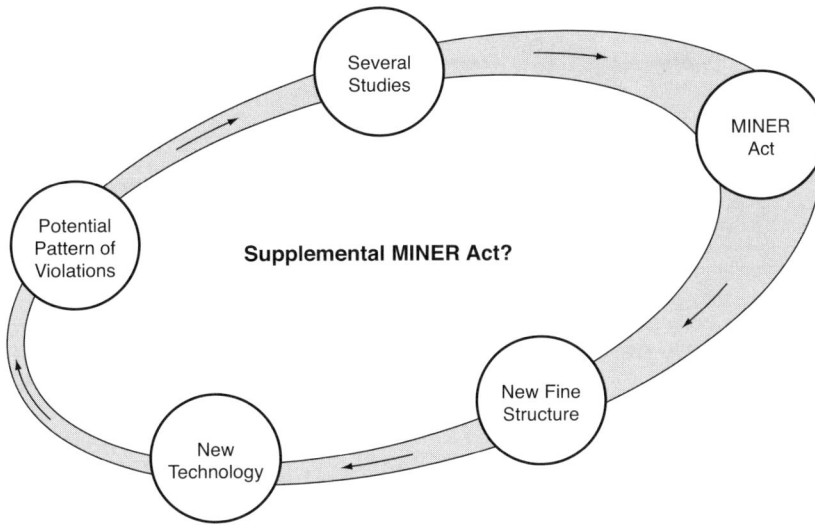

FIGURE 6.4   Impacts from 2006 U.S. mine disasters

One of the commission members, Consol Energy chief executive officer J. Brett Harvey, reinforced the study's position that the goal of the industry should be zero fatalities and zero lost-time accidents.[4] At the 2007 meeting of the Utah Mining Association, he embraced the new safety paradigm by saying:

> We are in the process of instituting a new approach to safety awareness and training that we believe will accelerate our drive to zero accidents throughout the company. We will start with the premise that our normal state of operation is no accidents. An accident is an abnormality that is unacceptable. Accidents are an exception to our core values.

Building a Mine Safety Culture of Prevention

Beyond the studies mentioned previously, the aftermath of these highly publicized mine disasters brought major new mine safety and health legislation passed by the U.S. Congress (including the Mine Improvement and New Emergency Response [MINER] Act of 2006),[5] a significant increase in the fine structure of citations for violations of regulations, and strong initiatives to develop new technology for mine communications and life-saving devices. It also led to the development by the Mine Safety and Health Administration (MSHA) of a methodology and algorithm—pattern of violations—to target poor safety performances at mine operations (Figure 6.4). A second mine safety initiative is now being considered in Congress called the Supplemental MINER Act. These activities and initiatives raised the stakes for poor safety performances to a much higher level, as follows:

- Having a fatality is a prohibitive stigma,
- Having lost-time accidents targets operations,
- Having a pattern of violations is costly, and
- In either situation, MSHA and public scrutiny will also be very undesirable.

Paramount in the new perspective developed following 2006 and 2007 is a new emphasis on systematic safety improvement. Focusing on the bottom line, major companies' operating plans

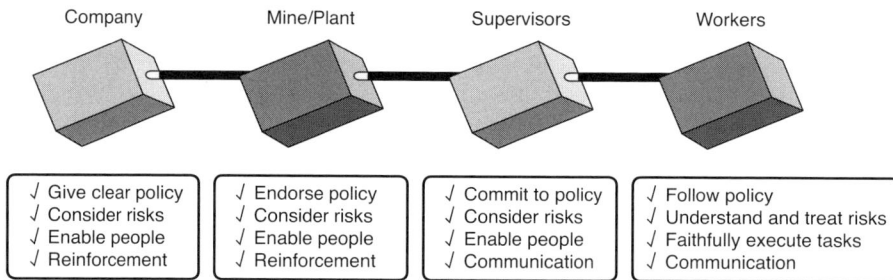

Company	Mine/Plant	Supervisors	Workers
√ Give clear policy √ Consider risks √ Enable people √ Reinforcement	√ Endorse policy √ Consider risks √ Enable people √ Reinforcement	√ Commit to policy √ Consider risks √ Enable people √ Communication	√ Follow policy √ Understand and treat risks √ Faithfully execute tasks √ Communication

Source: After L. Grayson 2001

FIGURE 6.5   Roles of managers, supervisors, and workers in building a safety culture of prevention

do not sanction losses of any type, adopt a continuous improvement mindset, expect good execution of all work tasks (by management and labor), and expect that losses of people, time, money, equipment, injuries, energy, and so forth will be avoided. The role of each manager, supervisor, and worker becomes critical as losses are perceived as exceptions to business—or the plan. For success, it is imperative that a culture of prevention be introduced, reinforced, and cultivated.

The roles of managers, supervisors, and workers are briefly described in Figure 6.5. In essence, the key facets in building a mine safety culture of prevention relate to setting, committing to, and following policy; considering, understanding, analyzing, and treating identified risks; enabling and empowering workers toward faithful task execution; and maintaining good, interactive communication among supervisors and workers, which is reinforced by higher levels of management.

Concerning the building of a culture of prevention, the NMA commission report noted the following:

> A critical action to ensure success of the process for any company is the creation of a "culture of prevention" that focuses all employees on the prevention of all accidents and injuries. In order to achieve this culture, operators, employees, the inspectorate, etc., share a fundamental commitment to it as a value. In essence the process moves the organization from a culture of reaction to a culture of prevention. Rather than responding to an accident or injury that has occurred, the company proactively addresses perceived potential problem areas before they occur.

> The tenets of the core business and personal values are critical links to achieving success. Founded on these aspects, a systematic and comprehensive process depends on other keys for success as well, which include the following:

> • Focus on ZERO Accidents, Injuries and Occupational Illnesses

> • Use a Holistic Approach

> • Identify, Disseminate, and Adopt Best Practices

> • Implementation of a Risk Management Process

> • Going Beyond Compliance

> • Minimizing the Footprint

> In the end, the industry must strive toward instilling a paradigm of prevention. There is no single path or approach to fashioning the safety culture. Rather the industry must

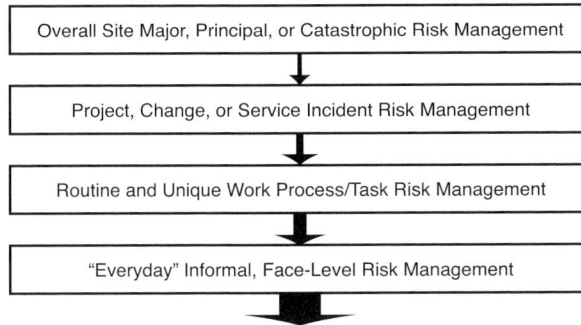

FIGURE 6.6   Evolutional journey toward risk management and building a culture of prevention

call upon engineering solutions, education of the workforce, and enforcement as tools to help create this culture. It must also enlist the commitment of all employees to be an integral part of the process aimed at zero incidents. Training is undertaken as a preventative measure, especially for honing critical skill sets for hazard awareness and control for every employee at every level of an operation. This safety commitment is the foundation on which the industry must build to fortify the protection for all employees from incidents or injuries, and not just from fires and explosions.

Culture building is not, however, easy to do. Companies that have applied risk analysis and management for many years also recognize that the change to a "culture of prevention" via "systematic and comprehensive risk management" involves a journey.[6] Professor Hudson depicted the journey graphically for Shell Oil in Figure 6.6,[7] where the process is shown as a series of steps. Importantly, moving through each of these steps is believed to take several years—each contributing to the building of the culture.

Corporate or division leaders set the stage and play a critical role in challenging all employees to seek accident-free performances, insisting on building a culture of prevention. Serious transfer of accountability then must permeate downward to the next level of responsibility. At this stage, the mine/plant manager plays a critical role in challenging supervisors to seek accident-free performances, which further builds the culture of prevention. In the end, supervisors must transfer accountability for accident-free performances to the workers. Ultimately, each worker plays a critical role in changing the culture permanently by (1) executing tasks faithfully according to best practice, (2) not taking shortcuts, (3) examining the workplace, (4) performing proper pre-op checks, and (5) using good judgment.

Risk Management

Risk management is a well-known loss control methodology that has been applied by many industries, including chemical, oil and natural gas, nuclear, military, aviation, environment, and space.[8] These industries consider risk management as an integral part of their daily business. A number of generic risk assessment and management standards and guidelines are available. The NMA commission report noted the development by the University of Queensland, Minerals Industry Safety and Health Centre (MISHC), of a guideline document aimed to provide advice on risk assessment within the Australian mining industry.[9] Based on several standards, risk management is a process comprised of several steps:[10]

1. Establish the context.

2. Identify risks.

3. Analyze risks.

4. Evaluate risks.

5. Treat risks.

During this process, there is regular communication and consultation as efforts progress, as well as regular monitoring and review of activities.

There are many different ways to assess risk relative to mine safety performances. Common methods include trend plots of incidents (violations, near misses, downtime, etc.), tabling data for prioritization of action plans, using matrix plots to prioritize multiple risks (major hazards, injury causes, violations), and quantitative risk analysis using advanced methods such as fault tree analysis, failure modes effects analysis, preliminary hazard analysis, and so forth. An example of a trend plot for a specific violation of the MINER Act is shown in Figure 6.7.

Table 6.1 presents lost-time accident data listed in descending order in column 2, for frequency of occurrence, and in column 4, for lost workdays. It is interesting to note that based on frequency, 79.1% of total reportable accidents are accounted for by five categories of injury while 92.8% of the lost workdays are accounted for by the same five categories, in a different order.

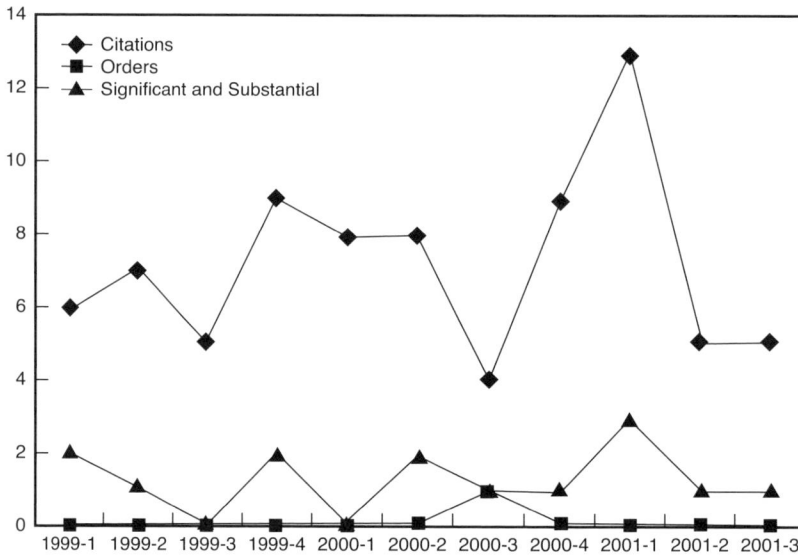

FIGURE 6.7   Trend plot of citations for violation of correction-of-hazardous-conditions regulation

TABLE 6.1   Prioritization of risk mitigation for lost-time accidents

Accident Class	Number	Accident Class	Workdays Lost
Material Handling	52	Material Handling	2,213
Handtool	23	Machinery	913
Slip/Fall	20	Slip/Fall	681
Machinery	17	Powered Haulage	910
Ignition/Explosion	9	Handtool	336

This approach shows the value of the 80-20 rule (Pareto principle) in prioritizing which lost-time accident categories should be addressed first.

Using probability and severity categories, a risk analysis matrix can be developed for any type of events and used to identify unacceptable risks for an operation (Figure 6.8).[11] It can also be used to prioritize which risks will be addressed (i.e., action or actions taken to eliminate or reduce risk, and in which order). An example risk analysis matrix is given in Figure 6.8 using some common mine hazards. The highest priority cells are located in the upper left part of the matrix, while the lowest priority cells are in the lower right corner. The approach could be used to compare the impact of many different events, and both quantitative and qualitative risks can be represented.

The Bottom Line

In a complex physical system, a weak link can cause a failure or loss. Similarly, in a human system, weak links in human performance, such as taking a shortcut, misjudgment, poor task performance, and so on, can cause a failure or loss. Ultimately, each person at an operation plays a role in safe, efficient, cost-effective production—whether that person is a corporate or division manager, the mine/plant manager, a supervisor in production or maintenance, a technical staff person, or a worker. As managers and workers alike realize their roles, a commitment to a safety culture of prevention and executing that commitment systematically reaps the following paybacks:

- The majority of excursions from plan are eliminated—lost-time accidents, citations for violations of the MINER Act, downtime, untimely progress on projects, problems with contractors, and so forth.

- The industry will strive for continuous improvement across the board as excellent performers—always looking for better ways of doing work and sustaining business.

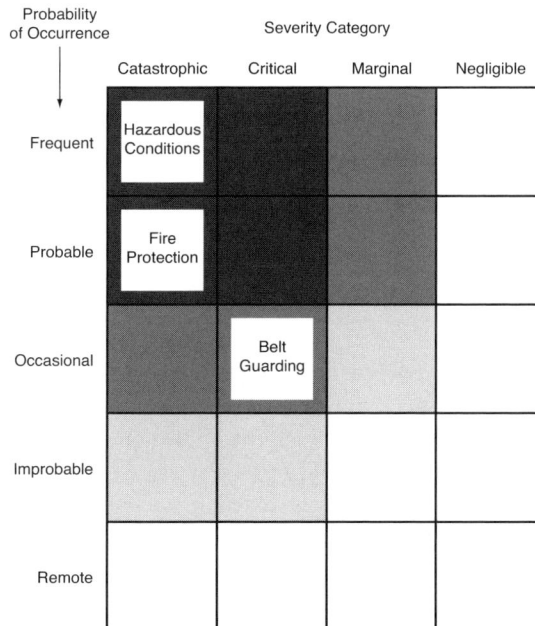

FIGURE 6.8   Example qualitative risk analysis matrix for three mine hazards

Acknowledgments

The material and ideas presented in this section were contributed by Dr. R. L. Grayson of the Pennsylvania State University, Department of Energy and Mineral Engineering, University Park, Pennsylvania. The chapter editors extend their sincere thanks to Dr. Grayson for his contribution and participation in the development of this section.

EMPLOYEES AS PORTALS

Depending on the specific circumstances, the triple bottom line (TBL) puts employees in a unique position of influence. In effect, these employees represent the interests of both parties participating in the social license (i.e., the company and the community). This represents a radical departure from traditional business philosophies, where large companies often viewed employees as simply the means through which they could achieve corporate objectives and create wealth. This is the traditional language of business. However, communities possess a very different perspective (language) that revolves around quality-of-life issues that are based on values systems. Values systems are attested to and measured by the actions of companies and their representatives (i.e., employees and contractors). Inside this intricate relationship, employees are both members of the company and the community and are, therefore, portals through which each party to the social license can communicate.

Carrying this analogy further, a portal (window) is only as good as the quality of the glass used in its construction. Any imperfections in the glass result in a distorted or obscured view of the events occurring on either side of the portal, regardless of what the actual reality is. Similarly, it follows that employees represent the company to the community and the community to the company. The values and the actions of the company are no more correct and proper than how they are perceived by the community. Furthermore, the needs of the community are only understood to the degree that employees understand and can communicate these needs to the company in such a way that management can take action. As such, the ability of a company to mitigate the risk of losing its social license largely rests on the actions, knowledge, and perceptions of their employees.

Therefore, employees serve three primary purposes:

1. Complete necessary tasks associated with mine production and operations;

2. Serve as a vehicle to accomplish wealth transfer from the mine operation and the community via salaries; and

3. Serve as portals of communication to maintain the social license as a dynamic, evolving covenant between the community and the company, thereby mitigating risk to the company (loss of operation) and to the community (loss of livelihood and preventable social and environmental degradation).

This complex relationship between local communities (stakeholders) and the company (shareholders) is facilitated through employee interaction, as is illustrated in Figure 6.9. The figure emphasizes the differences and risks between the opposing groups, where employees serve as the interface that provides balance and stability to the system. That said, who ultimately has the responsibility for initiating and maintaining the social license—the company or the community? Many believe that the balance is systematically shifting from the former to the latter. Although companies still have the responsibility to facilitate social license and fulfill commitments related to their core values, they rarely have the ability to dictate the terms of the social license, nor can they operate effectively without the continued support of the community. For companies, this places an even greater dependency on their employees to mitigate potential social risk and

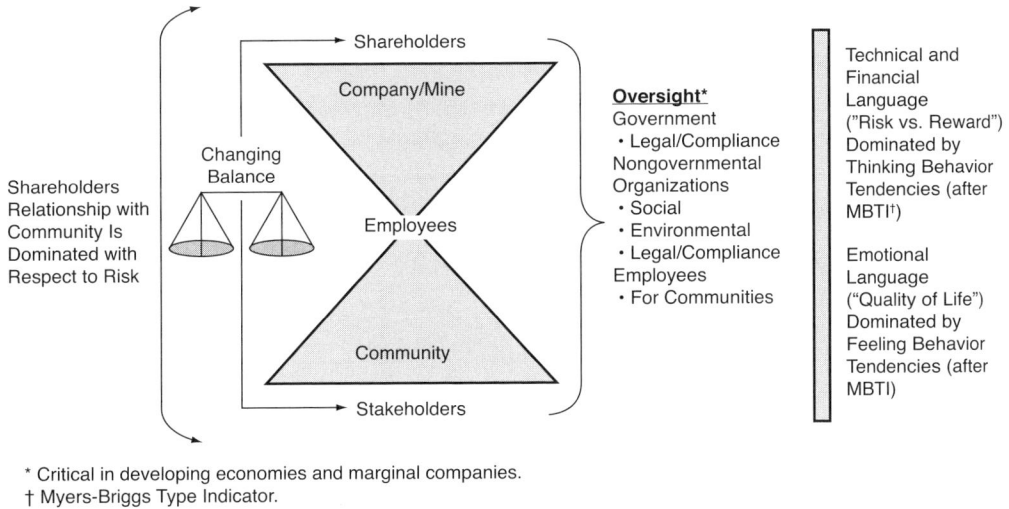

* Critical in developing economies and marginal companies.
† Myers-Briggs Type Indicator.

Source: After Freeman 2006–2007

FIGURE 6.9   Employee interface

represent their interests by advocating the merits and corporate values of a given operation. This has direct ramifications on the type and character of employees a company will recruit, as well as human resource policies related to employee training, development, and performance assessment.

THE QUIET REVOLUTION

As the minerals industry begins to experience the ramifications of a decade-long shortage of skilled labor as a consequence of demographics and tremendous growth, mining companies face an equally daunting challenge of finding managers with the types of leadership skills necessary to address the growing social responsibilities and demands they now face. Although many of the traditional character traits and skill sets associated with managers are still vitally important, the changing social expectations that stem from social license and the TBL necessitate that company leaders possess talents and capabilities very different from just a generation ago.

Managing social risk now requires the ability to interact effectively and influence a wide array of different groups and organizations external to the company, including those representing local communities, government and regulatory entities, media, environmental and social activists, and nongovernmental organizations (NGOs). Although managers have always needed the personal skills to influence and motivate labor, the difference is that managers rarely have substantive control over external stakeholder groups. In some situations, social license means delegating, or at least sharing, some element of control with these groups (e.g., the local community). As such, managerial responsibility for some activities that can directly impact mine and mill operations may be dependent on the participation, input, and/or approval of these stakeholders. To many managers in the minerals industry, this situation may be very unsettling. It recognizes that mitigating and managing social risk is largely dependent on influencing skills rather than through their authority as a representative and agent of the company, and requires a higher level of leadership in order to be successful. For those managers schooled and developed over the course of the last three decades, where authority tracked with responsibility, this may be difficult to accept and perhaps even more difficult to practice. This calls for no less than a revolution in the way management and future company leaders are recruited, trained, and mentored. In many progressive mining and resource companies, this process is well under way.

In most corporate settings, major changes of this type typically are met with resistance and are slow to materialize. Many human resources professionals, however, believe this won't be the case in the mining sector, given the extraordinarily high rate of managerial turnover expected in the near term. As projected, half of the current practicing professionals in the mining industry will retire in the coming decade. Similar trends are also expected for the skilled labor pool from which frontline supervisors are often identified and recruited. As a consequence, almost by default, a new generation of managers and supervisors will quickly be taking the helm of these operations. Fortunately, these individuals are likely to be more accomplished in the use of influence as a leadership style due to current trends in education. Regardless, company management must pick and train their potential successors carefully to ensure a successful transition with leadership capable of operating under the social terms and expectations of today's employees and stakeholders.

Shortage of People

A number of studies commissioned in recent years have identified several ominous trends that have and will continue to impact nearly every sector of the mining industry. Nearly 50% of the current workforce will retire in the next 6 to 8 years, and finding replacements with the necessary skill sets will become an extremely difficult challenge.[12]

Traditionally, those retiring from managerial positions in the mining industry fell into a demographic group ranging from 60 to 65 years of age. In most instances, these retirees were replaced by young, mid-career professionals from 30 to 45 years of age. Owing to the demographics associated with the baby-boom generation and long-term trends in the graduation rates of U.S. mining schools, there currently exists a distinct shortage of 30-to-45-year-olds in the industry.[13] Consequently, there are few successors to replace the managers who are currently retiring. Based on available data, it appears that this shortage will persist for at least the coming decade.

To illustrate this point, Figure 6.10 shows the number of mining engineering graduates from U.S. universities as compared to the average spot market price for copper. As the figure indicates, the dynamic associated with the shortage of mining engineers is not specifically related to the recent surge in commodity prices. Rather, the primary driver appears to be the attrition rate caused by retirement and the demographics associated with the baby-boom generation. That said, the impact of market prices in stimulating new projects and mine expansions is undoubtedly having an effect in tightening an already grossly labor-deficient market. The continued economic growth and activity in China, India, and other developing countries should promise to exacerbate this labor demand into the foreseeable future.

The x-axis of Figure 6.10 is delineated by both the year of graduation and average copper price. Also shown is the approximate age of the graduate in 2006. The y-axis is a percent normalized to the year 2000. As can be seen from the figure, the mining engineering graduation rate peaked in the early 1980s at more than 700 graduates. Since the late 1980s, the graduation rate has varied from 150 to less than 90, with an average rate of approximately 125 graduates per year.

Research conducted by the Society for Mining, Metallurgy, and Exploration, Inc.,[14] suggests that the sustaining rate for new mining engineers entering the industry is somewhere between 300 to 350 per year. This implies that the U.S. mining industry has been running at a deficit of some 200 graduate mining engineers per year for more than 20 years. It is, therefore, theorized that the mining industry has been "living off" the relative surplus of graduates from the early

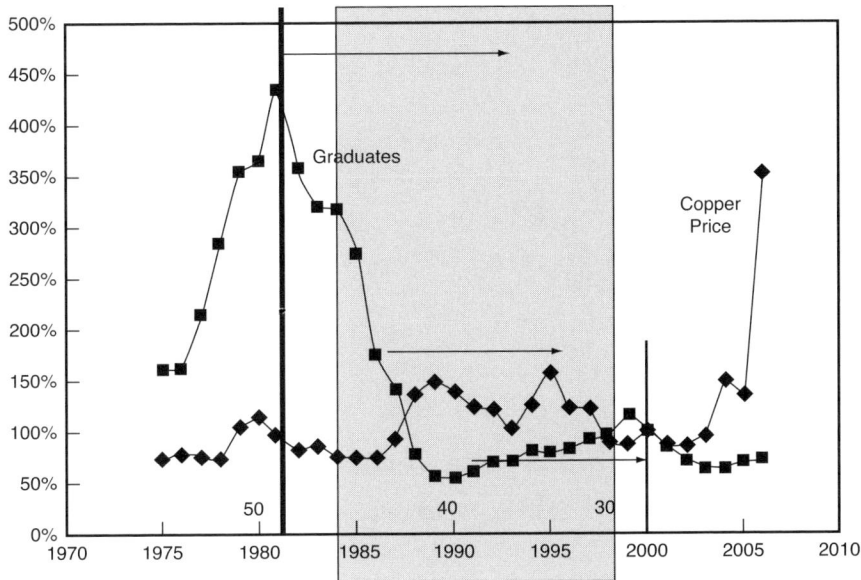

FIGURE 6.10   Total U.S. mining graduates versus average copper price

1980s in mining engineering, as well as other critical disciplines, including extractive metallurgy and economic geology. Virtually all other technical disciplines, in all other developed economies, exhibit similar statistics.[15]

The talent shortage is acute and not easily remedied. Although there is currently tremendous competition for new graduates, these inexperienced engineers simply cannot replace the vast majority of those who are retiring from the industry because they do not possess the required skills and competencies. In most cases, it takes time (experience), training, and mentoring to acquire and develop these attributes. This point is clearly illustrated in Figure 6.11.[16] This figure was developed by analyzing the career paths of hundreds of the most successful people in the mining industry in order to determine the time necessary to develop the competencies required for increasing levels of responsibility.

The x-axis of this figure represents the years of experience since graduating from college with a professional degree. The y-axis establishes the individual's work-related title, as well as his or her relative level of responsibility, ranging from new graduate (1), specialist (2), senior specialist (3), manager (4), general manager (5), and executive (6). The most successful professionals are represented by the left-most line on the graph.

This research indicates that it took 10 years for the best of these graduates to develop the competencies necessary to become managers. Consequently, it seems logical to conclude that most new graduates are not likely to be capable of replacing manager-level retirees in a period less than this (e.g., 10 years). The loss of management as well as leadership will leave a large gap in the talent pool of the mining industry.

Shortage of Competent People

Well before companies run out of people to fill positions, they will run out of competent people to fill positions. As such, employees and new recruits will be asked to assume the responsibilities of positions for which they are ill-suited or ill-prepared. As described previously, despite shortages

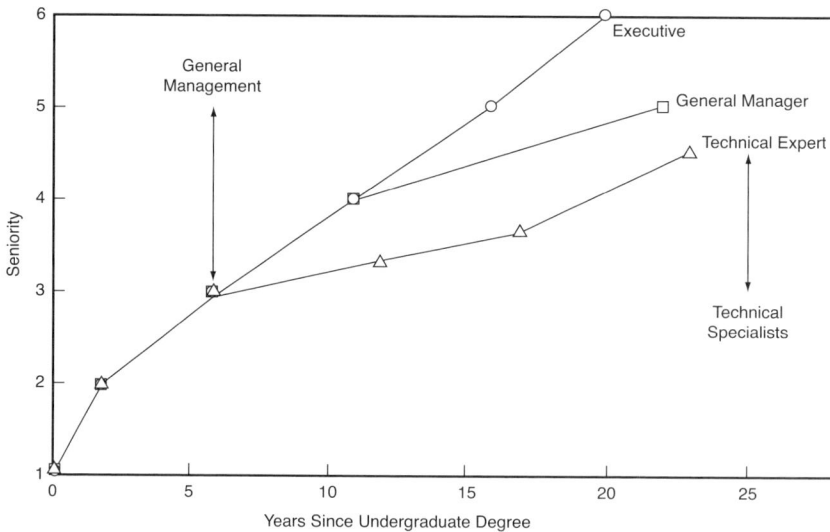

Source: After Freeman 2006–2007

FIGURE 6.11   Professional competency development in the mining industry

of skilled labor, there is particular concern over the pending shortfall of managers and supervisors with social skills sufficient to meet the demands of these positions successfully. High school graduates with particularly strong math and science skills are usually drawn to engineering when they reach college. In most university curriculums related to mining, economic geology, and mineral processing, education is strongly focused on math and science. Therefore, virtually all of the graduates from accredited universities are likely to be technically competent. This, unfortunately, cannot be said about their social skills. The typical student entering a university mining program in the United States is likely to be more interested in and display a higher degree of competency in technical skills than in those dealing with social issues. In fact, these students generally refrain from taking courses in communications, leadership, and management unless required to do so. In addition, the structure and emphasis of most engineering curriculums, along with accreditation requirements, further contribute to the relatively low priority placed on social and leadership skills development.

Figure 6.12[17] illustrates this point, where a normal curve represents the overall competency of people within the workforce. In this figure, the x-axis corresponds to different levels of competency and the y-axis denotes the percentage of the population of professionals. Relative technical competency is illustrated as a straight line connecting the upper right corner of the figure (representing the highest possible level of technical competency) to a mid-point on the y-axis (representing professionals at the lowest acceptable level of overall competency). Relative social competency is illustrated by a second straight line varying from the highest level on the right to virtually zero on the left. The statistical model of Figure 6.12 recognizes that these professionals were neither drawn to mining disciplines on the basis of social skills nor were their social skills specifically advanced during the education process.

Ideally, managers and leaders are drawn from the upper portions of the competency normal curve. As shortages of people emerge, leaders and managers with less competency will be called on to fill critical positions. Although their technical skills are likely adequate to fulfill the requirements of the job, they may fall substantially short with respect to social skills.

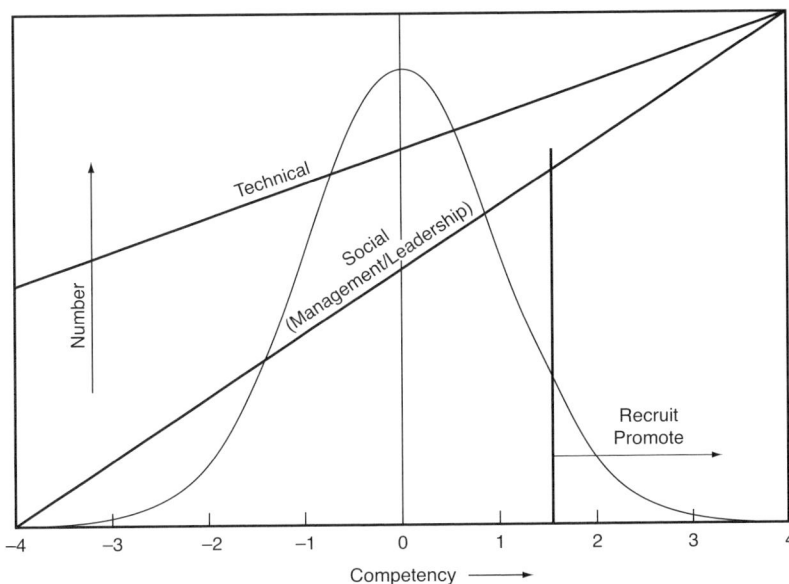

Source: After Freeman 2006–2007

FIGURE 6.12 Technical and social competencies

Thus, the industry is likely to run out of competent people before it runs out of people to hire. Demographics dictate that these unfortunate phenomena promise to become more acute over the next decade, even while the need for a substantially higher degree of social competency is necessary for mines to maintain a social license and operate successfully.

TALENT TOOLBOX

The concept of the talent toolbox is based on the practical application of fundamental management principles with respect to developing a corporate culture that nurtures employees to succeed, maximize their potential, and play a pivotal role as a portal to the community.

Leadership

In the context used here, social skills refer to competencies that are crucial for an individual to be successful as a manager as well as a leader. Management is simply defined as responsibility that corresponds with authority. Leadership, however, can be characterized as the ability to meet responsibilities without using authority. As such, leadership is fundamentally different than management in that it is often viewed as the ability to achieve specific corporate goals and objectives by influencing others over whom they have no authority.

Management implies a relatively high position on an organizational chart while leadership can exist anywhere within the organization. Leaders can influence those above and below them in the organizational ladder, as well as their peers. Most important, however, is the ability to influence parties not part of the formal organization. This might include local communities, regulators, investors, potential employees, and special interest groups. For example, it is usually ill-advised for mine managers to attempt to "manage" a community that is party to the mine's social license. They can, however, offer leadership and thereby provide some measure of influence on the community.

A fundamental concept of the philosophy of "employees as portals" discussed in this section is that it provides each employee the opportunity to be a leader.

Just as all employees have the potential to serve as leaders, all employees also have an opportunity and an inherent responsibility to serve as portals to the community. For example, a truck driver from a mine who coaches his or her son in Little League baseball has countless opportunities to represent the values of the company to the parents of other children on the team, as well as the parents of opposing teams and league officials. Similarly, this coach has a unique opportunity to understand the needs of the community on which the dynamic social license is crafted and influence how the resulting business practices (sustainable) are developed and implemented.

Fairly or unfairly, the truck driver coach can often be much more effective in communicating the values and culture of the company than even the general manager. The participation of mine management in civic and charitable organizations and community issues is often perceived by the community as stemming from the economic and vested interests of the mine rather than from a sincere concern over a community's well-being or quality of life.

Developed Versus Developing Economies

Employees of mines in both developed and developing economies can and do play an important role in serving as portals to the community. Although there are fundamental differences in the capacities of typical employees and communities that are dependent on the unique economic characteristics of the mine's location (developed versus developing countries), the overall dynamic remains fairly similar.

The capacity of the typical employee to understand and represent the circumstances of the mine and the community can vary significantly. Employees from developed countries are likely to be at least high-school-educated and have had an opportunity to encounter a number of professional and personal experiences outside their home community. Those from developing economies are less likely to be as well educated and to possess the same type of experiences in the outside world.

Although the role of employees to serve as portals of communication between the company and the community are important in both economies, this function is arguably more critical in developing countries where:

- The level of community understanding with respect to evaluating short-, medium-, and long-term benefits and risks associated with specific decisions related to project development are likely to be low;
- Governments are less likely to be capable of serving in a position of oversight; and
- The need for economic benefits is greater. This lack of capacity invites an overemphasis on short-term economic benefits by communities in developing communities, which in turn may be detrimental to sustainable and prudent decision making.

The deficiency in the capacity of communities in developing economies to represent themselves effectively in the process of developing a social license, as well as in the evolution of this relationship over time, is commonly filled by NGOs and special interest groups. Many NGOs aptly serve in this capacity, where mining and resource companies will often encourage their involvement. Unfortunately, many cases exist where some of these external groups and organizations possess ulterior motives and/or political and ideological agendas that destabilize this process and irreparably damage the relationship between the mine and the community stakeholders. In other cases, the NGO simply does not possess the expertise, knowledge, or resources to participate in this capacity properly and performs a disservice to the communities they are trying to help. As such, it behooves companies with mines in developing economies to build capacity and relationships with the potentially affected communities and stakeholders early in the process (e.g., exploration). This includes the NGOs and other groups that may participate on behalf of the community. In addition, the company must reach out and actively engage current and potential employees living in the community in order to convey trust and promote a corporate culture that will eventually enable these individuals to serve as effective portals to the community.

Information Management

Business strategy drives all activities in a company. Therefore, the human resources (HR) function requires defining the processes and tools around a specific business strategy. In HR, there are unique tools and levers a company can pull to get the outcomes necessary to drive strategy, but decisions need to be backed with data. This section is more about the data and pulling knowledge from that data than the systems themselves. Information management is one area where the HR function has something to learn from other functions of the business, such as the finance and commercial functions. Especially in engineering-led businesses such as mining and construction services, data is paramount in driving decisions. HR programs and initiatives that are in alignment with the strategy and backed up with a payback, NPV, IRR, or simply convincing data are easily sold as adding value. However, making those arguments with HR data is obviously easier said than done. Human resource information systems or human resources management systems (HRMS) are the databases and tools of HR-related data for driving decisions. HRMS capabilities have been enhanced into an integrated toolbox that can expand to the desired complexity of most organizations. Some of the most important HR decisions come down to what an organization will not use in a large and complex HRMS.

The Model

Information management is an integral part of the HR value proposition. The model in Figure 6.13 displays areas of HR capabilities that are demanded from the business. HR data and technology are an important part of the offering. There is a strong link from the HR technology and information systems to the ability to deliver. HR delivery has to do with responsiveness, staffing, development, and performance management, but without the processes and tools of an integrated HR information system, it cannot be delivered effectively.

Courtesy of Wayne Brockbank, University of Michigan

FIGURE 6.13   Model of human resources capabilities

Data-Driven Departments

HR departments that are data driven will be the most successful in helping the organization achieve business results. HR metrics are a way to drive these results. Whatever is measured and consistently reviewed is taken seriously in organizations. Understanding that the management team is dominated by managers who are data driven and quite competitive helps design effective communication to drive behavior in the business. For example, a large part of the HR strategy is built around retention of the right employees. Reviewing simple turnover rates and tracking of high-potential employees drive certain behaviors in the business. The regional managers understand the simple strategy and push the group to aspire to hold on to these valuable individuals and also add to the company's talent ranks by recruiting high potentials.

HR data is perhaps among the most difficult data to compile and retain. The complexity, volume, and continuous inputs stand out as reasons that records are not kept up to date. Employee self-service is perhaps the best methodology in certain areas, such as training records, development plans, and personal data. Every HRMS needs to be built for a purpose, as there are large amounts of data that have no value if kept. Only the data that is essential for driving business strategy should be a focus. Constant reviews of data are tedious but important. Any data tied to costs per department will usually be picked up immediately. Likewise, data linked to cost centers are usually kept up to date because of most organizations' focus on efficiency and costs. The greater the amount of HR data maintenance that can be tied to the business' bottom line (costs) or performed through employee self-service, the more efficient data maintenance will be.

High-Performance Organizations

A critical part of the HRMS is performance data on individual employees. Resource companies have generally kept and used performance ratings. Few have compiled and analyzed the data for use in driving performance in the future. A simple lag indicator analysis to run is the correlation between performance of employees, financial performance of the business, and management total rewards for the same time frame. It will give a good indication of how pay for performance is working in one's business. Taking that information and influencing behavior for the future can add value.

Another example of using HR data to optimize organizational performance is through what some companies label as "contestability" sessions. This is not the method of forcing performance data to fit a curve as a normal distribution; however, it is having a conversation with line managers and asking questions and understanding why managers have rated a certain way within a peer review. In the contestability session, before any employee is given a rating, managers at the appropriate level discuss the ratings they are giving their employees and discuss their reasons. The session is driven by data, and managers are encouraged to discuss specifics on progress of objectives. This process driven by data helps managers understand how the organization keeps people accountable within the culture of the business and places rigor in the process.

Focus List of High-Potential Employees—Key Retention Information

In addition to performance data, another critical area for information is in business succession plans and the resulting development plans of employees. There is no doubt that data on succession and key positions in the company is essential to strategic planning. This information is usually straightforward and simple to understand. Where it gets more complex is tailoring development plans with key individuals for particular roles in the future. With recruiting top talent becoming more difficult and an overall shortage in certain engineering areas, more in-depth time and energy should go toward sitting down with certain individuals and planning a career path. After the plan is discussed and agreed upon, placing the information in a system to track progress and help plan for the future is essential for this process to be successful. One competency that high-performing employees often lack is the ability to develop themselves continually and actively. This is a good way to drive behavior through information systems by tracking the progress of development plans.

Global Environment

Resource companies have always had a global aspect to them because of the orebodies they mine around the world. Given that fact, they are also becoming increasingly global in leadership and global organizational structures. Some companies have global matrix leadership teams and others make strong attempts to share best practices, synergies, and a common culture around the world. Never underestimate the "power of time zones" in regard to managing cross-border and cross-cultural work teams. HR information must take into consideration the complexities of language, foreign exchange (FX), country regulation on information, and culture. One example of complexity in global information is the FX issues with regard to short-term incentive targets. The question becomes a discussion on holding employees accountable for the risk of FX, a component over which they have no control. The only control they can take is possibly to mitigate against the downside. What happens when the company hits the target on a local level in local currency but they do not make the overall targets in the country where the shareholders sit because of FX rates? It has been determined that senior executives in the company must take on the additional challenge of FX rates because the shareholders are looking at the local market share price, wherever that might be. The stock market does not give the business a break because of FX rates. Deeper in the organization, short-term incentives are structured based on local currency. These decisions are strategic in nature and will reflect the culture of the business.

When it comes to information in a global organization, the simpler the process, the more successful it will be in organizing, updating, and holding people accountable. Translation issues in data and communication can widen the gaps of country borders. Having data in the HRMS in the local language is obviously very important; it's also mandated in certain parts of the world, such as Quebec in Canada.

Compensation—Total Rewards

Traditional HRMS have been able to collect and analyze employee compensation information. This is even more important given the higher demand for talent. The traditional 12-month review of compensation is being tested, with companies sometimes doing 6-month reviews, which are even more frequent for certain professions. HR information must be real time and proactive in keeping the basics, such as market data and local trends, correct. The reasons employees leave a company are often because of their managers and the corporate culture. But if a company does not get the basics correct, such as keeping salaries in line with the market, retention can be quite difficult.

Systems now attempt to compile total rewards or all parts of employee compensation in some form to help understand the costs of employment and also to educate the employees about all the other benefits. This is a bigger issue in the United States and other countries that do not have a government mandate on minimum health benefits, with employee benefits packages becoming a large part of total rewards. Employees often do not realize how much a company contributes to benefits packages and retirement plans. Employees often discount defined contribution and defined benefit plans unless companies take the time to educate their employees with data on how much the company contributes to the plans. However, the company needs to be competitive in the market before distributing a total rewards statement. Informing the workforce that they are underpaid in the marketplace does not inspire employee loyalty.

Knowledge Management

Many companies aspire to achieve a competitive advantage by sharing information, turning the information into knowledge, and using that knowledge to help with business issues. Large consulting companies have been doing this quite well for many years. Resource companies can use the same principles when experiencing issues at one site; by canvassing the issue with all operations and employees throughout the entire corporation, they may find experts who can help resolve these problems. There are many aspects to *knowledge management*, a term that is used frequently in consulting groups and universities around the globe. From an HR standpoint, one aspect of knowledge management is the use of systems and processes to leverage global knowledge. Another area where HR can add value to the business through systems and information is by attempting to place forums and knowledge groups together.

Employee Development

As the forces of globalization change the shape of industry at an accelerating rate, the area of employee development has struggled to keep pace. This is particularly evident in the mining industry, an industry that is faced with both the need for a high order of technical mastery to cope with a rapidly expanding global market and a widespread shortage of high-caliber young applicants. The result of these dynamics is a clear risk to the sustainability of companies within the industry who do not aggressively confront the need to develop future managers capable of executing company strategy across borders.

Globalization and the Next Generation of Industry Professionals

Tomorrow's leaders will be increasingly called on to operate in unfamiliar and uncomfortable settings and surroundings. The search for new markets, operational efficiencies, and resources poses a requirement for companies large and small to pursue opportunities in a virtually borderless context. That is not to say that the world has kept pace with the boundaryless charge by industry. Rather, important cultural, linguistic, and legal differences remain as barriers to organizational and operational efficiency. Only those managers capable of navigating the dynamics and idiosyncrasies of this global terrain will succeed and thrive in the future.

Despite its shortcomings and challenges, the mining industry is well positioned to answer the call and develop a new generation of global leaders. Mining companies have a long history of operations in some of the most far-flung locations around the globe. This early history of globalization has provided several extremely valuable points of strength to the industry. First, the established global footprint of many miners provides the ideal footprint for all levels of employees to be exposed to international and multicultural settings and projects. Next, as many of the operational and technical challenges are common across geographies, mining companies have the ability to send human resources on short-term and long-term assignments in a way rarely enjoyed in other industries. Finally, mining companies' vast experience in global operations has provided numerous trailblazers. As a result, mining companies generally feature location-flexible managers and are generally adept at promoting nontraditional career paths with the potential for changes on the spur of the moment.

There is one final catch—no matter what challenges the industry faces or its significant advantages, there must be a perpetual commitment to employee development, the key driver of organizational growth and effectiveness. Certainly one significant hindrance to the industry's ability to attract the best talent is the historical pattern of boom-and-bust cycles in many of the industry's sectors. This pattern has also often curtailed development opportunities during down cycles. The obvious result has been a curtailment of employee development during challenging times. Thus, it stands to reason that the only way to ensure effective employee development is consistent leadership and financial commitment to providing training and development opportunities for young professionals learning their trade.

The balance of this section is dedicated to exploring how to put this commitment into practice, using an examination of the levers of employee development, as well as a concrete example of mobilizing employees in the field.

Compensation

Compensation management[18] is the most visible cornerstone of talent management. Base annual compensation is considered to be a simple and universal metric for the value a company places on each individual employee. This metric, however, is more convenient than accurate.

Overview with Emphasis on Compensation for Salaried Employees

The actual value a company places on its employees is manifested by the degree to which they are developed to their full potential, nurtured in an environment (team) to which they are ideally suited, and tasked by the company to realize full value. As such, employees are clearly viewed as assets rather than costs. It follows naturally that an employee viewed and developed with this philosophy is worth more to the company than a peer employee in another company. Thus, the value of employees is demonstrated by how they are treated by the company. This applies to both salaried and nonsalaried employees.

This treatment defines the corporate culture. It defines a *good company*, an employer of choice. There is no shortage of talent for good companies because there are few good companies relative to the number of talented employees searching for such a company.

How does this relate to compensation? Compensation management aligns the financial benefits received by employees with corporate financial objectives. Employees that are worth more can be paid more. The principles upon which financial compensation are developed are simply based on a formula designed to share created value between the company's shareholders and its employees. Any number of formulas for sharing can be developed, but no one formula can be deemed either right or wrong. Ultimately, the formula adjusts the amount of remuneration relative to perceived risks and the ability to achieve corporate objectives. When viewed this way, compensation statistics drawn from peers of employees with similar titles and job descriptions become the metric of company performance with respect to employee recruitment, development, retention, and tasking; it is a direct reflection of the company's philosophy on sharing value creation.[19] As such, compensation statistics represent what value employees can create rather than the cost of employees. In this scenario, value links directly with the efficiency of a company's talent management program and the business model being used.

This view raises another important question: Are employees interchangeable between companies? The concepts of *Employer of Choice* and *Employer Branding*, when taken to extremes, suggest that employees are not readily interchangeable to the degree that each company's business plan, team of pre-existing employees, and culture is unique. In a perfect world, an employee plucked from one fully functional culture and dropped into another fully functional culture will result in an immediate and temporary drop in performance for both companies. Extrapolating this scenario further, the speed with which the company regains its value creation with the addition of a new employee is a measure of the robustness and quality of the company culture. And, after the new employee is fully integrated into the new culture, the company's business plan is richer by virtue of the diversity of background, development processes, and business approach introduced by the new employee. This concept is encapsulated in the phrase *Forming, Storming, Norming, and Performing*.

Carrying this perspective further, compensation statistics become inappropriate measures of employee value in as much as they suggest that employees are interchangeable between companies. Instead, the value of an employee varies with the culture of the company and how the employee is tasked. Nevertheless, companies need some measure of compensation with respect to peers. However, such measures should be used as guidelines and not as absolutes, where many companies present it with a "take it or leave it" attitude.

Ultimately, compensation is one of the simplest elements of talent management. A company can pay any amount relative to its peers so long as two criteria are maintained:

1. The business meets financial performance expectations in a sustainable way, and

2. The quality and quantity of talent is sufficient and sustainable.

These, of course, are not independent criteria. Figure 6.14[20] is a snapshot of base compensation in 2006 for mining engineers practicing in the western United States. It is described as an algorithm because the data points being used were drawn from divergent sources and are statistically invalid in as much as the sources were not exactly peers and the number of data points was substantially less than could be represented as meaningful. However, when simple mathematical trends are drawn through the unstable data points and weighted to a relatively small number of base cases, a usable compensation system emerges.

In Figure 6.14, rows represent "years of experience," starting with a new graduate and carried through 20 years. Other work demonstrates that employee advancement as measured by job title tends to level out after 20 years. The five columns represent "potential" in the vernacular used to describe new graduates. The row labeled *Statistics* refers to a hypothetical bell curve of potential, with 5% of the population on the two extremes representing the lowest and highest potential from left to right. Fifty percent of the population lies in the middle. The row labeled *Graduates* shows the statistical distribution for these five categories (columns) based on the approximate total number of graduates earning bachelor of science degrees in mining engineering annually from U.S. universities. Over the last few years, the average number of new graduates in mining engineering each year is close to 100. The top 5% of graduates, unfortunately, corresponds to

Statistics	5%	33%	50%	67%	95%
Graduates	5	20	50	20	5
Base	$45	$50	$55	$60	$65
Annual	2.5%	3.0%	4.0%	5.5%	7.5%

Years of Experience						
0	$45	$50	$55	$60	$65	Engineer
2	$47	$53	$59	$67	$75	
4	$50	$56	$64	$74	$87	Senior Engineer
6	$52	$60	$70	$83	$100	
8	$55	$63	$75	$92	$116	
10	$58	$67	$81	$102	$134	Manager
12	$61	$71	$88	$114	$155	
14	$64	$76	$95	$127	$179	General Manager
16	$67	$80	$103	$141	$207	
18	$70	$85	$111	$157	$239	Executive
20	$74	$90	$121	$175	$276	

NOTE: Salaries are given in thousands of U.S. dollars per year.

Source: After Freeman 2006–2007

FIGURE 6.14   Compensation algorithm

only five graduates. *Base* refers to starting base compensation for new graduates, where the value varies with *relative potential*. Graduates with the highest potential receive higher starting salaries. *Annual* represents the rate of salary increase compounded annually. This particular number was developed empirically to replicate observed compensation statistics. Along the right side of the table are simple titles (job classifications) ranging from *New Graduate* (not labeled, but represented in the upper right corner) to *Engineer* through *Executive*. Finally, the table is divided into sections with a series of diagonal lines representing a range of salaries for each title.

The data are self-evident. Young graduates with higher potential have higher starting salaries (US$65,000/year) and advance faster (manager after 10 years with a salary on the order of US$134,000/year). All professional jobs follow similar patterns. Algorithms can be developed for each group of disciplines. More importantly, this method allows a complete set of compensation statistics to be developed from a relatively small number of critical data points drawn from selected peers from which any particular employer would consider drawing its talent. This method is considerably more accurate than published and unpublished data for one simple reason: the mining industry employs a relatively small number of professionals. This reality is further affected by the fact that, in order to be valid, data points must be drawn from a population of peers. The already small number of professionals is further reduced when the concept of peers is introduced. For example, consider a senior planning engineer (8 to 10 years of experience) on a steady track of advancement for a large open-pit copper operation in the western United States. What is the appropriate base compensation? To start with, this individual would have to be a better-than-average professional to achieve this level in 8 to 10 years. How many qualified people at this level exist in the United States? How many such jobs are there in which these specific individuals could be developed? As one would suspect, the number is very small. Specific statistics regarding compensation drawn from these small numbers would be unstable. Within a 3-year time window of this engineer's graduation, the United States produced approximately 300 mining engineers. If only those graduates in the upper third of the skills/potential pool are considered, the total number of individuals drops to 100. Assuming half of these engineers went to work in the west (75) and one-third of these started with metals companies (e.g., gold and copper), the number declines to 25. If we suppose that half of these engineers go to work in open-pit operations (12), and only half of these individuals receive the proper mentorship and training required to perform this job adequately, the number sinks to a mere six people. Can an analysis to determine the appropriate compensation be generated from this pool of six possible candidates? Because this is a statistically unstable data set, the answer should be no. Many companies adjust their statistics by choosing to hire at the 75th percentile of a broader statistical pool. As can be seen in this scenario, we are looking at 6 candidates in a pool of 300. To start with, these individuals are in an upper quartile. Thus, the 75th percentile as applied here is a significant departure from the concept that a company is paying for talent based on the upper 75th of a much larger pool of candidates. As shown in Figure 6.14, the compensation guideline from these individuals is US$92,000/year to US$102,000/year.

Virtually all studies used to address compensation for a senior engineer position, such as in this case, normally combine all senior engineers to achieve a statistically stable population, even though they may not be appropriate for the position. This represents the Catch-22 of this type of analysis for very small talent pools. However, if candidates are qualified on the basis of talent and their ability to develop into a role rather than into a rigid prescriptive job description, the pool of talent is broadened and general compensation statistics based on talented mining graduates in all sorts of roles are applicable. It follows that this talent, when recruited, needs to be developed.

A number of additional observations pertinent to this discussion are worthy of consideration:

- *Raising the bar:* Managers must encourage supervisors to take an active role in challenging employees to improve and broaden their skill sets and to task them in positions where they will grow professionally. Whenever possible, supervisors need to be evaluated relative to how they professionally develop and motivate employees. Managers should also investigate how each position can be staffed to raise the bar. This implies establishing programs that recruit, develop, and train employees in order to allow them to increase their "value" to the company if they so choose.

- *What people are worth rather than what they cost:* The employees-as-portals concept offers a different perspective on compensation, where it gives individuals the opportunity to add value to the company and compensates them accordingly. It enhances the view of employees in terms of what they are worth instead of what they cost (in the economic language of the mine). In many cases, how a person is tasked governs their ability to add value to the company and, ultimately, establishes their worth to the organization.

  As discussed previously, employees have the ability (if not the responsibility) to serve as portals. Total cost accounting allows a company to view compensation as well as recruitment and development of employees in terms of their responsibility at the mine and their responsibility to maintain the social license of the company.

- *Employer branding:* Good people seldom change jobs for money. Normally, they change for better employers, better supervisors, better culture, better opportunities to develop, and better opportunities to create value. In many cases, employees who change jobs based solely on economics should be held suspect because they are likely to change again. A few exceptions to this rule exist, particularly when projects are short-lived or involve international assignments of a specified duration or activity.

  Good employers pay appropriately. Accordingly, although compensation is important, it's about culture and employer branding—not lip service but reality. A stable, mature, well-managed culture can be measured in any number of ways. Safety, employee turnover, sustainability practices, employee development practices, and compensation all speak to a company's culture and brand.

Workforce Incentive Systems for Nonsalaried Employees

Although mining is extremely capital intensive, it is also inherently dependent on skilled labor. Labor is often the largest direct operating cost associated with production, and labor productivity is, in many cases, the swing variable between economic profitability and loss. This is particularly true in underground mines, where unit operations are typically labor intensive and physically demanding. In these applications, miners are often exposed to work conditions rarely found in other industries, including dust, noise, extreme temperatures, darkness, and the constant threat of potential hazards.

The impact of labor on the economic viability of mining operations is the essence of developing techniques that influence the productivity of a given workforce. Traditionally, these systems were structured in the form of economic incentives and based on standards that gauged individual or group performance. The primary intent was to increase the efficiency of labor-intensive activities by constructing financial and psychological inducements that would motivate employees to achieve higher-than-normal productivities. Historically, these systems have been built on the premise that money is a motivating factor and that, psychologically, an employee will work more efficiently for himself than for an employer.[21] With the complexities of today's labor force, the use of incentive programs is no longer solely intended to enhance employee productivity. Rather, most systems are introduced as a managerial tool to achieve specific objectives and priorities, as well as to influence employee attitudes and behavior. Common examples include safety, absenteeism, participation in mine rescue and voluntary training activities, and the retention of skilled labor. In addition, these programs have also been used effectively to influence parameters generally viewed as independent of labor, such as dilution and grade control, equipment utilization and availability, and resource recovery.

The following paragraphs provide a brief introduction to the use of incentive systems in the mining industry and the primary factors that govern their success for a given workforce.

Compensation programs. Compensation, as it is commonly defined, refers to all forms of financial remuneration received as a consequence of employment. This includes earnings dispersed through direct payments, like wages and cash disbursements, and indirect compensation associated with employee benefits and tangible services (Figure 6.15).[22,23] The structure of a compensation program used at any given mine is influenced by a number of site-specific factors, including regulatory requirements, historical wage practices, labor requirements and availability, mining methods and systems, working conditions, and labor relations. Direct compensation is normally comprised of three distinct components: base wages, objective performance programs, and subjective performance programs. The most fundamental element of any compensation program is base wages. They represent the cornerstone of employee earnings and are defined relative to an employment agreement, established company wage structure, or provisions of a labor agreement. Objective-based programs are synonymous with incentive programs, where there is a direct correlation between financial inducements and employee performance objectives that are both measurable and quantifiable. Conversely, subjective performance programs are often used where job output is not quantifiable or the outcomes that can be measured are not representative of employee performance and skill. As the name implies, these programs require management to make subjective judgments about an employee's value to the organization in terms of performance and the contributions made to an operation. In these systems, there is not a concrete relationship between employee compensation and their accomplishments, where the distribution of monetary rewards is solely dependent on management's perception of an employee's merit and his or her contribution toward company goals. It is important to note, the monetary rewards earned through both objective and subjective performance programs are supplemental to base wages and assume that an employee–employer relationship exists.

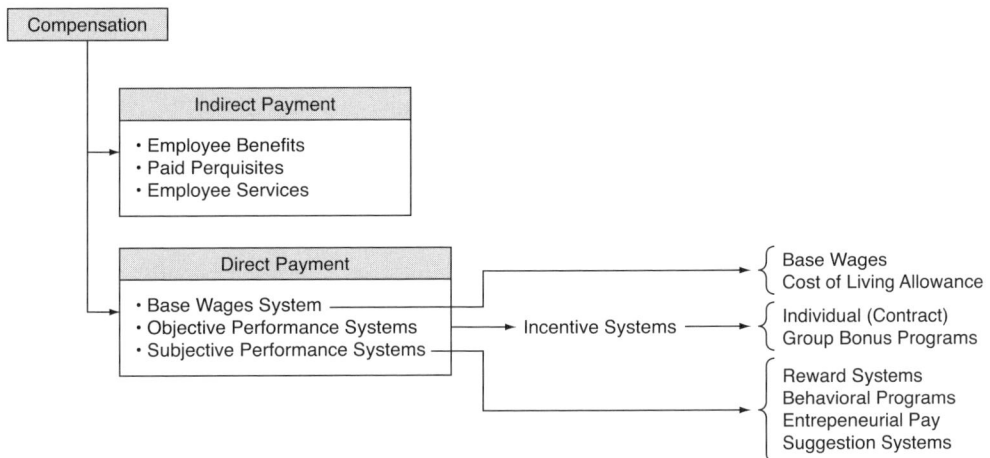

Source: After Miller 1991 and Milovich 1987

FIGURE 6.15 Schematic of typical compensation program

The philosophy of incentive systems.   There are many different philosophies of how financial inducements function as a motivational tool and the role of incentives with respect to today's workforce. Most incentive programs use monetary rewards to encourage and motivate employees to achieve higher-than-normal productivities. These systems are adopted on the premise that potential earnings have an inherent motivational quality. It also stems from the traditional belief that employee productivity is merely a reflection of effort and that monetary compensation is of such importance to employees that it can influence behavior.[24]

Most managers, however, recognize that these relationships are not as simple as they appear. Although money is generally valued as important to most individuals, its role as a motivating force is largely dependent on the need and desire of employees to increase their income. In many situations, income fulfills only a component of an employee's total psychological and physical needs. Recognition, self-realization, and other forms of achievement often possess a motivational impact equivalent to any monetary incentive. It is, therefore, imperative that mine management understand the basic needs of their employees and tailor incentive practices that are responsive to these needs. In addition, operations should be cognizant of how financial incentives function as a motivator and how they relate to nonmonetary forms of motivation. These relationships, as with employee needs, are dynamic and change with time. Unfortunately, these needs and relationships sometimes conflict with corporate goals and objectives. Successfully designed and implemented incentive programs allow companies to adapt to these changes and continually appraise the needs of their workers relative to managerial objectives.

During a survey of operating mines in the early 1990s,[25] nearly every company stated that the objective of their incentive program embodied more than just financial rewards to increase employee productivity; rather, that these systems provided numerous opportunities for an operation to optimize conditions deemed important by management. These systems have been used effectively to control specific operating parameters, influence employee attitudes, encourage quality workmanship, and promote safe work practices. Furthermore, incentive programs can be implemented to increase employee interest in activities that are often tedious, promote efficient and innovative methods of working in hostile environments, and enable supervisors to evaluate the performance of individual employees effectively.

Another primary benefit of these systems is that they can sometimes mitigate the adversarial relationship that can develop between a company and its employees. If designed properly, these systems can integrate the interests associated with both the company and labor through cooperative, win–win arrangements. Employees receive higher wages and improved working conditions, and fulfill other psychological and physical needs, while the company gains higher productivity, lower unit operating costs, improved employee morale, and other managerial objectives that have been designed into the program.[26] These mutually beneficial systems that link organizational objectives with the individual needs of labor can only be established through an environment of trust and communication. As such, it requires assurances that both parties will be treated equitably and that they believe in the integrity of the system, as well as each other.

It should be stressed that increases in labor productivity and other benefits derived from these systems do not necessarily result from employees being driven to work harder. Rather, it is believed that these improvements commonly arise because workers are more attentive, use more efficient means of working, and are dedicated to the activities that they are performing.

Incentive programs, however, should not be construed as a panacea for all operational woes. Often, mining companies implement these programs in hopes of solving production-related problems, including everything from excess dilution to deteriorating employee relations. Successful systems require skilled and knowledgeable management that fully understands the complexities associated with the incentive program, employees, operation, and corporate objectives. These programs are designed to improve functioning operations and are not meant to act as a remedy for production and management problems. The major dilemma of implementing incentive programs in these situations is that they are more apt to compound the problem than to alleviate it.

Types of incentives.   Incentives can be viewed as merely a vehicle by which a company measures and distributes the benefits created by labor through improved productivity or performance or as a direct result of their actions. This approach may be oversimplified compared to traditional academic definitions; however, it accurately conveys the mechanism through which these systems work. Performance is gauged with respect to established standards to determine financial rewards. The level of aggregation incorporated within an incentive program (e.g., individuals, groups, production areas, work classifications, shifts, or the entire operation), is an important issue in the determination of an efficient and equitable system. The characteristics unique to an operation should dictate the size of this labor component, with the intent of optimizing the relationship between employee performance and monetary rewards. In addition, the ability to evaluate the performance of individual employees objectively plays a role in designing a program that is applicable for certain job classifications.

These considerations have produced two distinct categories of incentives widely used in the mining industry for nonsalaried employees: individual and group bonus systems. Each category has traits that make it amenable to certain types of operating conditions, mining systems, and corporate objectives. In some situations, both types of incentive systems can be employed to fulfill the same needs. It then becomes the responsibility of management to choose and modify a system that will optimize their particular goals. It is common for an operation to implement both systems simultaneously, each encompassing different classifications of employees. [27]

Individual incentive plans.   Individual incentive systems are perhaps the oldest form of compensation used in the U.S. mining industry. Since the statutory discontinuation of tributing systems in the 1920s, individual incentives that maintain a formal employer–employee relationship have become synonymous with what are called contract systems. Also commonly referred to as gypo contracts, these systems have been widely used in underground metal/nonmetal mining for decades and comprise an array of different individually based performance programs.

In essence, these incentive programs can be characterized as a supplemental wage agreement between mine management and specific employees to perform predetermined tasks or activities, where financial remuneration is contingent on production, the rate of project completion, and/ or any other objective form of measurement that is representative of employee performance. How these standards are formulated into a system and how these systems are implemented into an operation predicate their operational value in achieving the initial design objectives.

Financial rewards are distributed to individual employees in conjunction with measurable contributions made to completion of predetermined tasks or projects. In essence, these systems are an agreement between mine administrators and specific employees that directly correlate performance standards to monetary awards. In some union operations, performance standards are contractually negotiated and included within their collective bargaining agreement. The Fair Labor Standards Act of 1938 required that operations engaged in incentive systems pay a guaranteed base pay level, which the employee earns irrespective of performance. Often called "day's pay," this guaranteed rate can correspond to an established company compensation schedule, to a negotiated rate between the employee and the company, or to a government-legislated wage rate.

The individual incentive programs used by most mining operations are comprised of numerous different contracts and subcontracts that take into account variations in performance standards tied to different work areas and job assignments (e.g., stoping versus drift development), mining methods and systems, equipment to be used, geologic conditions and orientation (e.g., vein width and dip), and various other factors that influence labor productivity. It is not uncommon for large, labor-intensive operations to "run" 20 or more contracts simultaneously. Five principle categories of individual contracts are used in the underground metal/nonmetal mining industry:

- Production and Stope Contracts
- Development (Shaft and Drift) Contracts
- Construction and Repair Contracts
- Exploration and Drilling Contracts
- Haulage and Hoisting Contracts

Contrary to what the name implies, individual incentive systems can be applied effectively to groups of employees as readily as to individual workers. In fact, a survey of operations using individual incentives showed that a vast majority of all contracts were configured with respect to specific work areas with minimal consideration toward the number of people included in each contract.[28]

Individual contract programs are generally associated with less mechanized operations, where production and operating efficiency are largely influenced by the skill and physical contributions of employees and less by the operating capacity of the equipment. While the perceived benefits of incentive programs have been previously discussed, there are advantages uniquely affiliated with contract systems. They tend to promote competition among miners, heighten pressure on individuals to perform, spawn innovation on how to work more efficiently, improve workplace organization, and force employees to accept responsibility for their own actions. Although these attributes may not be advantageous in all situations, they are usually highly valued by most mine managers.

Despite these benefits, there are several major challenges associated with implementing individual incentives at many operations. The first is establishing performance standards that are representative of employee effort, where earnings are proportional to increases in productivity or some other gauge of measure. Operators designing these types of systems must be cognizant of rewarding performance based on extra effort and not productivity. Although there is a general perception that the two directly correlate, this is not always the case. Under this scenario, there is a real possibility that employees can be rewarded for complicating their job, performing unnecessary activities, or be penalized for developing safe shortcuts. The level of aggregation also influences how performance standards are determined and, ultimately, evaluated.

Once performance standards have been determined, it becomes necessary to develop an objective means of measuring employee performance. This is actually more difficult than it appears. The outcomes that gauge the collective performance of larger groups, such as an entire operation, are normally available by measuring any number of factors, including overall production, unit cost, profitability, or metal yield. However, unbiased performance measurements of smaller groups or individuals can be a formidable task in that there are fewer concrete forms of output to measure. In addition, objectively evaluating the productivity and effort associated with a specific employee in relationship to the collective performance of a group is difficult at best. Economic considerations also play an important role in devising methods of measuring employee performance. Labor requirements associated with measuring and calculating contracts translate to a real cost incurred by the company. This becomes particularly evident for operations using shorter pay periods or where individual incentive systems are applied to large workforces.

In these cases, the labor required to perform these performance appraisals quickly becomes economically prohibitive. As should be obvious, the tangible benefits of using these contract programs must offset these and related costs before their use should be justified.

Another essential ingredient for a successful contract system is the acceptance of the program by the labor force, as well as by immediate supervisory personnel. To achieve this "buy-in," employees must believe that they are being treated equitably by the system, and have confidence in the integrity of their immediate supervisors and contract engineers. Similarly, management must also have confidence that the program is "not giving anything away" and that employees are earning the financial rewards that they receive. In order to work, these systems must have the complete support of shifters and production supervisors and should incorporate their input. Furthermore, the implementation of new incentive contracts, productivity standards, or measuring techniques must be designed properly and installed correctly on the first try if they are to gain the support of the employees. Lastly, as a consequence of more efficient work practices, performance standards must be periodically adjusted in order for the program to remain effective. This needs to be done with forethought and the utmost care; otherwise, it can have a significant detrimental impact on labor relations as well as employee productivity.

Group bonus plans. Group bonus plans are the most prevalent form of incentive used in the mining industry. In general terms, these programs distribute monetary rewards in recognition of achieving some level of success in obtaining specific organizational goals or objectives. In many respects, these programs reflect the same structural characteristics associated with individual incentive systems. For example, the intent of these programs is to establish objective standards through which employee performance is gauged and monetary rewards distributed in recognition of achievement. Similarly, performance can be judged with respect to productivity or in terms of any other combination of operational objectives. These systems differ from conventional contract programs, both in application and in reference to establishing performance standards.

Group bonus systems, often referred to as group-sharing plans, are highly amenable to certain applications where objective means of determining individual performance are not possible. Examples include individual job assignments that are so interrelated that determining the contribution of a single employee is difficult, in job activities where performance cannot be quantified, and in situations where performance is not representative of employee effort and contribution. These conditions are indicative of many activities performed in both surface and underground mines, particularly in highly mechanized operations that are generally more dependent on equipment capacity and economies of scale than employee productivity, skill level, and performance. Bonus systems can be adapted to fit any duration and, because they supplement base incomes, can be dispensed in a multitude of ways (e.g., lump sums, during regular pay periods, or deferred until some future date, like the end of fiscal quarters, holidays, or year-end).

The principal advantage affiliated with group incentives is that they encourage group cooperation toward achieving specific objectives. Because supplemental earnings are dependent on the cumulative performance of a group, most employees realize that cooperation is imperative. This relationship usually eliminates some of the problems normally associated with individual contract programs, such as the hoarding of equipment and resources, conflicts with reassigning workers to different job assignments, and the competition for specific work areas and partners. In addition, many believe these systems force employees to become more productive and develop skills quicker because of the peer pressure to perform being exerted by fellow workers. There is also an incentive for experienced miners to teach inexperienced employees skills and efficient operating techniques.

Another benefit of conventional group incentives are that they have been adopted success-fully in conjunction with programs associated with safety, absenteeism, and training. Further-more, in certain types of underground mines, these systems can be used to complement individual incentive programs. This is particularly true in mines where employees who are partic-ipating in individual contract incentives are dependent on the activities of personnel who are not. In these cases, it is highly advantageous to incorporate these noncontract workers into a group bonus plan. Dependence on one another promotes a relationship of mutual cooperation and possesses psychological benefits associated with promoting a "team" concept. It also unifies the priorities of all employees and establishes common objectives.

Potential negative aspects of group incentives are that they rarely possess the motivational influence affiliated with individual incentives. In large operations, individuals may feel their efforts have little effect or impact on the overall performance of the mine; therefore, the motivat-ing force behind the incentive program is greatly diminished or entirely lost. In addition, it is often more difficult to design a group bonus system that is responsive to secondary managerial objectives (e.g., dilution, equipment availability, direct cost reductions) because the relationship between individual actions and rewards is sometimes not readily apparent. Group bonus pro-grams are also sometimes viewed as being inflexible from the perspective of being able to meet changing strategic and managerial needs quickly because these programs often become entrenched within the compensation structure of an operation and intimately tied to employee job classification. These programs also do not provide an objective means for assessing the skills and performance of individual workers accurately and are occasionally plagued by problems with establishing equitable standards that are universally acceptable to all employee classifications throughout an operation.

Other managerial considerations.   The intent of most incentive programs is to influence, encourage, and compensate employees who exert an effort toward something deemed important to management. It is a structured process that alters the performance of labor through employee motivation. Although most systems use financial inducements as a means of motivating workers, numerous nonmonetary methods have also proven to be effective, particularly when used in con-junction with monetary incentives.[29] Examples include job status, recognition, competition, peer pressure, obligation, and self-respect. Although not all these factors may be desirable, each possesses a unique quality that potentially enables an operation to capitalize on its application in specific operating situations. Many times, these nonmonetary incentives have proven to be more influential than the motivation created by the lure of increased earnings. A common practice used to enhance a conventional incentive program is to incorporate small groups of employees under a single incentive contract or bonus system in order to promote greater interdependency between workers. In many cases, this creates peer pressure on the least productive members of the group to be more efficient. Another widely used approach is to post contract settlement sheets publicly in the miner's dry, so that every employee knows on the basis of contract earnings who the most and least productive miners are for any given pay period. This often incites competition for status and recognition, as well as putting pressure on miners to perform at or above certain levels or risk being criticized by their peers. Despite these advantages, it is vitally important that an operation understands how its particular labor force will respond to these types of motivating factors or it risks inciting employee conflict and resentment.

Safety is often cited as a major concern related to the use of incentive programs and, in par-ticular, individual incentives such as contract systems. Critics of contract mining have historically opposed their use on the grounds that they are inherently dangerous. The principal contention is that miners, in an attempt to maximize production, are more apt to take risks and expose themselves to hazards that pose a greater potential for injury or death. This claim has never been quantitatively substantiated in underground metal/nonmetal mines operating in the United States. In fact, it is the contention of many operators that the application of contract incentives actually contributes to safer work practices because miners have a financial inducement to be organized, practice good housekeeping in production areas, and concentrate on the activities they are performing, thereby minimizing apathy.[30] This is supported by research conducted by Chouinard and Billette,[31] who concluded in 1986 that "no direct and strong correlation exists" between incentive programs and accidents. Regardless of the claims, statistical evidence would seem to indicate that if a correlation between contract mining and accidents does exist, it is small.

Performance Management

Like many industries, mining is facing and will continue to face a shortage of skilled workers. It is estimated that more than 40% of baby-boomer workers will retire during the next 10 years. In mining, this number may be even higher because most workers were hired during the minerals boom of the 1970s and are now reaching retirement age. Recruiting new employees is only one facet of solving this talent shortage. Retaining and retraining incumbent employees is another critical part of the solution. Additional salary, benefits, and signing bonuses can help to attract new employees, but in the long term, compensation alone will not retain high-performing employees. Talented employees want challenging work, recognition for what they do, and development of their skills.

A well-designed and successfully implemented performance management process (PMP) can help employers retain key talent. A successful PMP can present recognition and development opportunities for employees while providing management with a tool for assessing current skills and competencies and identifying gaps that require additional training, development, or hiring.

What Is Performance Management and Why Is It Important?

Performance management encompasses managing individual and team performance as well as managing underperformers. The key to successful performance management is ongoing feedback, and communication. A successful PMP is based on the organization's mission, values, goals, and, most importantly, results. Organizations must communicate expectations clearly and consistently. Employees need to know what is expected of them, and how to prioritize where to focus their time and energy while at work. Communicating corporate expectations may be time-consuming for management but is critical to achieving positive performance results. When a company measures and reports its success in earnings to the marketplace, it is actually reporting the combined success of each of its employees. When viewed from this perspective, it becomes obvious that a company's PMP is a key tool to its economic success.

Traditional Performance Management

Employers today are rethinking the how and why of performance management. Often, PMPs are not seen as a value-added process but as a necessary evil enforced by human resources and unrelated to the day-to-day business. Traditional PMPs are annual processes wherein managers rate an employee's performance over the past year using a standard rating form. The manager rates the employee's performance in a number of areas and assigns a grade that is used to determine a salary increase. Often, the salary increase is mandated using a matrix that provides the manager with a factored percentage rate of increase. Under this program, the manager is not afforded much flexibility. It is therefore not surprising that when surveyed about their company's PMP, both managers and employees express a dislike of the process and a lack of understanding as to how it can positively affect an individual's or the company's success.

Many employees say their annual PMP meeting is the one time per year they have their manager's undivided attention, but that the experience is not always positive and often contains considerable criticism. Managers think the PMP takes too much of their time, and, often untrained in its use, may misuse it. Some managers give all their employees high ratings because it's the only way to give their employees a pay increase. Other managers may rate all their employees low because they see everything as just part of the employee's job description and not something to be rewarded. These are just some of the ways PMPs can be misused.

In many cases, employees and managers go through a traditional PMP and come out feeling negative and unmotivated. The question then becomes how can organizations change this process and transform their PMP into a positive, value-added experience that serves both employees and the company?

New Ways of Thinking About Performance Management

As many of the skilled and professional mining staff members retire, it's imperative to find a way to recoup the skills and expertise that are exiting. Performance management is a key to meeting this challenge.

One of the ways employers can improve the PMP experience is to tailor the process to their organization's business and culture and to apply a holistic approach to the process. In mining, this means focusing not only on achievement of production and safety goals but also on development of hard (technical, engineering, and mining) and soft (people and relationship) skills.

A successful PMP for the mining industry should therefore include measurements of a variety of goals:

- Goals tied to production (individual and that of the mine as a whole);
- Goals tied to learning new skills and skill development;
- Goals tied to safety (individual and that of the mine as a whole); and
- Behaviors exhibited and observed (how the employee does his/her job).

Developing a process that encompasses all of these areas is not difficult. It does require an understanding of the methods by which performance is typically measured and deciding which method or combination of methods will work best in a given company's culture. Whether you are the manager of mining operations or an HR manager, it's a good idea to include some line managers in developing a PMP. They bring daily experience and knowledge of the workforce and the work performed, and they can be instrumental in successfully rolling out the process. It might also be useful to include both a manager of unskilled labor and a manager of skilled and professional employees on the team. In union operations, a union representative must be involved throughout the process and the union contract referred to for guidance in evaluating union employees. The concept of "management by objectives" (MBO) was first popularized by Peter Drucker in 1954 and is a process where a company's objectives are established through mutual agreement of management and the employees, where both parties possess an intimate understanding of how the objectives are established, measured, and quantified.

Managers may ultimately develop different methods of measurement for different employee groups. For example, one mine recently surveyed (L. Sliman, personal communication) had a primarily MBO and skills-based PMP, but different performance measures were applied to hourly and professional employees. Performance management for hourly employees was more team based. Hourly employees were rated by their supervisors and their MBO performance was measured on the basis of mine-wide outcomes. These employees were also eligible for production and safety bonuses. Production bonuses were linked to safety performance, and no production bonuses were awarded if targeted safety goals were not met. Professional employees were rated on goals based on a combination of their individual performance and mine-wide performance, and the acquisition of higher levels of skills deemed critical to the mining operation.

Though measuring three of the four critical PMP areas, the above-mentioned mine was overlooking one of the primary keys to successful performance management: how the employee does his or her job. How employees do their jobs is as important over the long term as what they do. Consider this: is it acceptable to keep a manager on staff who gets results by bullying and threatening employees? Of course not; in such a case, it is the responsibility of senior management to coach and educate the manager on how to get results using appropriate management techniques. A successful PMP includes a tool to measure the *how* as well as the *what*.

Over the years, the PMP has become a process owned by managers with little input from employees. A new way of managing performance appraisals is to shift primary responsibility from the manager to the employee. This type of process is based on the premise that each employee is responsible for his or her own performance and should take charge of it. It also assumes that each employee can and wants to make a positive difference through the work he or she performs. If a company's culture can support this kind of thinking, an employee-managed PMP can be extremely effective.

Performance Appraisal Methods

There are many ways to measure performance; most methods can be categorized as one of the following:

- Category rating methods that typically use a standard form or checklist,
- Comparative methods in which employees are compared to each other, and
- Narrative methods that include written descriptions of employee performance.

Each of these methods has advantages and disadvantages, and when used alone may not achieve the desired goals. Table 6.2 lists some standard appraisal methods widely employed in industry.[32]

As is evident, several of the methods in Table 6.2 involve some type of rating. After a specific method has been chosen and incorporated into the PMP, a decision must be made on whether employees will be graded and on how feedback will be administered. In regard to PMP design, one size does not fit all.

If managers choose to link the annual performance appraisal to pay or merit increases, it will likely be necessary to convert employee feedback into some type of numerical score or translate it into some type of relative, categorical rating. If numerical scores are used, it is important to allow sufficient disparity in the rating levels to account for variations in performance. For example, a range of 1–5 is often used.

Instead of numbers, phrases such as "meets requirements," "exceeds requirements," and "does not meet requirements" are often used. It is the opinion of many that this approach is a less desirable technique for assessing performance and is open to a great deal of interpretation. It also limits a manager's ability to rank an employee effectively relative to others in a specific category. For example, how does one differentiate the low and high performers in the "meets requirements" category? In essence, there is no distinction because these employees would be lumped together in this method.

A better method may be to separate the annual performance appraisal from pay increases entirely. This ensures that the conversation is all about performance and not about wages or

TABLE 6.2   Standard appraisal methods

Graphic Scales: a category rating method where managers rate employees numerically on a designated form in common categories.

- • Advantages: Simple to use and provide a quantitative rating for each employee.

- • Disadvantages: Standards may be unclear and may be interpreted differently by different managers.

Ranking: a comparative method where managers list all employees numerically from highest performers to lowest performers.

- • Advantages: Simple to use but not as simple as graphic scales.

- • Disadvantages: May cause disagreements among employees and may be unfair if all employees are excellent.

Forced Distribution: a comparative method where managers use a bell curve to rate employees.

- • Advantages: Forces a predetermined number of employees into each group.

- • Disadvantages: Results depend on the adequacy of the original choice of cutoff points.

Critical Incidents: a narrative method where managers keep a record of employee actions, both good and bad, daily or weekly.

- • Advantages: Tool helps specify what is "right" and "wrong" about an employee's performance. Forces ongoing evaluation.

- • Disadvantages: May be difficult to rate or rank employees relative to one another.

Management by Objective: a narrative method where managers and employees set goals and the employee is rated against achievement of them.

- • Advantages: Tool is tied to jointly agreed upon performance objectives.

- • Disadvantages: May be time-consuming, goals may be unachievable or immeasurable. Difficult to compare employees to one another.

Behaviorally Anchored Rating Scale: a category rating method where employees are rated on their behaviors against examples of desired behaviors.

- • Advantages: Behavioral anchors can very accurate in measuring how an employee performs.

- • Disadvantages: May be difficult and time-consuming to develop.

Source: Dessler 1997

compensation. From a managerial perspective, this represents a major decision that can have significant implications to a company. Unfortunately, there is usually insufficient information for managers and HR professionals to make a definitive determination on what approach is best for a given company. As with most issues dealing with labor and performance evaluation, it is critical to have an intimate understanding of the business and its employees.

Developing and Implementing a PMP

Without a background in human resources, the development of a PMP can seem daunting and fairly intimidating. However, it does not need to be. A good PMP is all about feedback and communication and not about the tools being used. In some cases, the process can be as simple as using a blank sheet of paper to write down the employee's accomplishments and development needs. Although this approach might work well in a three-person business, it would not be practical for a large mining operation. As a consequence, where does one start?

1. It is important to understand the goals and values that drive your company. Make sure these are communicated consistently and regularly to all employees. Every employee should understand that what he or she does on a daily basis contributes to the bottom line success of the company. In a complex mining operation with multiple business units, this can be much more difficult that it might appear.

2. The next step is to analyze the company's culture. Through this process, it is helpful to think about the following questions, as well as any others that will facilitate a greater understanding of the company's culture:

- Is the company managed hierarchically, from the top down, or via horizontal matrices?

- Are decisions made unilaterally or does management regularly seek input from employees?

- Does management solicit employee suggestions, and if so, how often are they implemented?

- How does the company treat its employees? (Look at policies and benefits.)

- Why do employees leave?

- What is the general perception of the employees toward the company? Do employees seem happy or is there constant grumbling?

3. Identify the competencies and skills required to do the various jobs in the company. This is where supervisors and line managers can be extremely helpful. In this process, each supervisor should make a list of the jobs they manage and what is required to perform each job. If the company already uses job descriptions, their use might greatly aid in performing this analysis.

4. At this point, managers should be able to decide what to measure and how to measure them. Based on the results of the earlier steps, managers can begin to design a tool and a process that works for a particular company's supervisors and employees. Keep in mind that PMP is a process and, as such, is subject to change. It is important to understand that changes might be required in the PMP after it is first implemented. Consequently, managers must be flexible and keep an open mind.

5. Once a company has developed a PMP, be sure to get the approvals needed, including unions (if appropriate) and company executives, and a legal review.

6. Train supervisors and production managers on the skills necessary to manage performance effectively. Instead of providing only process-oriented training, consider training supervisors on soft skills such as listening, giving positive feedback, delegating, facilitating goal-setting, and having difficult conversations. These are skills all managers should have and should be measured on in their performance appraisals. During the training sessions, also get feedback from supervisory personnel on the process and tools and incorporate any changes that are appropriate before formally implementing the process.

7. Roll out the process to your employees and train them on how to use it. Employees need to know how to set measurable goals and how they will be evaluated on them.

8. Before a company implements a PMP, they must decide whether all employees will go through the PMP during the same time period (focal point reviews) or whether appraisals will be done on the employee's anniversary date with the company or relative to some other schedule (e.g., office workers one month, mine workers another). It is usually prudent to discuss and carefully weigh the advantages and disadvantages of each schedule before a specific approach is formally adopted.

9. Begin using the process. Whether a company tests its PMP against a small group or rolls it out to the whole organization is a management decision. After it is in use, however, it's important to get feedback on how well it is working from multiple points of view, including supervisors, line managers, and the employees. From this feedback, an opportunity exists either to tweak the process for the next cycle or to keep it as is.

10. Just as managers will regularly use the PMP to measure whether an employee is meeting his/her goals, it is important to assess whether the PMP is meeting its objectives and is remaining effective over time. Some things that a company might want to measure and compare on at least an annual basis include

- Turnover—Is it decreasing, remaining the same, or increasing?

- Skills—Are employees becoming more skilled in the areas designated as critical? Are the programs being implemented closing the skills gaps that have been identified?

- Satisfaction—If the company conducts employee surveys, are employees expressing more or less satisfaction?

Writing a Performance Improvement Plan

Another aspect of performance management is dealing with undesired employee behaviors or performance that does not meet company standards or values. Before a performance improvement plan (PIP) can be written, it must be determined whether the appropriate actions and the plan outlined are achievable. An employee should not be placed on a PIP just to go through the process if management sincerely thinks termination or some other discipline is warranted. PIPs should be goal oriented and measurable within a specific time period. When meeting with an employee to discuss a PIP, the manager should clearly explain what performance and/or type of behavior must be improved and within what time frame. The PIP should be in writing and include how the performance will be measured, how often, and the consequences if the PIP is not successfully completed. Both manager and employee should sign the PIP and date it.

Some managers may prefer to write the PIP informally during the employee meeting, but it is a best practice to write the PIP prior to meeting with the employee, and to have it reviewed by Human Resources, the company's attorney, or an executive before presenting it to the employee. This will help avoid misunderstanding and possible legal action in the event the employee is eventually terminated.

Sample Process and Forms for an Employee-Managed PMP

Figure 6.16 through 6.19 and the following paragraphs illustrate a sample employee-managed performance management process adapted from the Brady Corporation's performance leadership system.[33]

Acknowledgments

Several authors contributed materials and ideas presented in this section, including M. Conti (employee development), L. Fisher (information management), and L. Sliman (performance management). The section editors extend their sincere thanks to these individuals for their contributions and participation in making this section possible.

Sample Form:
Goals and Results Summary

Review Period: _____   Employee Name: _____

Date of Review: _____   Manager Name: _____

List your three to five most important business goals:

These are the priority tasks that allow you to add the most value to the company. At least one goal should be development related. Make each goal SMART (Specific, Measurable, Achievable, Realistic, and Time Specific.)

Summarize your results:

In narrative form, explain how successful you were in meeting each goal you described above.

Manager comments:

_____   _____
Employee Signature - Date   **Manager Signature - Date**

Source: After Brady Corporation 1998.

FIGURE 6.16   Sample form: Goals and Results Summary

Sample Form:
Development Goals and Results Summary

Review Period: _____   Employee Name: _____

Date of Review: _____   Manager Name: _____

Professional growth goals:

Identify one to three of the most important goals for developing skills and/or knowledge for your professional growth. Include development opportunities you want to pursue in the next year including educational, training, professional certifications or credentials, internal cross training, mentoring, or other pursuits.

```



```

Summarize your results:

In narrative form, explain how successful you were in meeting each goal you described above. Include the weight you assigned to each goal above, and determine the percent of completion you achieved since your last annual summary.

```



```

Personal growth goals:

Identify any personal development goal that you believe will help add value in your role at the company.

```



```

Summarize your results:

In narrative form, explain how successful you were in meeting your personal development goal.

```



```

_____
Employee Signature - Date

Source: After Brady Corporation 1998.

FIGURE 6.17   Sample form: Development Goals and Results Summary

Values Assessment Survey
(sample based on one company's values)

Employee: _____ Relationship: Self _____ Manager _____ Other _____

The company's values are listed below followed by a list of behaviors that demonstrate each of the values. Based on your experience with this person, please indicate whether you think each behavior is a Strength or Needs Development. If a person demonstrates the behavior more than 75% of the time, rate it a Strength. If the person demonstrates the behavior less than 75% of the time, rate it an area that Needs Development. If you feel you are unable to give feedback on five or more of the areas, you may not know the person well enough to give good feedback and should not complete this assessment.

	Strength	Needs Development
1. Integrity—We meet the highest standards of ethical behavior.		
a. Practices honest and open communication	_____	_____
b. Treats others with respect	_____	_____
c. Inspires trust	_____	_____
2. Judgement		
a. Acts and behaves professionally	_____	_____
b. Keeps the overarching goals of the organization in mind	_____	_____
3. Creativity		
a. Offers new ideas to further goals or solve problems	_____	_____
b. Constructively challenges ideas and processes	_____	_____
4. Initiative		
a. Accepts responsibility and develops self through growth opportunities	_____	_____
b. Seeks opportunities to contribute wherever he or she can	_____	_____
c. Is personally committed and actively works to continuously improve	_____	_____
5. Ownership		
a. Is accountable for own work and team outcomes	_____	_____
b. Takes responsibility for actions and behaviors	_____	_____
c. Doesn't blame or shame others	_____	_____
6. Communication		
a. Ensures that information flows to those who need to know	_____	_____
b. Encourages opposing points of view	_____	_____
c. Promotes understanding of all perspectives	_____	_____
d. Gives and receives feedback constructively	_____	_____
7. Teamwork—We work together to achieve our vision		
a. Creates strong morale or spirit on his or her team	_____	_____
b. Works to find common ground and solve problems to the good of all	_____	_____
c. Deals effectively with diverse groups of people	_____	_____

Additional comments: _____

Source: After Brady Corporation 1998.

FIGURE 6.18   Sample form: Values Assessment Survey

Process Summary

The goal of the performance management process is a simple, consistent way to set performance standards and measure results at all levels of the company. We want each of you to focus on the value you add to the organization and to be accountable for your actions and results.

The process includes: What You Contribute (SMART goals), How Others See You (values and behaviors), and What You Learn (skills development). These are the three dimensions of performance we want to measure.

Your review will include an evaluation of the goals you set with your manager at the beginning of the last year. If you are a new employee or just haven't set goals yet this year, you and your manager will need to take time before the review meeting to draft goals based on the work you have done this year.

Our intent is to begin using this new format, and to continue reviewing and revising it until it works for our company. Shortly after everyone has used the new process at least once, we will hold employee focus groups to get feedback on how the process worked.

How the Performance Management Process Works

Step 1: Understanding the Big Picture. You need to know the company's overall mission, values, strategies, and goals. Your manager can help you develop a clear line of sight from the company's mission and overall strategic goals to your individual performance objectives. As you begin to understand business expectations more clearly, the tasks that add the most value will become easier to identify.

Step 2: Goals and Objectives. Collaborate with your manager to set team and individual goals for the coming year. Your goals should consist of three to five objectives for you to focus on to bring the most value to the company. Work with your manager to set SMART (Specific, Measurable, Achievable, Realistic, and Time-Specific) goals.

Once SMART goals are set, you and your manager should agree on the right mix of direction and support for each goal. Each of the goals may get a different blend of direction and support. The manager provides the agreed-upon balance over the course of the year. The purpose is to have each employee become an expert in setting and reaching value-added goals.

Step 3: Tracking Results. Part of your job as an effective contributor to the company's success is to demonstrate results. By tracking specific measures of your performance and feedback from your customers and others, you can assess the impact of your goals and actions to the organization.

Tracking results need not be difficult or time-consuming. With agreement from your manager on what to measure and how to measure it, SMART goals should require just a few minutes each time you do an assessment. The payoff for you is that you will be aware of the day-to-day value you add to the organization. You should assess your results regularly and discuss them with your manager throughout the year.

Step 3 is important as it sets the right level of performance communication for each goal. It prevents performance "surprises" at your annual review and helps you build a great working relationship with your manager. Step 3 also sets the foundation for step 4.

Step 4: Annual Performance Summary Meeting. In preparation for the meeting, you must complete (1) the Values Assessment Survey for yourself and request that your manager complete it (If you and your manager decide it's appropriate to collect feedback from others you work with closely, send the form to them as well), and (2) a performance summary form listing your performance and development goals and your results.

Completed Values Assessment Surveys are returned to your manager. This will ensure that responses are kept anonymous and encourage honest and open feedback. If the giver wishes to give you feedback directly, he or she may do that as well. Your manager will summarize the values responses and provide you with a feedback report during your annual performance summary meeting.

Once you complete your portion of the performance summary and development plan and your manager has received and summarized the values feedback, you and your manager meet to discuss the results. By bringing together the performance results you have been tracking over the year and the feedback you have received from others, you will be able to assess how well you are performing. Because you and your manager have been talking regularly over the course of the year, there should be no surprises for either of you during your performance conversations.

Your manager may add his or her own comments to your summary. Together, you will reach agreement on the results you achieved during the year and assign a percentage of completion to each goal. When you have agreed to the goals summary, you and your manager should each sign and date the form and it will be included in your personnel file.

It is important to focus the performance conversations on the value you add to the organization through your results and behaviors. The discussion should be built on what you did to achieve your goals and the impact of the results and your actions to the organization. Finally, it includes what you learned and the skills you have acquired to add even more value in the future.

Step 5: Merit Increases. If the company ties pay increases to performance reviews, include the process here. Consider separating pay increases from performance appraisals for more effective performance management.

Average merit pay increases have been in the 3%–4% range for the past several years, and this amount is not an effective motivator. More and more successful companies are separating pay increases from the annual review and looking at profit sharing, bonus plans, and variable pay as ways of rewarding performance.

Ongoing Feedback

Throughout the year, you and your manager are encouraged to meet informally to make sure you are: (1) working on the right business priorities, (2) reporting results in a way that demonstrates added value, and (3) developing critical skills.

Source: After Brady Corporation 1998.

FIGURE 6.19   Performance management process instructions

NOTES

1. R. L. Grayson, "Hazard Identification, Risk Management, and Hazard Control," in *Concepts and Processes in Mine Health and Safety Management*, ed. M. Karmis (Littleton, CO: Society for Mining, Metallurgy, and Exploration, 2001), 247–261.

2. BHP Billiton, *Working for a Sustainable Future: BHP Billiton Health Safety Environment and Community Report 2004*, http://sustainability.bhpbilliton.com/2004/index.asp.

3. Mine Safety Technology and Training Commission, *Improving Mine Safety Technology and Training: Establishing U.S. Global Leadership* (December 2006), www.coalminingsafety.org/documents/msttc_report.pdf (accessed May 2008).

4. J. B. Harvey, *Keynote address* at Utah Mining Association's 92nd Annual Meeting, Park City, Utah, 2007, www.nma.org/pdf/misc/083007_harvey.pdf.

5. *Mine Improvement and New Emergency Response Act of 2006 (The MINER Act)*, Public Law 109-236, 109th Congress, effective June 15, 2006.

6. P. Hudson, "Becoming a Safety Culture: What Does It Take?" (PowerPoint presentation at OHS Conference, New South Wales Minerals Council Ltd., June 19, 2006), www.nswmin.com.au/about_nswmc/publications/2006_ohs_conference).

7. Ibid.

8. V. Kecojevic, Z. Md-Nor, D. Komljenovic, W. Groves, and R. L. Grayson, "Risk Assessment for Continuous Miner-Related Fatal Incidents in the U.S. Underground Mining," in Accident Analysis and Prevention, under review for potential publication with Elsevier.

9. J. Joy and D. Griffiths, *National Minerals Industry Safety and Health Risk Assessment Guideline*, Version 4 (Queensland: Minerals Council of Australia and Minerals Industry Safety and Health Centre, 2005), www.mishc.uq.edu.au/NMIRAG/NMISHRAG.asp.

10. Standards Australia/Standards New Zealand, AS/NZS 4360:2004, Risk Management, (Sydney: Standards Australia International Ltd and Wellington, NZ: Standards New Zealand, 2004).

11. Grayson, "Hazard Identification."

12. L. W. Freeman, *Internal Surveys and Analysis* (Denver, CO: Downing Teal, 2006–2007).

13. Ibid.

14. Steve Kral, ed., *SME Leadership Forum Addresses Shortage of Mining Professionals* (Littleton, CO: Society for Mining, Metallurgy, and Exploration, 2006).

15. Ibid.

16. Ibid.

17. Ibid.

18. G. T. Milkovich and J. M. Newman, *Compensation*, 2nd ed. (Homewood, IL: BPI/Irwin, 1987).

19. MRC Corp., *Recruitment and Retention Challenges in the Mining Industry*, 2007, www.miningsearch.com/recruitment-retention-challenges-in-mining-industry.asp.

20. Freeman, *Internal Surveys and Analysis*.

21. R. Peele and E. K. Judd, "Day's Pay, Contract Work, Leasing," Section 22, Article 1–7, in *Mining Engineers' Handbook*, 2nd ed. (New York: John Wiley & Sons, 1919), 1527–1533.

22. H. B. Miller, "A Comprehensive Analysis of Underground Contract Mining Systems" (master's thesis, Colorado School of Mines, 1991), Chapter 2, Compensation.

23. Milkovich, *Compensation*.

24. Miller, "A Comprehensive Analysis."

25. J. B. Leinart, *Mining Survey Sheds Light on Wages and Benefits Plans* (Littleton, CO: Society for Mining, Metallurgy, and Exploration, 1990).

26. Ibid.

27. Ibid.

28. H. B. Miller and M. J. Hrebar, "Contract Incentive Systems in Underground Metal Mining Industry," in *1994 SME Conference Proceedings*, Mining & Exploration Division, Underground Mining Session, Albuquerque, NM, February 1994.

29. Miller, "A Comprehensive Analysis."

30. Ibid.

31. J. L. Chouinard and N. R. Billette, "Bonus and Accidents: Is There a Link?" *CIM Bulletin* (October 1986): 31–35.

32. G. Dessler, *Human Resource Management*, 7th ed. (Upper Saddle River, NJ: Prentice-Hall, 1997).

33. Brady Corporation. *Performance Leadership Guide*, part of Certificate in Human Resource Management, Volume 2: Human Resource Best Practices, 1998.

CHAPTER SEVEN

Management of Exploration

J. A. Espí

INTRODUCTION

As defined in this chapter, mineral exploration is comprised of the technical and management activities and processes leading to the discovery, definition, and technical evaluation of mineral deposits.

Mining is necessary to sustain economic prosperity and quality of life, and this requires continued exploration for new mineral deposits. As easily discoverable near-surface orebodies are exploited, exploration needs to target deposits located at greater depths and more remote locations.

Mining and mineral exploration companies are faced with increased government regulations and voluntary adherence to sustainable development frameworks. This chapter discusses a mining regulatory framework and some voluntary "best practices" dealing with issues of corporate social responsibility and social license in the context of mineral exploration activities.

Regarding strategic trends, Heffernan[1] pointed out that in 2000, the exploration strategy was following three major trends: (1) a decrease in greenfields or "grassroots" exploration and increasing activity around existing operations (brownfield), (2) an increase in mergers and acquisitions, and (3) an upsurge in strategic alliances, particularly between majors and juniors. Today, these trends are basically valid.

Probably the most significant strategic change in this decade is the dramatic increase in the role of listed junior exploration companies. In the next section, Hudon refers to a study by Canadian-based Metals Economics Group[2] and reports it is estimated that in 2007, total world expenditures in exploration for nonferrous metals reached US$10.5 billion, whereas the junior mining sector has accounted for more than half of global exploration spending.

Regarding sustainable management issues, the exploration process does not require any alteration of the land uses where it takes place or any major development; therefore, its social and environmental impacts are rather limited. However, exploration represents the initial interaction between a mining company and the local communities, so building a trusting relationship with stakeholders at the exploration stages is essential to the achievement of social license. In this regard, the business perspective of social license as a strategic advantage in gaining access to mineral resources is gaining the attention of mining companies. In this context, an excellent sustainable development draft framework for exploration activities has been developed by the Prospectors and Developers Association of Canada.[3]

Also highlighted in this chapter, the application of international standards to the evaluation and reporting of exploration results is essential for transparency and ethics considerations when reporting to shareholders and when dealing with the hosting communities and other stakeholders. The balance of this chapter is composed of four sections.

"Regulatory Framework for Mineral Exploration" discusses the mining regulatory framework and voluntary "best practices" dealing with issues of sustainability development and corporate social responsibility in the context of mining exploration activities.

"Exploration Strategy of Mining Companies" reviews the different exploration strategies of the minerals industry, focusing on the links between exploration strategy and sustainability.

"Exploration Management and Sustainability" describes the exploration management activities and processes, focusing on corporate social responsibility issues. It also discusses several sustainable development frameworks as applied to mineral exploration.

"Ore Resources Inventory Management" focuses on key criteria for sustainable management of ore reserves inventory, the use of resource management systems, the classification of resources and reserves, and the concepts of grade control as it applies to ore resource management.

REGULATORY FRAMEWORK FOR MINERAL EXPLORATION*

Mining is a nonrenewable resource industry and may therefore at first appear to be on a collision course with sustainable development. In order to reconcile mining activities with sustainable development objectives, the mining regulatory framework, including the securities regulatory framework, must provide for guarantees that mining companies must be required to follow.

The World Business Council for Sustainable Development states: "Corporate social responsibility is the continuing commitment by business to behave ethically and contribute to economic development while improving the quality of life of the workforce and their families as well as the local community and society at large."[4]

This section discusses the mining regulatory framework and some voluntary "best practices" dealing with issues of corporate social responsibility (CSR) in the context of mining exploration activities.

The Concept of Mineral Exploration

It is generally accepted that mining is a process that begins with the exploration for and the discovery of mineral deposits that continues through ore extraction and processing to the closure and remediation of worked-out sites. Exploration is thus at the very heart of mining. Mineral exploration basically consists of a number of interlinked and sequential stages, which involve material expenditures, financing, and risks. Each successive stage involves more time and more money.

With increasing environmental restrictions and administrative hurdles, the search for minerals has extended to more hostile geographical environments. Exploration investment is further affected by the metals demand and supply, the prices of metals, the ability to raise capital, and shareholders' satisfaction.

The mining industry is capital intensive. From exploration to mine closure, mining operations may have a serious impact on the environment and communities where the mine is located. The sum of all the risk factors determines why mining is unique and different from other economic activities. According to some, it takes one thousand grassroots prospects to make a discovery and it takes one hundred discoveries to make a mine. According to a study by Canadian-based Metals Economics Group (MEG), it is estimated that, at the end of 2007, total world exploration expenditures for nonferrous metals companies listed on the stock exchange had reached US$10.5 billion.[5]

---

* This section was written by M. G. Hudon.

Regulatory Framework

Policy and regulatory regimes affect the attractiveness of a country for exploration investment; the timelines, predictability, and certainty are important factors. The exploration companies are faced with an increased general regulatory framework including land use, project environmental assessment reviews, aboriginal issues, abandoned mines, and regulatory issues from the regulatory securities authorities. There is a feeling that the mining industry is increasingly the subject of complicated regulations. Some say that it is more difficult now to be a mineral exploration company than it was 15–20 years ago.

In some regions, regulations regarding security, health and safety, land tenure, and the environment have tripled. According to a PricewaterhouseCoopers report, mining companies are now spending upwards of 20% of their annual expenditures meeting regulatory requirements.[6] A whole new set of regulations now deal with social license to operate and the duty to consult and accommodate stakeholders.

The Mining Code

In almost all cases, ownership of rights in or over mineral substances forms part of the domain of the state. No person may therefore prospect or carry out exploration or mining activities without a permit or license issued by the governmental authority having jurisdiction.

Exploration licenses or permits. In some instances, exploration permits or licenses are issued upon paying the prescribed fees and completing and delivering the form prescribed by regulations. In these cases, the required work program is prescribed by regulation. In other instances, the exploration permit or license will be issued subject to the applicant satisfying the responsible governmental authorities that it has the financial and technical capabilities to carry out a work program negotiated with the responsible governmental authorities.

It is recommended that in all cases, a qualified independent technical report be filed upon the application for an exploration permit or license. Such a report, prepared by an independent qualified person, should include social and community factors, as well as technical aspects.

The holder of a mining exploration permit or license is generally subject to general applicable laws dealing with, among other things, the general regulatory environmental framework, including the forest, water, wildlife conservation, and agriculture regulatory frameworks.

Disclosure—notices. The holder of an exploration permit or license must generally be required to transmit to the mining authorities a report on exploration work performed and the results of the work, the whole subject to prescribed confidentiality rules.

Before commencing grassroots or surface exploration work, no notice, under the Mining Code, is generally required to be given to the mining authorities. However, in the case of underground exploration, a provision should provide for required written notices before commencing operations.

Protective measures. The mining regulatory framework should generally require of the holder that protective measures be taken to prevent any damage that may result from the exploration work or the discontinuance of the exploration work. The mining regulatory framework should provide that if the holder fails to take such measures, the mining authorities have the right to suspend the holder's right until such time as the default is remedied or have the right to cause such measures to be taken at the expense of the holder.

Land rehabilitation and restoration work. The principle provided by the Mining Code should be that every holder of an exploration permit or license must submit a rehabilitation and restoration plan approved by the mining authorities before commencing exploration work, the whole as determined by regulation. Usually, such plan is approved after consultation with the governmental authority responsible for sustainable development.

The contents of the plan should be spelled out in the mining regulatory framework; the plan should provide for rehabilitation and restoration work to be performed during the duration of the permit or license. The plan must include a description of the guarantee serving to ensure performance of the work required by the plan.

The mining regulatory framework will generally provide for provisions in cases of default and more particularly will award the mining authorities the right to cause the work required by the rehabilitation and restoration plan to be performed at the holder's expense and to recover the cost thereof out of the guarantee.

The mining regulatory framework should also give to the mining authorities the right to suspend or revoke exploration permits or licenses when the prescribed work has not been executed or when the work report submitted has not been accepted. Before suspending or revoking an exploration permit or license, the principles respecting administrative justice must apply.

Guarantees. The rehabilitation and restoration plan should include a description of the guarantee serving to ensure the performance of the work required by the plan.

The amount of the guarantee should correspond to a prescribed percentage of the anticipated cost of carrying out that part of the work required under the plan that relates to the rehabilitation and restoration of the site. The forms of the guarantee may include

- Bonds issued or guaranteed by a recognized government and having a market value at least equal to the amount of the guarantee,

- Guaranteed investment certificates or term deposit certificates issued by a recognized bank or a trust company,

- An irrevocable and unconditional letter of credit issued by a recognized bank or a trust company,

- A security or a guarantee policy issued by a legal person legally empowered to act in that quality, or

- A guarantee provided by a third party secured by an acceptable collateral.

The mining regulatory framework may, in either circumstance, also allow for the right of the mining authorities to subject the approval of a rehabilitation and restoration plan to the advance payment of all or part of the guarantee. Powers should be given to the mining authorities: (1) to increase the amount of the guarantee or reduce it depending on the circumstances, and (2) to request the payment in full of the guarantee if the financial situation of the holder could prevent the payment of all or part of the guarantee.

Environment Quality Legislation

In general, the environmental legislation requires the mining company to perform an Environmental Impact Statement (EIS), aiming to evaluate the environmental impacts and risks associated with mining operations and determine the monitoring, control, and remediation actions that should be implemented in the project to minimize the environmental risks.

Conflicts. The mining regulatory framework should provide that none of its provisions should affect or restrict the application of the general legislation dealing with the quality of the environment.

Authorization. The environmental quality legislation framework will generally provide that no person may undertake any construction, work, activity, or operation, or carry out work as provided for by regulation, without following the rules for Environmental Impact Assessments (EIAs) and, as the case may be, without obtaining an authorization certificate from the responsible environmental authority.

Generally, the regulatory framework will

- Determine the classes of construction, works, plans, programs, operations, works, or activities to which such rule applies;
- Determine the parameters of an EIA with regard, namely, to the impact of a project on nature, the biophysical milieu, the underwater milieu, human communities, the balance of ecosystems, archaeological and historical sites, and cultural property;
- Prescribe the terms and conditions of the information and of the public consultation pertaining to any application for an authorization certificate or for an EIA for all or some of the classes of projects contemplated; and
- Define types of EIAs and the terms and conditions of presentation.

The rules should clearly state which projects are automatically subject to the EIA or notice procedure and those that may be or are automatically exempt from said measures. For example, all mining site developments, including the additions to, alterations, or modifications of existing mining site developments should automatically be subject to assessment, while exploration projects, except for below-ground projects, should not automatically be subject to the assessment and notice procedure contemplated in the environmental legislation framework.

No project not automatically subject to the EIA and public notice procedure should be undertaken unless a certificate of authorization or an exemption from the EIA and review procedure has been issued by the responsible environmental authority.

Every person intending to undertake a project that is not automatically subject to the EIA and public notice procedure must give prior written notice of his intention to the responsible governmental authority and briefly indicate the nature of the project, the place where the project is to be undertaken, and the date foreseen for the start of the work, the whole as prescribed by regulation. The said notice must deal with technical, economic, and social implications of the project.

The proponent of a project that is automatically subject to EIAs shall prepare an EIS, either preliminary or detailed, or both, according to the prescribed directions and recommendations of the responsible governmental authority and in conformity with the regulations made under the law. Such statement should deal namely with a qualitative and quantitative inventory of the aspects of the physical and social environment that could be affected by the project, such as fauna; flora; human communities; the cultural, archaeological, and historical heritage of the area; agricultural resources; and the use made of resources of the area.

The content of the notices and statements should be prepared by a qualified independent person.

The Capital Markets and Securities Regulatory Framework

A mining exploration company will finance its operations and growth through the capital markets. The main source of financing for mining exploration companies comes from the issuance of new shares in their capital stock, be it by way of public or private financing. Access to venture capital is essential given the nature of mining exploration and that stock exchange listed companies account for the bulk of worldwide exploration budgets.

Financing

According to MEG, which tracks only junior financings for more than 2,000 companies, nearly 90% of the companies covered by its studies are based in Canada (1,002), Australia (439), the United States (89) and the United Kingdom (82); 60% of the world's exploration and mining companies are listed on the Toronto Stock Exchange or the Toronto Stock Exchange Venture in Canada; and 50% of all equity financing is done in Canada.[7]

Disclosure

Once a company becomes public and/or gets listed on a stock exchange, it is subject to certain disclosure requirements and other good governance requirements. Transparency and good governance are the underlying basic principles of stock exchange rules, securities laws, and regulations.

The policies governing such a framework do not generally require from listed companies the obligation to disclose, in annual reports, environmental or sustainable corporate policies. Such policies sometimes require listed companies to adopt and disclose their good governance policy, which mainly deals with internal management issues.

There appears to be a tendency to regulate financial disclosures and to forget disclosure of CSR. Some securities commissions that regulate companies that have issued shares to the public (and which are listed on stock exchanges) are urging such companies to improve reporting of risks and liabilities regarding the physical environment.

Investments

Institutional investors appear to be requesting improved disclosure of environmental risks; some investment firms' portfolio managers and pension funds analyze, before an investment decision is made, the past record on how mining companies treat the environment and the communities in which they operate. As a result, some mining companies are now considering the interests of their shareholders, those of their other stakeholders including the communities they operate in, and the protection of the environment as part of their duty.

A number of banks have adopted a set of environmental and social guidelines, the Equator Principles. Under these principles, companies or countries that apply to the member banks for certain infrastructure projects must show that these projects would not have negative effects on people and the environment.

Technical Report

For publicly listed mining exploration companies that intend to obtain financing, a technical report often must be filed describing the geology of the project, prior programs, if any, the details of the exploration program to be executed, and the social and economic implications. This report is generally prepared by an independent competent person, a geologist, or an engineer duly registered with a recognized professional corporation.

Voluntary Frameworks

Mining companies interested in improving their social and environmental performance as part of their business have a wide range of voluntary tools available to them. Such tools can vary widely in terms of objectives, scope, costs, level of formality, partnerships, extent of stakeholder involvement, and many other characteristics. These tools can be applied to one or more of the planning and implementation stages of corporate operations.

There are a variety of principles, guidelines, and codes of conduct that mining companies can use to develop their commitments:

- The OECD (Organisation for Economic Co-Operation and Development) Guidelines for Multinational Enterprises and for Corporate Governance
- United Nations Global Compact
- Canadian Institute of Mining, Metallurgy and Petroleum (CIM) Guidelines
- The Prospectors and Developers Association of Canada (PDAC) Guidelines
- Global Sullivan Principles
- Association of British Insurers (Corporate Governance and Investment Guidelines)
- Canadian Coalition for Good Governance
- Global Reporting Initiative (GRI) Guidelines[8]

In its voluntary "Exploration Best Practices Guidelines," CIM has adopted the following principle regarding environment, safety, and community relations: "All field work should be conducted in a safe, professional manner with due regard for the environment, the concerns of local communities and with regulatory requirements. An environmental program, including baseline studies, appropriate to the stage of the project should be carried out (...)." The guidelines add: "This set of broad guidelines or 'best practices' has been drawn up to ensure a consistently high quality of work that will maintain public confidence and assist securities regulators. The guidelines are not intended to inhibit the original thinking or application of new approaches that are fundamental to successful mineral exploration."[9]

GRI is the leading international standard for sustainability reporting. More than 1,000 organizations from 60 countries use the GRI Guidelines to produce their sustainability reports. All sorts of organizations report using the GRI Guidelines, such as corporate businesses, public agencies, smaller enterprises, nongovernmental organizations, industry groups, and others.

PDAC has developed online tools to assist companies in developing their own sustainability policies in exploration. This organization's "e3—Environmental Excellence in Exploration" online manual launched in 2003, is a comprehensive Internet-based toolkit that offers leading examples of environmental and social responsibility in the minerals industry.[10] Such a framework should namely provide good practice guidelines, performance indicators, and reporting criteria. The draft principles and management essentials proposed by the PDAC are summarized in the "Exploration Management and Sustainability" section.

Challenges and Conclusions

Access to lands for mineral exploration is critical for a healthy mining industry, and the interests of affected communities must be considered. Permitting and consultation issues should be dealt with by the regulatory framework. Issues in exploration that involve people must be addressed at the outset, including community relations.

A large portion of the world is open to mining companies, which are offered numerous exploration investment options. Key factors—some controlled by the government, others beyond its control—have an impact on the ability of a country to attract investments. Given that mining exploration is a high-cost, high-risk business, investors will be attracted by those countries with prospective geology, a reliable mining code, a reasonable tax regime, political stability, and reasonable sustainable development best practices.

Today, there appears to be growing pressure on mining companies around the globe to adopt CSR guidelines as part of their business plan. Social sustainable development guarantees do not generally appear to be a requirement prescribed by law or regulation. It is left to the mining companies to subscribe voluntarily to various codes and standards, which are being promoted by various companies and seldom without material consequences.

In order to promote sustainable development, governments should adopt a regulatory framework that would provide for consequences in cases of default. The mining companies should be required to adopt a sustainability development policy in which their business and social responsibility is spelled out; it should be clearly stated that business goals will be achieved in a safe, transparent, environmentally and socially responsible way. An effort should be made by governments to promote a better balance between the competing interests of mining companies and communities.

Companies requesting an exploration permit or license should be required to file their corporate governance policy with the mining authorities at the very start, including sustainability development and corporate responsibility issues; the application for listing on a stock exchange should also include a similar policy. Such corporate governance policy could (in the absence of an applicable regulatory framework) make reference to international guidelines, disclosure, and transparency.

The regulatory framework should also provide for the filing of a prescribed technical and social report regarding all mineral exploration projects, or at least for below-ground projects. Such disclosure requirements would be a condition precedent to the delivery of exploration permits or for the authorization required for the commencement of work. The content of such a technical report would be prescribed by regulation. The report would be based on information prepared by or under the supervision of a qualified independent person.

The regulatory framework could refer to the permitted use of foreign or international codes, standards, or policies regarding such reports. Given that mining guidelines, industry practices, and standards are evolving, the regulatory framework should adopt such forms (regulations, guidelines, policies, etc.) that are flexible and that may be adapted quickly to developments regarding best practices.

Every person engaged in exploration should be requested to forward annually or quarterly to the mining and, as the case may be, to the securities authorities, a report and a forecast for the following year, showing namely the nature and cost of the exploration work, the rehabilitation and restoration work performed or to be performed, and community expenses and involvement.

Wherever impact assessment reports or notices are required for an exploration project, the mining regulatory framework should prescribe the contents thereof and provide for specific CSR requirements as to its content. Table 7.1 shows sample content for CSR.

Once the exploration permit is awarded or renewed, title holders should be required each year to describe completely their practices with specific reference to the prescribed guidelines or rules. Such disclosure should be filed with the responsible mining and securities regulatory authorities. In their disclosure, the mining companies should describe their practices against each guideline or rule and, if their practices differ, reasons should be provided for the discrepancies. This mandatory disclosure against guidelines or rules should provide investors with useful information and enhance the quality of the capital markets and the environment. Mining companies listed on stock exchanges should be required to issue yearly comparative sustainability reports outlining the social, environmental, and economic performance of all operations.

In the event such guidelines or rules are not fully addressed by the mining companies, the mining and securities regulatory framework should provide for measures dealing with sanctions ranging from daily penalties to the suspension of the title holder's rights up and until such defaults are remedied. The capital markets' regulatory bodies should provide for the suspension or delisting of a company's securities where it continuously fails to meet such disclosure requirements.

TABLE 7.1 Sample regulatory framework for corporate social responsibility

Business Practices
• The policies for corporate governance, ethics, and sustainable development
• The budget for environmental and social aspects and human resources
• The required compliance of contractors and subcontractors in contracts to the company's policies regarding social, environmental, health and safety issues, and human rights
• All and any information that is relevant to their activities, subject to confidentiality constraints
• The company's transparency policy
• Project-related social and environmental matters and risks
• The policy dealing with employment matters and its public disclosure
• The policies and procedures for the management of environmental issues, including remediation and reclamation of lands, health and safety, and public disclosure
Due Diligence
• The nature and intent of the due diligence activities regarding social, cultural, environmental, and human rights issues
• The policy and procedures for prior and informed consultation with groups affected by the project
• The policy and procedures for update of informed consultations
Communities
• The policy concerned with communities and public disclosure including indigenous peoples, lands, and resources
• The policy and procedure for community relations; land access, compensation, and dispute resolution
• The procedure for the mutual exchange of information
• The execution of land access agreement
• The information about the company and exploration program
• The implementation of monitoring and reporting information procedure as to the social and environmental aspects of the project
• The consultation process, the adoption and public disclosure of policies and procedures for employment, use of local suppliers and services, and community development

Regarding the role of the regulatory framework in the promotion of sustainable development, the tax aspects related thereto should be examined. The costs of community consultations, baseline environmental studies, and feasibility studies should qualify as exploration expenses, giving the investors the right to deduct such expenses; the taxation framework should encourage the cleanup of abandoned sites. Under Canada's Income Tax Act,[11] contributions to a qualified environmental trust (QET) are deductible in the year of contribution. Such a QET is defined as "a trust resident in a province and maintained at that time for the sole purpose of funding the reclamation of a site in the province that had been used primarily for or for any combination of, the operation of a mine or a waste dump, where the maintenance of the trust is or may become required under the terms of a contract entered into with Her Majesty in Right of Canada or the province, or is or may become required under a law of Canada or the province (...)."

EXPLORATION STRATEGY OF MINING COMPANIES

Exploration is the most strategic activity of any mining company. Therefore, the efficient management of the exploration portfolio is essential to sustainability. By and large, the success of a mining company is determined by its capacity to transform today's exploration targets into tomorrow's cash-flow streams.

Sustainable exploration strategies should be focused on controlling decision risks, considering risk as chance of failure or loss. Singer and Kouda[12] consider three ways to control failure risk:

1. Increasing the number of prospects,

2. Increasing the economic potential of prospects, or

3. Improving prospect management.

Management aspects are discussed in the "Exploration Management and Sustainability" section, where successful exploration management is mainly linked to three management aspects: (1) in-depth knowledge of the legal framework and organizational structure of local, regional and national administrations where the company operates, (2) high ethics, and (3) environmental and social responsibility.

MEG economist Michael Chender[13] considers the traditional drivers of exploration strategy to be the need for corporate growth, high metal prices, availability of risk capital, and new discoveries that open up new areas to exploration. Some "internal" drivers are also important, such as: what is in our pipeline now, the potential of discovery near our mines, and how much should we spend just to keep up with our present exploration rights.

From the strategic management perspective, resource growth can be achieved by direct investment in exploration or by relying on others (e.g., junior companies) and subsequently acquiring the resources. As previously highlighted by Hudon, listed junior mining companies accounted for almost 50% of the world exploration spending of US$10.5 billion in 2007.

David Timms[14] highlights that the tremendous success of the junior exploration model is related to seven competitive advantages:

1. Operate with cost efficiency,

2. Select the best people (mine finders),

3. Develop a multidisciplinary team,

4. Devise meaningful incentive schemes,

5. Explore in ore-permissive areas,

6. Design long-term programs, and

7. Drill more holes into selected targets.

Other important strategic decisions are the distribution of capital budget among exploration targets and the use of contractors (drilling, geophysics, and logistics).

Strategic Alternatives for Exploration

What shareholders and other investors expect from a mining company[15] is that it can deliver a growth in reserves and production by whatever means, either by organic growth through the success of company's own exploration division or, alternatively, through acquisitions or joint-venturing. Another factor is the investor's perception of the quality of company's management and the technical expertise of the exploration group.

An important strategic question is deciding between achieving mineral resource growth through in-house exploration or by focusing on acquisitions of exploration assets with high upside potential. It's also important to find out how these two options are perceived by investors. From a strategic perspective, this question will determine the size and the expertise required for the company exploration group, as well as its role in the organization.

Gunthorpe[16] proposes a structural model (Figure 7.1) describing the functions of the mineral resource management group of a mining company. The model defines four management sections, the functions and responsibilities of which are related to the risk, and the uncertainty of the decisions to be made.

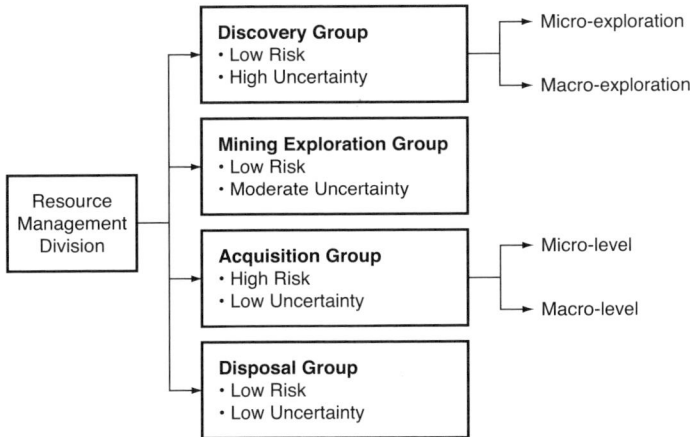

Source: After Gunthorpe 2001

FIGURE 7.1   Four functions of resource management and growth

The model considers two discovery strategies: greenfield and brownfield. The greenfield exploration strategy focuses on the discovery of new orebodies in areas where other orebodies have already been discovered or in areas with no previous discoveries but showing favorable geologic conditions. Brownfield exploration focuses on the environs of existing mining operations, aiming for the discovery of extensions and repetitions of known orebodies, and a lower probability of a major new discovery.

The strategy of growth via "discovery" is considered as low risk/high uncertainty because capital intensity may be controlled, but the probability of success may be low. Similarly, brownfield exploration risk is low because it can be controlled and uncertainty is somewhat lower (higher probability of success). Therefore, growth through exploration should be considered a conservative, low-risk strategy.

As an example, AngloGold's exploration strategy[17] is to discover gold reserves via brownfield exploration at a cost of less than US$9/oz Au reserves, and greenfield discoveries at an exploration cost of less than US$30/oz Au reserves.

In contrast, strategies of growth via acquisition are uncertainty-controlled (the acquisition decision may be taken under low uncertainty), but the financial exposure is associated with the decision and therefore is maximum.

Asset disposal is an important part of exploration strategy that carries a substantial risk. Here, risk is associated with disposing of an asset without having fully assessed its economic potential.

As previously stated, the exploration strategy should be focused on managing for risk. However, risk management strategies vary depending on the size and the financial capacity of the company. Only a few global players have the financial capacity to run high-risk exploration strategies involving both acquisitions and discovery-focused exploration.

As an example, two leading gold companies, Newmont Corporation and Barrick Gold Corporation, with similar sizes and business purposes, have different growth strategies. While Newmont focuses on growth through exploration as its main strategy, considering acquisitions on an opportunistic basis only, Barrick Gold focuses on growth through acquisition of assets. Both

---

Exploration Strategy of Rio Tinto

✓ Our prime focus is on mining large, long-life, low-cost orebodies. We recognize that medium term growth will come largely from further development of assets we already own.

✓ Today's mining industry is highly competitive and Rio Tinto needs to improve continually in order to keep ahead.

✓ Our fundamental principles will not change, but we do contemplate some changes in emphasis.

✓ Capital spending on new projects has been rising steadily in the last few years as we have developed a number of high-quality greenfield projects and some major brownfield extensions. In addition, we have development studies progressing on 15 more projects. Over the past year we have transferred five projects from exploration to the product groups.

✓ Our level of success reflects efforts we have made over a number of years and is a tangible result of our commitment to exploration.

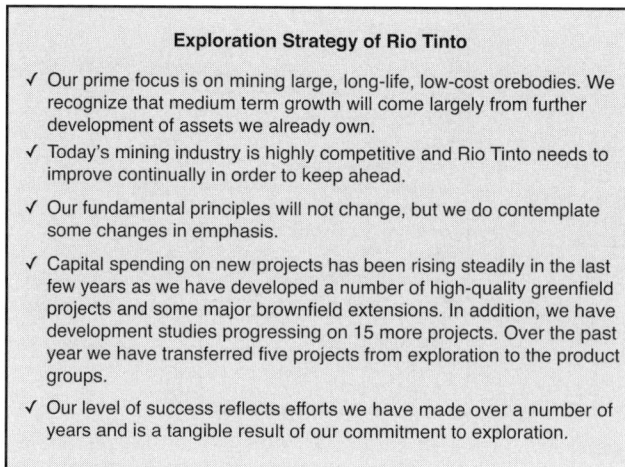

Source: Data from Rio Tinto Ltd.'s Web site

FIGURE 7.2   Exploration strategy: Rio Tinto

strategies are valid, though probably Newmont's seeks economic efficiency while Barrick Gold seeks faster growth.

Other examples of exploration strategies for three global mining companies, Rio Tinto, AngloGold Ashanti, and Barrick Gold, are described below.

Rio Tinto's corporate strategy[18] (Figure 7.2) focuses on developing large orebodies that can be mined at a low operating cost. This strategy allows for basing medium-term growth on resources already discovered and evaluated.

AngloGold's exploration strategy[19] (Figure 7.3) gives maximum priority to growth from expansion of ore reserves nearby existing orebodies (brownfield exploration), considering high-risk greenfield exploration as a second priority.

Barrick Gold's exploration strategy[20] (Figure 7.4) is based on expanding its resource portfolio through acquisition and focusing its own exploration effort on the expansion of ore reserves from existing operations.

In general, the exploration strategies of most gold and base metal companies seek a balance between both exploration and acquisition strategies by operating their own exploration program while following the market for acquisition opportunities.

Also, most medium/large mining companies carry out exploration through joint ventures with junior companies, seeking competitive advantages in case of a discovery. Exploration joint-venture programs are also carried out between large mining companies, in this case, aiming to reduce financial risk.

The present business cycle of high prices is significantly influencing the strategy of mining companies. The process of globalization of the mining sector is taking place at a much higher rate, with mergers and acquisitions becoming the main strategy for survival. The structure of the mining sector has changed dramatically through a rapid process of business concentration, with a handful of global giants dominating the sector.

Exploration Strategy of AngloGold Ashanti

✓ The replacement of production ounces through near-mine (brownfields) exploration continued to remain a high priority for AngloGold Ashanti in 2006. During the year, brownfields exploration activities continued around most of the group's current operations.

✓ In 2006, exploration activities in new areas (greenfields exploration) were primarily focused on the Tropicana joint venture project in Western Australia, in Colombia, and in the Democratic Republic of Congo (DRC). Joint ventures and partnerships with other companies facilitated additional greenfields exploration activities in Russia, China, Laos, Colombia, and the Philippines, while the company divested its exploration assets in both Alaska and Mongolia during the year. The discovery of new long-life, low-cost mines remains the principle objective of the greenfields exploration program, although AngloGold Ashanti is also committed to maximizing shareholder value by exiting or selling those exploration assets that do not meet its internal growth criteria and also by opportunistically investing in prospective junior exploration companies.

✓ During 2006, total exploration expenditure amounted to $103 million, of which $52 million was spent on brownfields exploration. The remaining $51 million was primarily invested in three key greenfields areas (the Tropicana joint venture in Western Australia, in Colombia, and in the DRC), with the remainder being spent in Russia, China, the Philippines, and Laos. Exploration expenditure is expected to increase to $163 million in 2007, with $77 million to be spent on brownfields exploration and $86 million to be spent on greenfields exploration.

Source: Data from AngloGold Ashanti's Web site

FIGURE 7.3   Exploration strategy: AngloGold Ashanti

Exploration Strategy of Barrick Gold

✓ Barrick has a motivated, discovery-driven team looking for gold in many countries around the world and plans to spend about $170 million on exploration in 2007. Reserve development and replacement of production is a major priority at all sites. The Company consistently funds its exploration programs throughout all gold cycles, and has a proven track record of finding ounces at both greenfield and brown-field projects.

✓ Exploration is focused on highly gold-endowed districts where Barrick controls large land positions, the primary ones being the Goldstrike and Cortez districts in Nevada, the Frontera District in Chile/Argentina, the Lake Victoria District in Tanzania, and at Porgera in Papua New Guinea. In addition, the Company is exploring earlier-stage projects and evaluating exploration opportunites in emerging districts around the world.

✓ Three key factors drive the Company's exploration success: the motivation and technical excellence of its exploration team, the policy of consistently investing in exploration, and the robust and balanced pipeline of exploration projects. The Company's disciplined exploration strategy maximizes the chance of near-term discovery by putting the best people on the best projects and advancing the best projects more quickly up the exploration pipeline.

Source: Data from Barrick Gold's Web site

FIGURE 7.4   Exploration strategy: Barrick Gold

Focused Strategies

Strategic management of exploration relies on three main factors: the people, the processes, and the quality of the projects. All mining companies aim to excel with respect to these management principles (people, processes, and projects) but may differ in focus and the standards of quality of the projects.

As an example, Gold Fields[21] focuses on high geology standards and exploration databases. In Gold Fields, any geological knowledge should be included in a geographic information system (GIS) and exploration data should pass a quality assurance process. To this aim, the company runs high-quality technical support systems and employs highly qualified exploration geologists and quality assurance experts who control all stages in the exploration process.

Exploration strategies may also aim for specialization by focus, substance, geography, geological environment, and even technology. This is illustrated in Table 7.2.

Another important strategy aspect is the fact that exploration investments tend to go to countries and regions where they are well received (i.e., those with a stable and favorable legal framework) and to areas with higher potential for discovery.

However, some of the recent "world class" orebodies had a metalogenesis unknown at the time of discovery. This is where no experience on similar orebodies was available, as was the case of Neves Corvo, Candelaria, and Century. New discoveries have confirmed the mineral "fertility" of the circum-Pacific belt and the potential of the belt in areas like Pakistan, Iran, Turkey, and the Balkans. Furthermore, the giant orebodies of Escondida and Grasberg were found by using geological criteria only.

EXPLORATION MANAGEMENT AND SUSTAINABILITY

Mineral exploration is the first phase of the mine life cycle and is comprised of the technical and management activities and processes leading to the discovery, definition, and technical evaluation of mineral deposits. Usually, mineral exploration starts with the definition of large areas where mineral prospecting may be carried out to identify one or more exploration targets of smaller size within it. In a second stage, each of these targets is subject to more detailed and cost-intensive exploration activities, aiming to define mineral deposits with economic potential, where detail drilling may lead to the definition of orebodies of economic interest.

Mineral exploration generally has a low environmental and social impact, and, in most cases, it does not succeed in its objective of developing a new mine. However, being the initial environmental and social footprint of mining, managing exploration in a responsible and transparent manner is critical to achieve social license to operate should exploration lead to project development and mining.

Stages of the Exploration Process

Exploration is normally carried out in three stages:

1. Area selection,
2. Target generation, and
3. Resource evaluation.

Area Selection

Area selection is the process whereby the area having the greatest potential for obtaining economic ore is selected. Area selection is based on the application and testing of orebody genesis models from previously discovered orebodies on the basis of the knowledge on ore occurrences

TABLE 7.2 Typical differences in exploration strategy of mining companies

Strategy Focus	Company	Priorities
Substance	Orezone	Au
	Northern Shield	Platinum group elements
	Anvil	High-grade Cu, Au
Orebody—Genesis	Anvil	In mineral belts
	Compass	Regional experts
Orebody—Placement	Anvil	Shallow orebodies
	Newcrest	Deep orebodies
Region, exploration type...	Minvita	United States
	Anvil	GIS data storage
	Albidon	Districts well known
	Newcrest	Grass roots—good databases
	Teal	Known orebody models
	Northern Shield	Good data, politically stable areas
Environmental impact	Anvil	Low impact and good local contacts
Exploration teams	Albidon	Technical expertise and experience of team
Know-how	Compass	Rare metals
	Albidon	State-of-the-art exploration technology
Portfolio	Anvil	"Balanced pipeline of products"
Financial objective	Aur Resources	ROI > 15% Life > 10 years
	Nippon Mining	High value—downside cost potential
Stage and business structure	Silver Gray	Minority share
	Teal	Existing projects
	Alba Mineral Resources	Advance project stages
	Minvita	Joint venture in "grass roots"
Country risk	Teal	High country risk
	Kinbauri	Political stability

in the area and the mechanisms of their formation. The required knowledge is acquired via the study of geological maps and visual inspection of the area by experienced geologists, using available databases and publications.

The exploration activities performed in this stage, often referred to as reconnaissance exploration, include the compilation of previously existing information, aerial photography, and remote sensing investigation. Also, limited geophysical work (magnetic, electromagnetic, and radiometric methods) and some geochemical methods may be performed in this stage. The reconnaissance may sometimes affect areas from hundreds to several thousand square kilometers wide.

Target Generation

The target generation phase involves investigations of the geology via mapping, geophysics, and geochemical or intensive geophysical testing of the surface and subsurface geology. In some cases, for instance in areas covered by soil, alluvium, and platform cover, drilling may be performed directly as a mechanism for generating targets.

This process applies the disciplines of mineral deposit modeling, geology and structural geology, geochronology, petrology, and a host of geophysical and geochemical disciplines. The process is used to make predictions and draw parallels between the known ore deposits and their physical form and the unknown potential of finding a "look-alike" within the selected area.

Geological mapping is performed over the selected area in the previous stage and geological information is digitized into databases, and geological maps of the selected area are produced. Mapping in this stage includes outcrop lithology, stratigraphy, and structural geology, mineralogical alterations, and structural data, such as intrusions, faults, folds, and so on. This information is essential to deduce the location of hidden mineral deposits. It is usual to take samples for the mineralogical, grain size, and textural study under the microscope.

Geochemical exploration is performed by means of a systematic sampling and a chemical analysis of rocks, stream sediments, soil, water, and vegetation to determine the location of dispersion halos of chemical elements. Geochemical exploration would start with the reconnaissance of wide areas in order to reduce the survey areas gradually, and it would end up with the selection of "targets" to be considered for the resource evaluation stage. Thus, the geochemical studies must be completed with the chemical and mineralogical knowledge of the deposit.

The type of study carried out will depend on the degree of geochemical and geological knowledge of the surveyed area, on the nature of the explored land, and on the company operative strategies.

Geophysical exploration is performed to measure the variation of physical properties of the ground, which may result from the existence of a hidden orebody. The most common geophysical methods are gravimetric, magnetic, electrical, forced current, seismic, and radiometric. The environmental impact associated with modern techniques of geophysical exploration is considered minimal and not permanent.

Resource Evaluation

Resource evaluation is undertaken to quantify the grade and tonnage of a mineral occurrence. This is achieved primarily by drilling to sample the prospective horizon, lode, or strata where the minerals of interest occur. The ultimate aim is to generate a sufficient density of drilling to satisfy the economic and statutory standards of an ore resource.

Drilling is performed to obtain a direct knowledge of subsurface mineral deposits. The first drilling stage is performed on a very wide drilling pattern and aims to locate possible mineral deposits at an acceptable drilling cost. If results are satisfactory, a closer pattern of drilling may be required to determine the geometry and grade distribution of potentially economic orebodies within the mineral deposit. The drilling method may be auger, rotary drilling, diamond drilling, and others, depending on the type of ground, the depth, and the drilling objectives. Additionally, this stage may require metallurgical testing in order to define the possible mineral behavior in a technical and commercial view.

Following the exploration process, a stage of "reserve definition" is undertaken to convert a mineral resource into an ore reserve, which is an economic asset. The process is similar to resource evaluation, except that it is more intensive and technical, and is aimed at statistically quantifying the grade continuity and mass of ore. The end of this process opens the project feasibility stage to carry out detailed studies to determine if the deposit may be mined at a profit.

Mineral Exploration Players

The following paragraphs discuss the main actors and stakeholders involved in the exploration process.

Governments and Government Agencies

Governments are in charge of the public administration of the mining laws and regulations. They also issue exploration and mining permits, and enforce compliance with safety and environmental regulations.

In many countries, the geological service and other government agencies make available the basic data on geography, geology, and airborne geophysics to the exploration companies at a moderate price. Moreover, some countries provide exploration right holders with the exploration information developed by companies that have previously held the exploration rights for a given area.

Within the European Union countries, the role of the government in supporting mining and exploration is defined in a report by the Enterprise Directorate General,[22] as follows:

- Coordinating or undertaking geological or geophysical surveys to provide general data on the location and nature of mineral reserves,
- Contributing to the funding of a national geological institute or survey,
- Providing financial assistance to private companies involved in exploration, and
- Providing information for land use planning and issuing licenses/permits to allow exploration.

Individual Prospectors

These are freelance geologists, landowners, or exploration consultants who hold the mineral exploration rights in areas with exploration interest. In most cases, their limited financial capacity only allows prospectors to carry out the area definition stage and offer the rights to junior or senior exploration and mining companies.

Junior Exploration Companies

"Juniors" are stock exchange listed companies that only operate in the exploration segment of the mineral industry. Typically, they become highly specialized in exploration as a business and have a small but highly specialized professional structure. The business objective of a junior is to add value to their exploration targets, through raising funds in the stock markets and making efficient use of these funds by bringing exploration to a more advanced stage with a lower risk of failure. Eventually, they hope to bring exploration targets to the feasibility stage and then sell them for a profit.

As previously highlighted by Hudon, juniors play a key role in today's exploration. According to a study by MEG,[23] in 2007 junior companies accounted for more than half of global exploration spending.

Senior Mining Companies

"Seniors" are companies that operate in both the mining exploration and mining operations segments of the minerals industry. These companies may choose to gain access to new ore resources by carrying out exploration by themselves or in joint ventures with other companies or alternatively by acquiring resources through purchase.

Contractors and Services Suppliers

Exploration companies contract out some services, such as drilling, geophysical surveys, geochemical exploration, geological services, engineering services, geostatistical evaluation, land acquisition, information technology (IT) systems and data management, labor recruitment services, legal services, sampling campaigns, auxiliary labor, and vehicle and air transportation.

Social and Economic Stakeholders

In addition, the exploration activity involves several social stakeholders and interested parties:

- The landowners, occupiers, and users of the land, with whom the exploration company must negotiate for the rights of access and/or temporary occupation;

TABLE 7.3   Land access requirements

Activity	Access Requirements
Basic exploration	Exploration license
Airborne surveys	Generally, only Civil Aviation flight approval required
Exploration rights	Record of the rights depending on the legal requirements
Drilling campaigns	Landowner permission (direct negotiation with landowner)
Tree cutting	Permit from local environmental authority
Construction/excavation	Regional/local government permit and agreement with landowners

- The communities hosting the exploration activity with which the exploration company must maintain a relation based on trust and transparency, aiming to earn the social license; and

- Potential investors and financial institutions in the controlling and auditing role.

Land Access to Mineral Exploration

The access to lands with mineral potential for mineral exploration and development purposes and certainty of tenure for viable deposits is a major component of maintaining a viable mining industry. Single or conflicting land use designations create a climate of uncertainty and are a serious impediment to attracting and retaining mineral exploration.

Land use planning is done at the government level (national or regional). It is done through an integrative process in which different claims of utilization are subjected to an evaluation process on the basis of which the land use planning authority identifies areas where, in principle, no minerals extraction will be allowed, areas where extraction may be allowed but is subject to certain conditions, and areas where, in principle, extraction will be permitted.

The mining industry faces restrictions when accessing the area to be explored and exploited, going into direct competition with other users of the territory, who can see in mining activities an impediment to the development of other activities.

The required licenses in the different exploration stages vary, depending on the territories to explore. Table 7.3 summarizes the access requirements for different exploration activities.

For land use planning to be an effective tool, it is essential that it be based on a solid and well-substantiated database that includes all necessary information. From a minerals development point of view, it is crucial that the information concerning mineral deposits be entered into the land use databases to ensure that minerals are considered in all land use planning decisions. Incorporation of minerals in land use planning decisions is considered good practice and essential for a sustainable minerals supply in Europe.[24]

Environmental, Safety, and Health Issues

In general, environmental and social impacts of mineral exploration activities are associated with short- and long-term changes in land use. Exploration requires longer-term infrastructures, such as campsites, roads, and fuel storage; it also generates short-term changes in land use in relation to exploration excavations (pits, trenches, etc.) and access and support infrastructure for exploration drilling. Other environmental impacts are related to the effects of exploration activity on water quality and wildlife. Environmental impacts of exploration activities are summarized as follows:

- Temporary infrastructure campsites:
  - Access roads

TABLE 7.4   Information requirements for a Declaration of Environmental Factors

General information	The general details of the proposal, as licensee or operator name and address
Physical environment	Landform and topography, soil and surface units, surface cover, drainage, hydrogeology
Biological environment	Vegetation cover, fauna, and habitat
Environmentally sensitive locations	Areas having particular ecological, cultural, or conservation value
Human environment	Uses of land, roads, and tracks
Exploration program	Access tracks, drillilng, water supply arrangements, camp, excavation, and drill site clean-up procedures
Potential environmental impacts	Disturbance of native vegetation, soil disturbance, disturbance to scientific and cultural sites, fauna disturbance, visual disturbance, fire, groundwater contamination, surface drainage interference, introduced weeds, and rubbish and waste

Source: After Minerals and Energy Resources South Australia, 2002

- Landing runways
- Fuel storage
- Ditches, pits, trenches
- Drilling site excavation and spills
- Impacts on water quality by uncontrolled effluent discharge to environment
- Impacts on wildlife:
  - Approach of animals attracted by food waste
  - Effects on migrations derived from human presence
  - Damage to wildlife

Regarding land reclamation and impact mitigation, the exploration company must comply with the conditions imposed in their required licenses or permits, as well as with the conditions agreed upon with landowners and the local authority. Furthermore, the exploration activity is subject to inspection and audits by the different government agencies. As an example, the Minerals and Energy Authority of South Australia[25] requires exploration companies to submit a Declaration of Environmental Factors (DEF) in situations where field exploration involves the use of heavy earthmoving equipment (e.g., bulldozers, backhoes, excavators) or when drill rigs are proposed in areas deemed environmentally sensitive.

By applying the DEF procedure, the company is required to identify elements of the environment that may be at risk from the proposed exploration activities and the ways in which potential impacts can be prevented or managed, as summarized in Table 7.4.

Few occupations expose individuals to such a variety of hazards as mineral exploration. Several characteristics are somewhat unique to the industry and affect safety considerations and monitoring. The workplace encompasses wilderness areas ranging from alpine to near-desert conditions and from arctic to temperate environments. The unwary could succumb to any one of many potentially fatal hazards, including falls in crevasses or on precipitous ground, avalanches or falling rock, hypothermia, hyperthermia, asphyxiation, exposure, drowning, lightning strikes, tree falls, animal attacks, insect stings, and injuries resulting from aircraft, vehicle, and boat travel.

Additional health and safety hazards, which some may consider more conventional, may also be encountered in the workplace and must be addressed. These hazards can include noise, ergonomics, working in confined spaces, working around heavy industrial machinery, working in and around excavations, and working at heights, which requires fall protection.

The exploration managers, field supervisors, and technical staff should be thoroughly familiar with safety procedures. Particular attention must be directed to contractors' personnel and
new labor. Appropriate safety and first aid equipment, and suitably trained personnel should be
available at working locations. In this regard, many companies have developed safety manuals
and guidelines with the continuing objective of reducing accidents.

The Association for Mineral Exploration British Columbia[26] has recently updated their
health and safety document, *Safety Guidelines for Mineral Exploration in Western Canada* with
the objectives of ensuring that

- Exploration sites are equipped with appropriate first aid kits, attendants, and access to
  emergency communication;
- All persons employed at an exploration site are trained in safe working practices specific
  to site conditions;
- Any pits, trenches, and excavations are made safe;
- Exposure to uranium and thorium is limited;
- People are protected from electrical hazards, such as those potentially posed by the use of
  induced polarization geophysical survey systems; and
- Explosives are used and stored safely.

The manual includes detailed procedures for first aid, training, pits, trenches and excavations,
induced polarization geophysical survey systems, use and storage of explosives, working in and
around fixed wing and rotary aircraft, and legal responsibilities.

Social Impact

Although the largest social impacts of mining activities are linked to the development and production stages, the exploration activities may have a social impact associated with increased
expectations for wealth growth as a result of a discovery, which would lead to the creation of new
jobs and the opening up of new businesses. This early social impact is more significant when
exploration takes place in poor and underdeveloped regions or in areas where First Nations have
been recognized as having special rights on mineral wealth. In these regions, mineral exploration
is perceived as an opportunity for human development and better living standards; therefore, any
exploration activity generates great expectation. Some business opportunities associated with
mineral exploration may generate new jobs in hosting communities; the construction of a camp
site and other infrastructure, food and lodging, vehicle rental, drilling and earthmoving operators, and so forth.

Exploration Management Essentials

Probably the best and the most complete sustainable development framework for exploration
activities has been developed by the PDAC.[27] The PDAC framework provides information at
two levels:

1. Exploration Principles and guidelines (vision and mission statements)
2. The e-3 Environmental Excellence in Exploration online manual[28]

The Exploration Principles and guidelines aim to "implement and maintain ethical business
practices and sound management systems that include sustainable development as a factor in
business decision making." The guidelines are specific recommendations for responsible exploration and development activities, which are classified into seven groups:[29]

1. Ethical business
2. Human rights
3. Project due diligence and pre-engagement
4. Community engagement
5. Community development
6. Environmental protection
7. Health and safety performance

The e-3 Environmental Excellence in Exploration manual is an exploration management manual focusing on community engagement and environmental practices. PDAC offers it in the form of an online toolkit for mineral exploration.

The management approach of the e-3 Manual is expressed by the "Management Essentials" section, which outlines key management issues to be addressed proactively by site project management. Management essentials are summarized into 13 areas:[30]

1. Exploration code of conduct
2. Environmental and socioeconomic challenges
3. Community relations
4. Legislation and permitting
5. Planning
6. Contractor selection and management
7. Health and safety
8. Wildlife
9. Fire prevention, policy, and response
10. Training
11. Reviews and audits
12. Record-keeping
13. Reporting

For example, the area of management essentials on planning is outlined as follows:[31]

- Take into account costs required to remediate or reclaim any environmental impact, and to address the concerns of local communities.
- Carry out exploration work in a thoroughly professional manner.
- If necessary, perform baseline study prior to environmental impacts.
- Assign individual responsibilities to each employee in relation to environmental performance and empower them with required decision capacity.
- Make sure that exploration imperatives do not take precedence over environmental issues.
- Involve environmental professionals at an early stage, so that their input can be considered and implemented where appropriate.
- Perform due diligence if potential liability may have been acquired with the purchase of mineral rights.

The technical content of the e3 Manual is laid out in six activities, which are (1) land acquisition, (2) surveys, (3) access, (4) camp and associated facilities, (5) stripping and trenching, and (6) drilling. The information for each of the six activities is assembled by management issues, such as planning needs, land disturbance, site management, air management, fish and wildlife management, water use and conservation, spill management, hazardous materials, waste management, and reclamation and closure.

Reporting of Exploration Results

Reporting exploration results in accordance with international standards is essential for listed mining companies and in general when reporting to government institutions and agencies. Furthermore, standard reporting of exploration results is essential for transparency and ethics consideration when reporting to shareholders and when dealing with the hosting communities and other stakeholders.

International standard definitions are required to improve the transparency and consistency of reporting mineral reserves and resources worldwide. They will also help to improve communication, aid in the reassessment of mineral inventories, reduce risks for investment by helping to avoid the nightmare of share price inflation, and, most importantly, "*keep the bears out of the beehive.*"[32] The international standards for classification and reporting of ore resources are described later in the "Ore Resources Inventory Management" section.

For a general sustainability reporting framework, readers should refer to the Global Reporting Initiative.[33] GRI is a multistakeholder initiative, formed in 1997 to develop and disseminate globally applicable sustainability reporting guidelines in collaboration with various United Nations agencies, notably the United Nations Environment Programme. GRI became independent in 2002 and published the second version of its *Sustainability Reporting Guidelines*. A third iteration, known as G3, was launched in October 2006.

GRI incorporated a Mining and Metals Sector Supplement, which was developed in cooperation with the International Council on Mining and Metals (ICMM). This was described in Chapter 5.

Sustainability Frameworks for Minerals Exploration

Nearly all governments of countries with important mining potential have already included guidelines for environmental management in mining exploration processes. Moreover, there are many easily accessible handbooks for the development of both mandatory and sustainability reports (GRI and others). An excellent example is *Guidelines for Environmental Management in Exploration and Mining—Exploration and Rehabilitation of Exploration Sites,* published by the State of Victoria (Australia) in 2002.[34]

Every mining exploration campaign must identify and foresee the possible environmental and social impact that the process can have. It is true that during the course of exploration campaigns, the environmental impact is not excessive, especially in the "grassroots" stage.

Other sustainability frameworks make reference to corporate social responsibility (some are described in other chapters of this book):

- Australian Minerals Industry Code for Environmental Management (1996), now Code 2000
- ICMM Sustainable Development Framework (2003)
- International Organization for Standardization (ISO) 14001 Environmental Management Systems (2002)
- Mining Certification Evaluation Project—Australian Regional Initiative (2002)
- United Nations Global Compact (2002)
- U.S.–U.K. Voluntary Principles on Human Rights and Security (2003)
- World Bank Operational Directive on Involuntary Resettlement (2003)

- The Equator Principles (2003)
- United Nations Universal Declaration of Human Rights (2001)
- The Voluntary Principles on Security and Human Rights (2000)
- The OECD Guidelines for Multinational Enterprises
- The International Finance Corporation (IFC) Performance Standards on Social and Environmental Sustainability
- The IFC Environmental Health and Safety Guidelines

Acknowledgments

The author of this section benefited from the ideas and suggestions of A. Arribas and F. Vazquez, professors at the Madrid School of Mines. The contributions of both are gratefully acknowledged.

ORE RESOURCES INVENTORY MANAGEMENT*

Mineral deposits are finite, either physically or economically, so efficient management of ore reserves is critical for sustainability. This section focuses on key criteria for sustainable management of ore reserves.

The classification of geological resources as ore reserves is a dynamic process; therefore, the management of ore resources requires resource management systems (RMSs) capable of calculation and updating whenever internal or external parameters change. Internal parameters, such as mining costs and metallurgical recovery, or external parameters, such as market prices, may cause geological resources to become ore reserves or vice versa.

An RMS capable of fast and reliable data management will enable a company to react quickly to any operational and market change affecting business economics. Furthermore, an efficient RMS will help in the identification and management of environmental risks associated with the varying characteristics of the ore.

The software market offers integrated software packages for resource management and mine planning; often, mining companies use several commercial and internally developed software applications to fit their needs.

Regardless of the number of software applications used, the key requirement is that data storage, manipulation, and backup are centralized in a single-server system, where users may be assigned specific levels of reading and editing authority as a function of their responsibilities in the project. Unfortunately, this is not always the case; too often, data management is split among several departmental databases, causing duplication, loss of information, and data processing problems.

Terrain Models and Digital Models

Most commercial software packages can import and export topography of terrain and excavation data in several formats, but DXF is the most commonly used format. DXF can use many types of drawing units, including triangulation. After the topography has been imported, the software is ready to work on mine planning and design, creating digital renderings, and so forth.

Modern mine design software uses triangulation to generate surface and three-dimensional (3-D) volume models. Considering the geometric complexity of geological and orebody models, very few users have the technical capacity and the spatial vision to work in three dimensions (Figures 7.5 and 7.6).

---

* This section was written by C. Castañon.

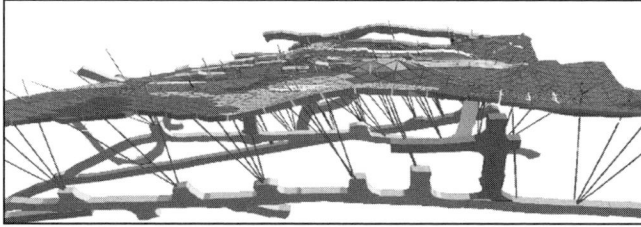

FIGURE 7.5   3-D model of a flat tabular orebody

FIGURE 7.6   3-D model of a steeply dipping tabular orebody

Definitions and Classifications of Resources and Reserves

The distinction between resources and reserves is not limited to geologic and mining aspects, but it extends into some ethical and economical implications. Also, public mining companies are subject to regulations by the Securities and Exchange Commission (SEC) for filing and public reporting of exploration information, resources, and reserves.

In any case, the estimation, classification, and reporting of ore resources requires certain flexibility because different criteria may be applicable, depending on each particular case. Also, mining companies may apply their own resource management methodology and use international reporting standards to comply with SEC requirements. The following paragraphs outline the CIM reserves and resources definitions.

A *mineral resource* is a concentration or occurrence of natural, solid, inorganic, or fossilized organic material in or on the earth's crust in such form and quantity and of such a grade or quality that it has reasonable prospects for economic extraction. The location, quantity, grade, geological characteristics, and continuity of a mineral resource are known, estimated, or interpreted from specific geological evidence and knowledge.

An *inferred mineral resource* is that part of a mineral resource for which quantity and grade or quality can be estimated on the basis of geological evidence and limited sampling and reasonably assumed, but not verified, geological and grade continuity. The estimate is based on limited information and sampling gathered through appropriate techniques from locations such as outcrops, trenches, pits, workings, and drill holes.

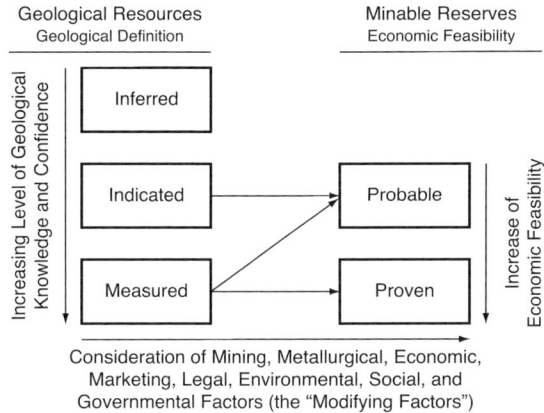

FIGURE 7.7   Relationship between resources and reserves

An *indicated mineral resource* is that part of a mineral resource for which quantity, grade or quality, densities, shape, and physical characteristics can be estimated with a level of confidence sufficient to allow the appropriate application of technical and economic parameters to support mine planning and evaluation of the economic viability of the deposit. The estimate is based on detailed and reliable exploration and testing information gathered through appropriate techniques from locations such as outcrops, trenches, pits, workings, and drill holes that are spaced closely enough for geological and grade continuity to be reasonably assumed.

A *measured mineral resource* is that part of a mineral resource for which quantity, grade or quality, densities, shape, and physical characteristics are so well established that they can be estimated with confidence sufficient to allow the appropriate application of technical and economic parameters to support production planning and evaluation of the economic viability of the deposit. The estimate is based on detailed and reliable exploration, sampling, and testing information gathered through appropriate techniques from locations such as outcrops, trenches, pits, workings, and drill holes that are spaced closely enough to confirm both geological and grade continuity.

A *mineral reserve* is the economically minable part of a measured or indicated mineral resource demonstrated by at least a preliminary feasibility study. This study must include adequate information on mining, processing, metallurgical, economic, and other relevant factors that demonstrate, at the time of reporting, that economic extraction can be justified. A mineral reserve includes diluting materials and allowances for losses that may occur when the material is mined.

A *probable mineral reserve* is the economically minable part of an indicated and, in some circumstances, a measured mineral resource demonstrated by at least a preliminary feasibility study. This study must include adequate information on mining, processing, metallurgical, economic, and other relevant factors that demonstrate, at the time of reporting, that economic extraction can be justified.

A *proven mineral reserve* is the economically minable part of a measured mineral resource demonstrated by at least a preliminary feasibility study. This study must include adequate information on mining, processing, metallurgical, economic, and other relevant factors that demonstrate, at the time of reporting, that economic extraction is justified. A graphical model of ore resources is shown in Figure 7.7.

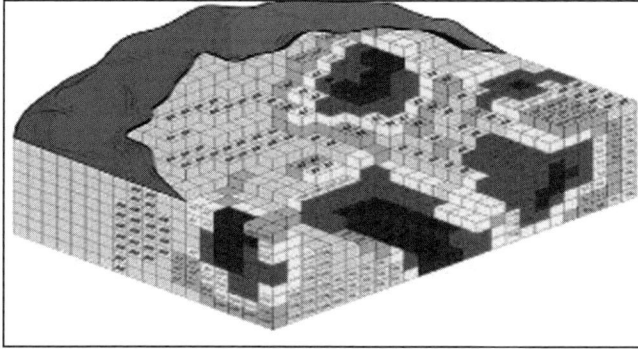

FIGURE 7.8   Block model of a massive orebody

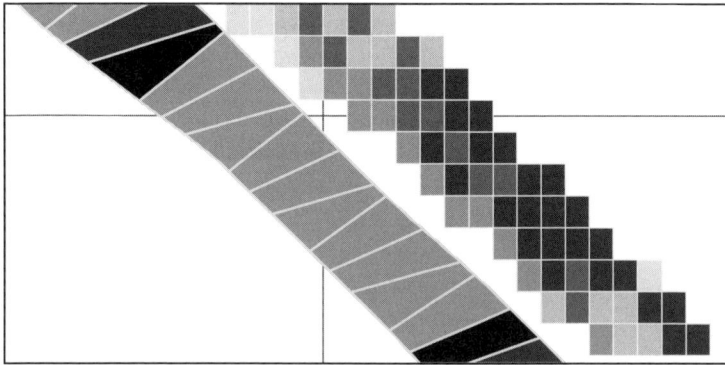

FIGURE 7.9   Block modeling of thin, tabular orebodies

Estimating Resources from Block Models

Several methods for estimating ore reserves have been extensively used in the past (sections, polygons, triangles, etc.). Today, the generalized use of commercial 3-D software, IT-assisted massive data management systems, and geostatistics has resulted in the generalization of the use of block model methods, reducing the use of all other methods to preliminary estimates or to double-checking calculations performed using block models.

In this section, we focus on the use of the "digital blocks" method, in which the orebody is modeled as a set of parallelepipeds (blocks), all with the same dimensions and where the grade (or property) estimated for each block is assigned to its center (Figure 7.8).

Almost any orebody may be modeled by using block model methods. In the case of thin tabular orebodies (Figure 7.9), block models have limitations. In this case, standard block modeling generates a notched outline requiring smoothing. To solve this problem, some commercial software has implemented methods for smoothing model boundaries. A second problem associated with the application of block models to tabular bodies is defining how the core sample assay data should be used, and in what direction. Block models, in general, interpolate drill samples rather than using the entire hole intersection. Regarding the direction of the mineralization, interpolation using drill samples—not hole intersection—generates the loss of directional information (bedding plane direction), which is important when intense folding is present.

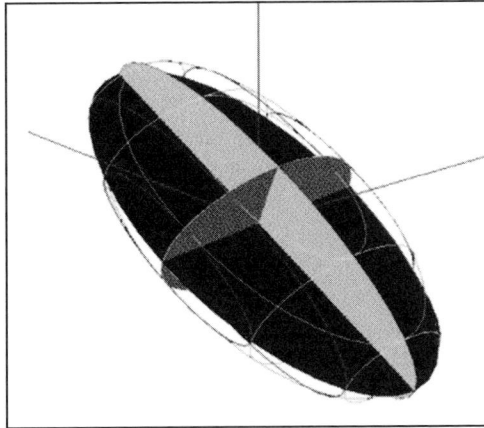

FIGURE 7.10   Search ellipsoid

Search Ellipsoid

The first step is the definition of the "ellipsoid of search." This is the ellipsoid that, when centered at a point of the orebody, will contain all data points that are close enough to be assigned a "weight" in the estimation by interpolation of a regionalized variable associated to the center point. As an example, the grade at a certain point is estimated by weighted interpolation of the grades measured at all data points located within the ellipsoid of search. Other data points outside the ellipsoid will not be used.

Normally, orebodies present "directional anisotropy," that is, different regularity in different directions. Anisotropy is modeled by orienting the search ellipsoid with its longest and shortest axes respectively oriented with the directions of maximum and minimum regularity, the intermediate axis being perpendicular to the other two axes. The length of the ellipsoid axes is a scale factor where the scale distance in the three axis directions is transformed by dividing actual distance by the scale factor.

For a given orebody, each element (e.g., Cu, Au, Zn), property (e.g., ore type, rock quality designation [RQD], color) or parameter (thickness, density, etc.) shows a different anisotropy field and, therefore, a different ellipsoid.

As an example, Figure 7.10 shows an ellipsoid with scale factors of 1, 2, and 3. When estimating the grade at its center, the same weight should be assigned to a sample located at 30 m along the principal axis (scale factor = 1 or 30/1) as to a sample located at 15 m along the secondary axis (scale factor = 2 or 30/2) or a sample at 10 m along the tertiary axis (scale factor = 3 or 30/3).

For tabular orebodies, it is common practice to choose the principal and secondary axis (axis of maximum and minimum regularity), respectively, in the bedding plane and perpendicular to it.

Block Model Structure

Following are the main criteria for the design of a block model database:

- All unit blocks are equal in size and geometry. Some commercial programs allow for subdivision of unit blocks to adapt better to the geometry of the orebody. Each block is identified in the database by its position indices (id_x, id_y, id_z).

- Position indices are integer numbers representing the relative Cartesian coordinates of the center of the block with respect to three orthogonal axes (x, y, z) or (East, North, Elevation). Normally, the origin is at the center of the block that has these coordinates: (1, 1, 1).

- The table of blocks (database) is defined by the said position fields (id_x, id_y, id_z), plus fields for all data and fields for the interpolation of the properties to be associated with the block (e.g., block grades, density, lithology, hydrology). In addition, the block registry includes other fields to record codes for important block properties (e.g., geology, rock mechanics, hydrology, resource type, planning codes, block monetary value).

International Standards for Ore Resource Reporting

The first mineral resource reporting standard was the *Australasian Code for Reporting of Exploration Results, Mineral Resources and Ore Reserves*.[35] It was developed by the Joint Ore Reserves Committee (JORC) in 1971 to establish minimum standards, recommendations, and guidelines for Public Reporting in Australasia of Exploration Results, Mineral Resources and Ore Reserves. In 1989, it developed into the "JORC Code" and was incorporated in the Australian Stock Exchange listing rules in the same year, with guidelines added in 1990.

In 1991, the Society for Mining, Metallurgy and Exploration, Inc., in the United States published its guide for reporting and, that same year, the Institution of Mining and Metallurgy in the United Kingdom also revised its standards, largely based on the JORC Code.[36]

The leading principles governing the operation and application of international reporting standards are transparency, materiality, and competence with focus on public reporting of exploration results, mineral resources, or ore reserves, prepared for the purpose of informing investors or potential investors and their advisors.

Commonly, the standards provide guidelines for the classification of mineral resource and mineral reserve estimates into various categories. The category of an estimate implies confidence in the geological information available on the mineral deposit, the quality and quantity of data available on the deposit, the level of detail of the technical and economic information that has been generated about the deposit, and the interpretation of the data and information.

Mineral resource and mineral reserve estimates and resulting technical reports must be prepared by or under the direction of, and dated and signed by, a qualified person. A "qualified person" means an individual who is an engineer or geoscientist with at least 5 years of experience in mineral exploration, mine development, production activities, and project assessment, or any combination thereof, including experience relevant to the subject matter of the project or report, and who is a member in good standing of a self-regulating organization.

Sustainable Management Considerations

Terrain and orebody block databases may be used as a tool for sustainable management of regionalized variables related to sustainability. When used in combination with commercial mine planning and design software, they provide exploration management the capacity to analyze different exploration scenarios.

The following data obtained from field geology, geochemistry, and exploration drilling should be stored in databases from the early exploration stages:

- Geological data: Stratigraphy, petrology, and structural data;

- Topography data: Position of holes, dip, direction, deviation measurements, ground surface survey points, and so forth;

- Geotechnical data: Core sample RQD and joint sets;

- Groundwater parameters: Water table, permeability rates, and water quality;
- Multielement assays: Performed on core samples, including elements that may be of economic importance or may have a negative impact on the environment or on metallurgical processes (e.g., As, Sb, Hg, Bi, Se);
- Land management data: Land ownership and uses, soil type, surface water quality; and
- Environmental baseline data: Data required for an environmental baseline study.

NOTES

1. V. Heffernan, *Growth Strategies for the Mining Industry: Acquisition or Exploration?* (London: Financial Times Energy, 2000).

2. Metals Economics Group, "Record-Setting Exploration Continues in 2007," press release, November 13, 2007, www.minesearch-usa.com/.%5Ccatalog%5Cpages%5C2007%20CES%20Press%20Release.pdf.

3. Patricia J. Dillon, "Continually Improving Performance—Providing Leadership and Tools for Global Exploration" (presentation, 6th Fennoscandian Exploration and Mining Conference, Rovaniemi, Finland November 27, 2007), www.pdac.ca/pdac/publications/papers/2007/dillon-finland.pdf.

4. Mike Wright, "Corporate Social Responsibility: What Stakeholders in Emerging Economies Had to Say" (presentation at the Corporate Citizenship Conference, The Royal Institute of International Affairs, Chatham House, London, November 8, 1999), www.wbcsd.org/DocRoot/OR515bqVKBeOFcmpogtT/CSRStakeholders.PDF.

5. Metals Economics Group, "Record-Setting Exploration."

6. PricewaterhouseCoopers, "Multinational Executives Expect Compliance Costs to Increase, PricewaterhouseCoopers Finds," press release, November 23, 2004, www.barometersurveys.com/production/barsurv.nsf/vwallnewsbydocid/fb079fa11b1829f885256f5400587e47.

7. Dillon, "Continually Improving Performance."

8. Organisation for Economic Co-Operation and Development, www.oecd.org; United Nations Global Compact, www.unglobalcompact.org; Leon H. Sullivan Foundation, www.thesullivanfoundation.org; Canadian Coalition for Good Governance, www.ccgg.ca; Institutional Voting Information Service, Guidelines, www.ivis.co.uk/Guidelines.aspx; Global Reporting Initiative, www.globalreporting.org/ReportingFramework/.

9. Canadian Institute of Mining, Metallurgy and Petroleum, "Exploration Best Practices Guidelines," www.cim.org/definitions/exploration/bestpractice.pdf.

10. Prospectors and Developers Association of Canada, "e3—Environmental Excellence in Exploration," online toolkit, www.e3mining.com and www.pdac.ca.

11. Canada's Income Tax Act (Part XCII, s. 248).

12. Donald A.Singer and Ryoichi Kouda, "Examining Risk in Mineral Exploration," *Natural Resources Research* 8, no. 2 (June 1999): 111–122.

13. Russell A. Carter, "PDAC Report: Explo Spending's Up, But Is It Enough?" *Engineering and Mining Journal* (April 2004), http://findarticles.com/p/articles/mi_qa5382/is_200404/ai_n21363644.

14. David Timms. "Multinational Major to Junior Explorer—Oh What a Feeling!" (presentation SMEDG–AIG 2001 Symposium, North Sydney, NSW, Australia, April 27, 2001).

15. Bob Gunthorp, "Achieving Normandy's Resources and Reserves Growth Objectives: Exploration Versus Acquisition" (presentation SMEDG–AIG 2001 Symposium, North Sydney, NSW, Australia, April 27, 2001).

16. Ibid.

17. AngleGold Ashanti, "Exploration," 2007, www.anglogold.com/About/Exploration.htm.

18. Rio Tinto Web site, www.riotinto.com, 2004.

19. AngloGold Ashanti Web site, www.anglogold.com/default.htm, 2006.

20. Barrick Gold Web site, www.barrick.com, 2007.

21. Gold Fields Web site, www.goldfields.co.za/operations_exploration.asp?navDisplay=Operations, 2007.

22. European Commission, *Study of Minerals Planning Policies in Europe—Extended Summary.* Enterprise Directorate General under Contract no. ETD/FIF 2003 07812003.

23. Metals Economics Group, "Record-Setting Exploration."

24. European Commission, *Study of Minerals Planning Policies in Europe.*

25. Office of Minerals and Energy Resources South Australia, "Mineral Exploration Guidelines for the Preparation of a Declaration of Environmental Factors (DEF)," 2002.

26. Association for Mineral Exploration British Columbia (AMEBC), *Safety Guidelines for Mineral Exploration in Western Canada*, 4th ed (AMEBC, 2006), www.amebc.ca/healthsafety.htm#safetymanual.

27. Prospectors and Developers Association of Canada, www.pdac.ca.

28. Environmental Excellence in Exploration (e3), Prospectors and Developers Association of Canada, www.e3mining.com.

29. OCG/PDAC Principles and Guidelines, Version 7.1.

30. E-3 Environmental Excellence in Exploration, www.e3mining.com.

31. Adapted from e-3 Environmental Excellence in Exploration, www.e3mining.com.

32. G. P. Riddler and N. Miskell, "An International Reporting Standard for Mineral Reserves and Mineral Resources—An Odyssey Nears Its End?" in *Mineral Resource Evaluation into the 21st Century*, edited by Simon Dominy (Cardiff University, 2000).

33. Global Reporting Initiative (GRI), *Sustainability Reporting Guidelines*, 1997, www.globalreporting.org.

34. The State of Victoria, Department of Primary Industries, *Guidelines for Environmental Management in Exploration and Mining—Exploration and Rehabilitation of Exploration Sites*, 2002.

35. The JORC Code and Guidelines, www.jorc.org.

36. Ibid.

CHAPTER EIGHT

Managing Project Feasibility and Construction

J. A. Botin

INTRODUCTION

Project management begins with the selection of areas for grassroots exploration or when entering into project acquisitions. As the project develops, building a trusting relationship with stakeholders is essential. A mining project will be more successful when guided by targeting the highest possible standards in technical, safety, and environmental issues, and in corporate social responsibility, and by openly sharing these core values with the wide variety of stakeholders.

Integrating sustainability into a mining project begins with the identification of stakeholders, their values, and objectives. Shields and Solar[1] visualize this process as follows:

> Alternative management approaches are developed that reflect those objectives. Social and environmental impacts are predicted for each alternative, technical aspects are considered, and costs estimated. Technically or economically infeasible, or unsustainable, alternatives are revised or rejected. Feasible alternatives that support sustainable outcomes are then presented to the public for debate and negotiation with the goal of choosing an alternative that is acceptable to the public. Once an acceptable management alternative has been identified, it is implemented, monitored, evaluated, and revised as needed. The process of revision once again requires public participation and the cycle is repeated.

This chapter presents an approach to the management of mining projects where the focus is placed on sustainability as the key to successful project management and to ensure long-term benefits throughout the mine life cycle.

In its content and objectives, the chapter compiles the personal experience and viewpoints of the authors as project managers, each with a different perspective. In its objectives, it aims to provide the reader with the authors' thinking without attempting a systematic approach to project management systems and techniques.

This chapter is composed of five sections. "Mining Project Feasibility and Construction: An Overview" presents the owner's vision of how a mining project should be managed from the advanced exploration stages and early scoping studies to construction and startup. The author focuses on the organization and structure of mining projects, project evaluation procedures, the key project management decision, and how financial risk relates to project assumptions and parameters. The author, Norm Anderson, a former chief executive officer (CEO) of a leading mining company, gives very valuable views, opinions, and guidelines for the best practices on the key tasks and problems to be dealt with when managing a mining project.

"Mining Project Management: The Sustainability Challenge" provides an overview on the importance of integrating sustainability in the project. It also addresses the management challenges related to environmental and social issues and other project aspects affecting the hosting communities, the permitting agencies, employees, contractors, and other stakeholders.

"Project Management and Stakeholders" offers an in-depth management view of the problems of identifying the stakeholders who might be present in a mining project, learning about their position and expectations, and deciding how the company should interact with stakeholders to achieve project objectives. It also focuses on the importance of proper management of external and internal communications and reporting.

"Project Feasibility Evaluation: New Trends" introduces readers to the economic side of sustainability, the importance of the project evaluation process in the optimization of the project returns, and the long-term benefits to stakeholders. It also describes how criteria for financial analysis are established and presents methods of evaluating investments.

"Case Study: Las Cruces Aquifer Protection System" presents a case study on the environmental risk caused by an important aquifer intersecting the pit in the Las Cruces mining project, and the remediation measures implemented to control this risk.

The chapter editor gratefully acknowledges the assistance of the mining companies that have contributed information and suggestions on the preparation of this material. Special appreciation is due to Cobre Las Cruces, S.A., for the information and assistance the company provided to the authors of the sections titled "Project Management and Stakeholders" and "Case Study: Las Cruces Aquifer Protection System."

MINING PROJECT FEASIBILITY AND CONSTRUCTION: AN OVERVIEW*

The management of the preproduction evaluation stages of a mining project can be seen as a "stepwise risk reduction process" (Figure 8.1), where increasingly larger amounts of capital are invested through time to reduce uncertainty and financial risk. This process can last anywhere between a few years and decades and is divided into stages of increasing capital intensity.

Scoping Studies Through Feasibilities

At the end of each stage, a drop/continue decision is made on the basis of existing information and, should the decision to continue be made, the project enters into a new evaluation stage of higher capital intensity. At the last stage, referred to as the final feasibility stage or simply feasibility stage, the project reaches a level of financial risk at which it is acceptable to stakeholders to proceed to construction and reach the production stage.

Although the number of evaluation stages and its scope are variable, three levels of evaluation studies are often required to lead a project to the production stage: (1) scoping studies, (2) prefeasibility study, and (3) final feasibility study.

Scoping Studies

After a project is identified, whether by discovery or acquisition, it is necessary to assess its worth. This might be a preliminary quick look, with rough numbers as to tonnage, grade, recovery, costs, and prices; if encouraging, this quick look should lead to a scoping study. This is the first formal level of feasibility work, and probably would have a 25%–40% accuracy . . . sometimes greater, sometimes less. As an example, a scoping study was done on the Red Dog Zinc

---

* This section was written by M. N. Anderson.

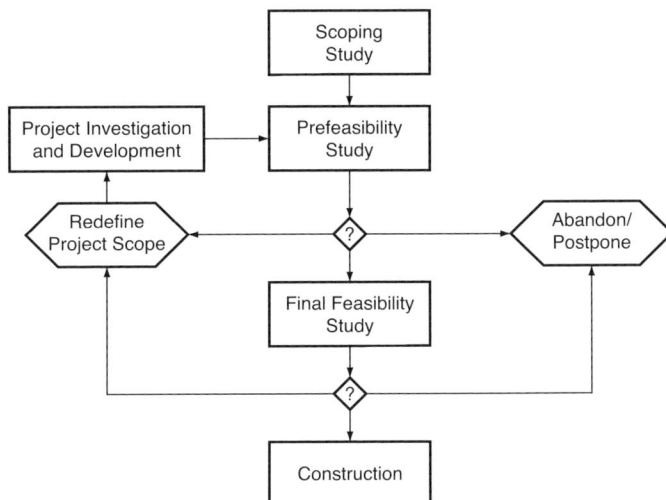

FIGURE 8.1   Typical legal and regulatory framework for mining projects

deposit in Alaska after only seven diamond drill holes that turned out within 10% accuracy. Good people and a little luck can do a lot in a short time at the start, and it helps set the strategy going forward!

Scoping studies would not likely be considered bankable but would cover all the subjects one would expect in a full-blown, final study (e.g., tonnage, grade, geology, mine plan, metallurgy, flow sheets, capital costs, operating cost, environmental considerations, economics). Most importantly, early studies also identify "what's missing," such as complete in-fill drilling, complete metallurgical testing, geotechnical work, environmental testing, a better understanding of the product markets, competition . . . and other risks.

Assuming the scoping study does not kill the project, the company is now ready to raise or appropriate the moneys to do the further drilling, tests, and studies.

Prefeasibility Study

If the scoping study indicates a very robust project, it is possible to bypass this study and proceed directly to the final feasibility study. The "what's missing" work is launched in any event.

Less certain projects are probably best covered with a prefeasibility study first, which will cost less than the final study (but is also less accurate); it can also provide time to seek solutions to the most vexing problems, whatever they might be.

Again, these studies will cover all the topics covered in a final feasibility study, only in less detail.

Final Feasibility Study

This one will be expensive, it will be thorough, and if it is well done, it will be conclusive and bankable if the results are positive. The company's banks will determine whether it's bankable. How competently the study is written, how well the risks and their management have been identified, how well the technical aspects have been solved, and so forth, all determine the project's bankability, but the banks will make the final determination.

It's crucial to keep in mind—it might not be a positive report; the project could fail . . . and it must be allowed to fail if that's where a good study leads. The author of the report is the proponent. Others will act as opponents looking for the weaknesses. The report should be written with this in mind. This subject will be discussed further later in this section.

Organizing and Managing a Feasibility Report

Hiring a competent engineering contractor to prepare the feasibility report is, of course, important. However, it's equally important or perhaps even more important to appoint senior, experienced people to be the owner's representatives, to manage the exercise, to help make the contractor selections, to work closely with them and make the day-to-day decisions in a timely and wise way. The owner's representative should know the history of the project, the company's management philosophy and goals, and the resources available (financial and technical) within the company; this representative should also have sufficient authority to move the project forward. Likewise, he or she also needs to know when and how to inform senior management of progress and to get them involved in decision making and approvals.

In today's buoyant business scene, finding competent, experienced engineers is challenging. Younger, less experienced people are being thrust into key roles. They can do good work, but this does underline the need to manage the process well, understand the limits, and to call for help when required.

The owner's representative therefore must interview the staff being nominated by the contractor, understand these limits, and know where to get help.

Environmental and Social Considerations

Permitting, social, and environmental considerations are probably the most significant changes that have taken place in feasibility work in the last 50 years. Good management has always called for reasonable care and attention to these matters, but since becoming law some 40 years ago, these concerns have added considerable time and cost to projects. Indeed, these considerations have created a whole new and growing industry.

These subjects are covered in more detail in other chapters of this book. Their importance in this section warrants these highlights:

- Background flora, fauna, and weather records and surveys begin immediately as soon as a new project is identified.
- Safe disposal sites for tailings and waste rock are identified and planned.
- Air emissions and water effluents are identified and, if necessary, plans for mitigation begin.
- Staffing, hiring, and accommodations can sometimes be a huge consideration, particularly in remote locations.
- Safety and security, on and off the job, is a growing concern and cost.

Safety, environmental, and social considerations are part of the concept of sustainability, one of the more recent issues that have affected the way companies plan projects. All of these concerns lead to permitting, which involves two or three levels of government, which has grown to be another impressive subset industry. It takes time, it takes money, and it takes talent and good laws to be successful.

Codes of Account

It's worth mentioning here how important it is that, once project cost estimating begins, a code of accounts be established from the beginning that can be used through to the end of construction.

In the past, when accounting programs were less sophisticated, changes in the code of accounts often made it difficult or impossible to compare how costs changed from scoping to feasibility to construction. It was frustrating to management and boards of directors and, equally importantly, it made a valuable cost-control mechanism useless.

Know When to Stop/Complete the Study

As with construction, as studies draw to a close, they often start to drag. There's the potential to start asking questions that should have been settled weeks or months earlier or perhaps to begin chasing minutia, trying to anticipate too much, trying to be too perfect. Delays at the end of the prefeasibility or final feasibility study cost money and may add little of value. Two examples can best illustrate this:

- A great (robust) project at the prefeasibility stages does not need all the i's dotted or t's crossed. It's time to finish it up and move forward to the bankable stage.

- At the final feasibility study stages, a weak project will need all the best people working together at the end to find the best technical input and reach the right conclusion.

Technical Considerations

A few of the more technical considerations necessary for a good report merit some mention: geology, metallurgy, and mining.

Geology. When a potentially good discovery or acquisition comes into play, it is often the geologists who bring it forward. The discovery has been drilled, mineralogy has been studied, reserves and resources have been cataloged, and areas requiring more exploration identified. It's ready for an initial scoping study, which, if successful, eventually leads to a final feasibility study.

Two or three problem can arise. The prospectors/geologists, who have been celebrated for their successes, now have competition for the spotlight as the mining engineers and metallurgists arrive. If not handled well, this sense of no longer being the center of attention can be heartbreaking for the discoverers, but it is inevitably part of the process.

In addition to this stress, the focus of drilling evolves, changing from drilling to find new ore to fill-in drilling to upgrade resources to reserves and prepare for initial mine planning, drilling metallurgical test holes and condemnation holes to ensure the shaft or the mill or the tailings pond will be safely located, and so forth. Further exploration can sometimes be stopped too soon and important upside opportunities can be missed. On the other hand, continually adding new ore while feasibility work and permitting are underway can create troublesome delays. An example of this occurred at the Red Dog mine in Alaska in the early 1980s. What appeared to be a thrust faulted offset of the main zone was discovered and tested briefly. It was 5 miles away from the main zone and appeared not to be of better grade, larger, or lower cost, so the exploration was stopped, thus locking in reserves and facilitating permitting. Twenty years later, it was treated with confidence as a new discovery, so it's important to know when to stop exploration.

Mining engineers and metallurgists sometimes do not know as much geology as they should nor do they always pay sufficient attention to the geologists with whom they work. This can lead to mistakes. Geologists, on the other hand, must inform themselves of potential downstream problems stemming from their reports.

In conclusion, it is important to include a well-written geology/mineralogy report in all feasibility studies. Likewise, it is important that the mining engineers and the metallurgists read and understand this section. Readers of these reports should ask for help if they encounter something they don't understand. Here are some simple rules of thumb to keep in mind when reading this section:

- Some secondary minerals like chalcocite, smithsonite, or franklinite, although high grade, can lead to metallurgical problems. Energite can carry too much arsenic to be easily marketed. Marcasite, which is not usually a marketable by-product, can also create metallurgical problems.

- Certain massive iron sulfides can create environmental problems. They can be easily oxidized to the point where secondary sulfide explosions can be encountered during underground mining. When brought to the surface, they can oxidize quickly, heating up and resulting in serious pollution and transportation problems.

- Certain host rock (waste rock) contaminants can also create problems. Small amounts of iron sulfide in country rock can lead to acid rock drainage problems. Gold contained interstitially in pyrite might not be dissolvable without first oxidizing the pyrite. Organics in host rocks may act as a price grabber. Clay in a mill or in a heap leach can be a very serious problem.

Representatives of each discipline working on these studies must develop "antennae" to search for potentially fatal flaws that, left uncaptured, can kill the project. If captured early enough, these flaws can often be solved or their effects mitigated. Such a mishap appears to have occurred at a Tonopah, Nevada, sulfide copper heap leach project 10 years ago. The basic geology described a mudstone host rock, with argillic alteration and fault gouge. Subsequent metallurgic column testing showed more degradation to fines (clay) during acid irrigation, but it appears to have been missed earlier. The project moved forward, operations began, only to fail soon, with clay preventing efficient irrigation and recoveries that were dismally low. This fatal flaw should have been identified—it was a geological problem that could have been solved.

Metallurgy.   Refractory ores have been cursed over the centuries, yet many great solutions to the problems they pose have been found and will continue to be found.

So, when one sees a new project with good tonnage but modest grades and bad metallurgy, it always warrants a second look-maybe there is some higher-grade, less refractory ore to be found to get the project launched . . . and to provide the time to find the right solutions for treating the tougher ores. A small oxide gold cap over a large refractory sulfide body is one example. Another might be a small, sweet chalcocite or bornite blanket overlying or underlying an arsenic-ladened energite body.

In these cases, the capital cost, the marketing, and the environmental/permitting problems to do the sulfides can be overwhelming and the project can stall . . . particularly in a down market.

If the tonnage and grade and location are inviting, and the bank account allows pursuing them, these projects can be worth the effort required to move them forward. Evolving technology also offers hope for their development in the future. One hundred years ago, flotation was only a dream before it unshackled an enormous number of previously refractory ores. Similarly, 50 years ago, heap leach technology was also only a dream until it changed the metallurgy textbooks on how to tackle low-grade ores. Autoclaves are tackling many of the high-grade refractory sulfides, but this process may not be affordable for the lower-grade ones. Arsenic continues to plague arsenopyrite-hosted gold and energite coppers, but for how long?

Some of the many innovators have dreams worthy of investment. However, some critics do not invest time and capital in prototype processes. Fortunately, dreamers have always been a persistent lot, and a new generation of solutions certainly will inevitably emerge.

Mining. Rarely does one see bad design decisions in open-pit mines that cannot be revised at a sensible cost. But developing a new underground mine incorrectly, then changing the mine plan, can lead to an expensive redevelopment problem.

Queue River (copper) in Australia was a small, high-grade project, conceived as a trackless, Avoca cut-and-fill project. The Avoca system was all the rage at the time and the building of a ramp access was begun, but it quickly ran into very bad ground.

The situation was reassessed. The ore and surrounding host rock were in good ground. The mining method was reviewed and the team chose instead to sink a small woodframe shaft, a long-hole open stoping scheme, and backfilling with cemented rock/gravel fill. Although this approach was old fashioned, it worked well.

So, new underground mine systems must be selected carefully. Any neighboring mines should be visited if possible or at least any mines using the system being considered. And it's not always wise to follow the herd when a new system evolves, whether it's designed for the mine, the mill, or the smelter.

The previous paragraphs assume a mine project, which is the start of the cycle, but similar disciplines apply to smelters and refineries, of course. Nothing really changes except the technology and people, but the rules are generally the same.

For example, in Trail, British Columbia, when the Kivcet Lead Continuous Smelting Process was being studied to replace the old sinter/blast furnace system to overcome age, rising costs, and new environmental, hygiene, and safety rules, there was close communication with the organization's Australian competitors, who were pursuing the same goals, to find the best way. It is quite legal to collaborate technically, and the joint effort was worthwhile.

Feasibility Reports

The authors of any feasibility report must understand and keep in mind who will read it. The readers will include the board of directors and the CEO . . . who might or might not be interested in or able to understand much of the technical "stuff." The bankers, brokers, and accountants might concentrate on the financials, evaluation, and risks. The regulators will focus on their end of the business in detail. Finally, those who follow, the construction team who will do the final design and build the new plant, will use the feasibility report as their starting point but might not reach exactly the same design conclusions. The need to answer the questions of a wide range of readers often leads to the creation of a multivolume report.

The board of directors might only read the executive summary, and some might not get past page 3 or 4, plus some other details. Some on the design and build team, however, should read every page. The bankers' consultants should, too . . . but might not. The bankers and brokers likewise might only read the executive summary, the financials, and the risks.

Therefore, it's important to write a thoughtful executive summary, as well as a complete report that covers all the bases, including thoughtful conclusions and recommendations sections that flag all the items that need follow-up, and a full appendix. All the sampling, testing, and security precautions that were done and by whom, what source material was used for capital and operating costs, and so forth should be shown.

These reports cost a lot of money and cover a lot of ground, so it's critical to make them complete and readable.

How Is a Project Evaluated?

Evaluating the worth of big-dollar investments, which require large amounts of upfront capital and the long payout times commonly encountered in mining projects, is a serious task. Although the topic is worthy of several full chapters in a book such as this, here it will be dealt with in a few pages.

Sometimes the CEO and board of directors will accept a less rigorous feasibility report—if it describes a robust project—and run with it. They must keep in mind, however, the report might have underestimated things by 15% or 20% or more . . . but if it's robust, it might be judged worthy of approval. Sometimes, on less robust projects, extra evaluation is warranted.

When acting as a bank consultant, one should look for a number of items, described in the following paragraphs.

Size Improves the Rate of Return

Sometimes, the company might see a large, low-grade copper porphyry mine, for example, that initially was conceived as producing 30,000 t/d, but which produced an unsatisfactory return. It is enlarged, perhaps several times, and ends up as a 90,000 t/d project with an internal rate of return (IRR) of 15%, which reduces the mine life from 30 years to 10 years. It might become a "bet the company" investment for a small company in a down market. Also, due to the now-shortened mine life, there's little time to recover from bad markets or mistakes. This scenario warrants a hard second look!

Blue Sky Resources

Some investors and bankers often won't consider any resource other than measured and indicated reserves. It's a conservative rule that they use and one that must be lived with. However, when evaluating new projects or acquisitions, there is a time to consider the "blue sky potential" of a property, as well as a time for sound geological thinking. This is a time, too, when the board of directors will benefit from this advice.

Perhaps the most obvious example of when resources other than reserves must be considered are deep, narrow-veined gold mines. These typically have only 2 or 3 years of reserves on their balance sheets, which is hardly sufficient to pay out a new investment. This is why these projects often start small and grow. It's too expensive and unnecessary to develop more than 2 or 3 years of reserves in such mines—if the geology is well understood.

So there is a time and a place for such resources to be used in evaluation, and it should be clearly described. Maybe the banker should even give it a peek.

Use of a Mine Contractor

When mine contractors are available, they can often come in, do the stripping and initial mining and crushing, and relieve the capital budget. Some mine contractors can be nothing but headaches. It's a valid strategy, but eventually the owners should consider self-mining, which can save anywhere from 10% to 30% on mine operating costs.

This is useful to keep this in mind when first engaging the mine contractor. He or she will properly amortize his or her equipment on the job, with the mining company's money. Mining companies should check out a rent-to-own arrangement whereby they might, in 5 years, earn a significant equity in this equipment at a modest extra cost. This is often more applicable to open pits than underground mines.

It is also notable that conversion to self-mining can be accomplished with many of the contractor's crew changing over.

The same kind of thinking is applicable during construction. Usually, some of the construction equipment (cranes and loaders) could be purchased through this sort of arrangement, so that at the end of the job, it belongs to the mine owner, who can choose to dispose of it.

Capital and Operating Costs

Contractors doing feasibility studies have data on capital and operating costs from around the world and are able to factor this data intelligently into new projects as they arise. Having said that, the owner and the banker's consultants must judge the validity of this factoring. If serious (i.e., damaging) discrepancies arise, with the owner saying costs were too high or the banker's consultant feeling they are too low, then a problem develops, which is usually solvable. There is also the matter of future liability that the contractors must consider. If the differences in opinion are not large or damaging, they can usually be bundled and analyzed in the sensitivity part of the report.

Tonnage and Grade

When the owner continues to drill after the official reserves have been set, new (hopefully more positive) results can best be addressed in the sensitivity or conclusions sections of the final report.

Economic Models

Owners, contractors, bankers, and investment houses will all have their own models and will also enjoy putting the mining company's numbers in their models.

At or near the start of the study, the owner and the contractor should decide on how they want the economics (spreadsheet) presented. They should each check their model against the others and resolve any significant differences. This can often be resolved in the footnotes, which would address how to indicate tax treatment, how payout is calculated (from the start of construction or from the start of commercial production), and so forth.

Price of the Product

Large companies might have in-house annual pricing committees to study and recommend long-term pricing to be used in budgeting and feasibility studies. Bankers, contractors, and smaller companies will depend on the myriad of "crystal ball gazers" in investment houses and commodity consulting houses, all with good people studying the same data used by the big mining houses. The crystal ball gazers will publish those results for all to see (for a price). This usually results in a good, healthy spread in the estimates.

These estimates can then be assembled to show this spread and calculate the means and the averages. Table 8.1 shows three such assemblies of estimates, gathered over a few months in early 2007.

TABLE 8.1   Assemblies of metal price estimates

Product	Assembly I			Assembly II			Assembly III		
	High	Low	Average	High	Low	Average	High	Low	Average
Copper 2007—US$/lb	3.25	2.20	2.61	3.06	2.27	2.85	3.50	2.27	2.88
Zinc 2007—US$/lb	1.95	1.00	1.65	2.06	1.46	1.72	2.15	1.15	1.65
Copper (long term)— After 2010	NA	NA	NA	1.50	1.00	1.19	1.30	1.10	1.20
Zinc (long term)— After 2010	NA	NA	NA	0.75	0.52	0.62	0.85	0.50	0.67
Gold (long term)— After 2010	NA	NA	NA	710	490	556	700	450	575

NA = not applicable

The assemblies are a mix of from 9 to 27 "guesses" on metal prices by various investment houses. The average is the arithmetic average of all the guesses for each assembly.

By using different assemblies, of course, it would be possible to come up with different numbers—and to know they are all likely to be wrong. The average numbers of the assemblies in Table 8.1 are "close enough," except for gold. And this is (likely) the most difficult and most volatile, and the guesses will have some wide swings at the top and bottom. Naturally, investors will have their ideas, too—all of which should be captured in a good sensitivity analysis in the reports.

Inflation and Leverage

In a down market, one might see financial engineers play with leverage and inflation and *not* make the game rules clear. It is certainly possible to change the return on investment significantly doing this.

The best recommendation is not to do this. In the evaluation, this game in sensitivity should be played only if necessary and the rules should always be made clear.

Compare the Project with That of the Competition

Several good firms are producing cost curves based on data they acquire by visiting mines worldwide for each of the principal commodities. These are *not* inexpensive. Although, of course, they cannot be 100% accurate, they can provide a well-thought-out guess, each calculated the same way, for a lot of mines. It's a good measure of competitiveness.

Cost curves allow a company to evaluate, albeit roughly, the cost position of its project compared to that of competitors. When interpreting cost curves, one should be cautious about the following:

- During a serious downturn, state-owned mines might not follow the same rules as in the private sector. They might not (ever) shut down, even though they have high costs.
- Some mines produce high grades in a downturn (they might have to do so to survive) and mine lower grades in good times. Or they might produce high grades in good times to maximize profit. This makes it difficult to judge supply and demand (forecast prices) and make good cost comparisons.

It's crucial to understand the pricing risk in a downturn *or* through a price cycle. This is complicated by the "super cycle" in which the industry now appears to be stuck. But even long cycle prices do come down eventually.

Risks

As the conclusion of work on the feasibility report approaches, it's wise to take a moment to consider the risks listed in Table 8.2.

TABLE 8.2   Site-specific risks

Country Risk	Geotechnical Risk
Permitting Risk	Geological Risk
Social Risk	Transportation Risk
Environmental Risk	Market Risk
Manpower Availability	Infrastructure Availability
Weather	Capital Costs
Competition	Operating Costs

These and many other site-specific risks will raise questions that the CEO, the board of directors, the bankers, and all the other stakeholders have a right to and will/should ask. *All of them should be covered in the feasibility report.*

What does a competent due diligence consultant look for when studying a new feasibility study? This is the frame of mind into which report authors must put themselves before issuing feasibility studies. So what exactly *do* due diligence consultants look for?

- **Economics**. The last page of the executive summary and the spreadsheet in the economics section of a good report will usually tell the banks the story. If the return on investments (IRR) is robust, the life of the mine is around 15 years or longer, the long-term price projections look reasonable, and the author of the report is reputable, chances are it will be a worthy project. However, they will read the rest of the report, looking for obvious flaws. But if no serious unanswered concerns arise, the company should be ready to negotiate financing.

  At the other end of the spectrum, if the IRR is showing around 12% to 15% (depending on location), the life of the mine is 10 years or less, the long-term pricing seems to be a stretch, and the report is obviously not well-written or understandable, then the hunt for the "fatal flaw" begins.

  Most studies will fall somewhere in between these two extremes. One might have an IRR of 15% to 16%, a mine life of 12 to 14 years, and a long-term price in the high-normal range. In this case, one might find that the geology has evolved during the study to be larger than originally thought, resulting in lower unit capital and operating costs, which has the delightful effect of improving the IRR to meet or exceed the "hurdle rate," which one likes to think of as being between 15% and 18%.

  One such (as yet undeveloped) project, looked at several years ago, was conceived as a 30,000 t/d copper open-pit/mill operation and was rather isolated. During the study, it grew to 60,000 t/d, then to 75,000 t/d, then to 90,000 t/d—and the mine life dropped, the grade dropped, and the strip ratio dropped as expected. The resulting project hit the magic IRR of 14.9%, but it was not financeable.

  Sometimes, projects like this need a completely new start. Is there a high-grade core with which a 5,000-t/d used mill can be justified? Bethlehem Copper in British Columbia started this way in 1955, on a large, low-grade orebody. Initially, they focused on the higher-grade areas, paid off their debt, then expanded the mill and lowered the cut-off grade. After several of these additions in 20 years, they were up to 30,000 t/d, the correct size for the reserves and with no debt, so the project was profitable. Perhaps other companies should occasionally do some retroactive thinking like this. Economics is obviously an important risk and a driving incentive. What else?

- **Geological risk**. The geology of a new project is centered on tonnage and grade, as it must be. Today's computer-oriented geologists have mastered what used to be tedious, time-consuming hand calculations. Today, with the press of a button, it's possible to change the cut-off grade and have a new higher-grade or lower-grade mine design in minutes. With a second click, one can see how the economics and mine life change with these new parameters. It's wonderful.

  But there is more to geology than that and sometimes fatal flaws appear, which might be ignored if a metallurgist, mining engineer, or bank consultant doesn't read, recognize, and reject them. Mining history is littered with such casualties, particularly where the grade has been overestimated, sometimes on purpose.

· **Mining and metallurgical risks**. Although these risks have already been addressed several times, it's worth mentioning again because it's critical to ensure that competent people do this planning well.

· **Cost risks**. Clients might push for and some report writers will yield to the pressure to include unachievably low capital and operating costs in their reports. Even in today's Sarbanes-Oxley- and 43-10-regulated environment, these lowball numbers can slip though the net. These incidents will continue to be part of the business, but they must be identified and dealt with appropriately.

· **Environmental risks**. Environmental risks are similar in many ways to cost risks. Permitting/regulatory oversight mechanisms are in place, but projects with likely environmental shortcomings can slip through. They too must be identified and dealt with appropriately.

· **Market risks**. This is an integral part of the economic risk exercise. Markets for products change, usage levels change, price cycle lengths change. Enormous changes in technology over many years have resulted in a reduction in the cost of a unit of production and, in so doing, have lowered the noninflated price of copper, zinc, aluminum, and many other metals. A big, high-grade discovery like the Escondida copper mine in Chile or the Red Dog zinc mine in Alaska can also contribute to this price deflation effect.

But the industry is now undergoing what some might call it a paradigm shift. China, India, and many third-world countries are having an impact on the business. Rising energy prices are making a serious impact. Government intervention, wars, and terrorism continue to affect the industry.

How will these situations affect the price of products going forward? It is necessary to ask oneself a question: is there a real paradigm shift that will keep this super cycle motoring on for several more years, resulting in a super-long up cycle in metal prices and a reversal of the long-term price erosion that the industry has witnessed for nearly a century? Or will the industry once again see copper prices of less than $1.00/lb, zinc at $0.40/lb and nickel at $2.50/lb? The answer is vague, but the risks are quite clear.

Construction Stage of a Project

For a new project that has been approved and has been financed, permitted, and so forth, it is often tempting to consider turning it over to an engineering, construction, and project management company, giving them a budget and telling them to build it. However, this approach is generally unworkable.

The feasibility work has accomplished less than 5% of the final design, and not all of that work might be acceptable. This is the last opportunity to reconsider *all* the major design decisions. A new revised budget will be developed; every few months after that, a new estimate (that is, a forecast) will arrive. Later, when the design reaches about 90% completion, construction might have reached 40% completion, and the project is now on a version 3 forecast. After 2 or 3 years, the project is at 100% completion and the plant starts. Only then will it be possible to see the final cost number and know how good the initial code of accounts was and how well the construction was managed.

The principles and criteria for good feasibility management apply to construction as well. A good owner's representative or project management team should be appointed and that person or team given sufficient authority and support to do the job. The project leader should participate in the hiring and contract negotiations with the engineering/construction contractor, hire his or her own staff, know what downstream construction permitting approvals are required, and be aware of what reports are expected and what startup responsibilities the owner has. The owner's representative should also understand project insurance and the insurance coverage that should be provided (by owners or contractors) during the construction phase of the project.

The owner's representative during the construction phase might or might not be the same person who was in place during the feasibility work, and the construction contractor might or might not be the same contractor as the one who performed the feasibility work. It is tempting to use the same personnel during both processes. The problems, challenges, and rewards in this engineering/construction phase are remarkably like those during the feasibility work.

Someone once proclaimed that any new project has three phases:

- **Phase 1—Discovery and Feasibility:** This is a period of thoughtful, well-considered plans and judgments, based on not too many details or final plans other than tonnage and grade. These are thoughtful times with one eye always on costs; time might not be a primary issue.

- **Phase 2—Construction and Startup:** This is a headier period during which details are worked out. Time is an issue (sooner is better) and cost prudence is generally set aside to speed the process—right through to 1 or even 2 years of production and on to full production rate.

- **Phase 3—Fine Tuning:** This is when the size of the capital expenditure (capex) is finally known. It's now possible to gain a good idea of what the achievable mine grade will be and there is probably still some way to go to achieve optimal operating performances. This is not an easy time. It takes a different management style and it is an important period!

After a successful feasibility study, it's time to start *all* over again as well as to take one last look at all the design decisions made earlier, particularly from new eyes on the owner's side, from people who will later bear the responsibility of starting up, maintaining, and operating the new infrastructure. Likewise, during the actual construction and development of a mine, a future general foreman will spot and improve on details that might have been missed in the final design. Techniques on how to manage water inflow (depth of the ditches), for example, or how to monitor effluent outflows from the mill can be incorporated. A lot of small but important details can be added to the as-built drawings and need not/should not go back to design.

The nitty-gritty of construction—union or open shop, housing for construction people, temporary offices for contractor and owner, buying versus leasing construction equipment, security, weather-related logistics, timing of orders for equipment with long lead times, selection of subcontractors—will go on for as long on the contractors stay. A sound owner's representative/project management team will assist in finding the best solutions to problems as they arise.

Relations with regulators, neighbors, and local political officers (and tax assessor) will need attention.

All the details—from design, to where the spare parts will come from, to the new code of accounts for operations, to hiring good staff (from superintendents to janitors)—will need to be completed before startup.

As construction nears the end, it is necessary to decide *how* and *when* to take over from the contractors and who from that group (if anyone) should be kept onboard for startup. There's no more hand-holding, no others to blame for shortcomings. This is when previous attention to detail pays off dramatically. It can also be a time of adrenaline rushes and sleepless nights, particularly if some details have been missed.

Despite all of everyone's efforts, some mistakes will undoubtedly emerge during the startup. Small mistakes can be corrected in minutes or during planned shutdowns. Big mistakes might call for a more careful understanding of the problem and the solution. If it's an expensive correction, it might mean another trip to headquarters for more money—not easy, particularly if the project is already over budget or behind schedule. But it must be corrected and the sooner the better.

The construction contracts and financing arrangements will both contain completion and performance guarantees and targets. It will be to everyone's advantage to achieve these targets quickly and well. This again leads back to all the details that are so important during final design and throughout construction.

Rather than dwelling further along these lines, it might be helpful to look at a few examples (case studies so to speak):

- There have been cases where the capital cost breakdown from prefeasibility to feasibility to the final project budget was different and therefore, not comparable. This unfortunate situation shouldn't be allowed to happen.

- There are cases in which good feasibility work was followed by faulty final design, without proper owner input, so the finished product was not what the owner wanted. It's important not to let this happen.

- Sometimes, new projects in new areas fail to gain local acceptance. At times, this is the result of neglect because someone didn't recognize the need to build local understanding or support or someone wasn't assigned to do the job. At any rate, it doesn't get done and major problems ensue. There are many examples of this happening, particularly when dealing with cross-cultural issues. It's often not easy to do and it can be *fatal* if it's not done right.

- One newly hired mill general foreman, who arrived early to help supervise equipment installation, spotted a +12-inch-diameter horizontal pipe connection leading from a slurry sump to a pump. He had it shortened to 8 inches to reduce the risks of sanding up (plugging) in the connection. It was a simple fix to do and likely prevented an untold number of shutdowns.

- The cost of shipping heavy or bulky equipment by rail (if available) is probably less than the cost of shipping by truck, but once the equipment arrives, it might have to be unloaded from the railcar onto a truck and moved another mile or 10 miles to a lay-down yard at the site. What does that cost and would it not be better to pay the extra trucking cost all the way? Good procurement advice is important.

- When one organization started a new underground mine in Missouri, it was thought that the ventilation had been well planned. However, after a year, during the summer, the ambient heat, high humidity, and diesel smoke in the underground air stream produced serious (dangerous) fogging problems. What to do? A bright young engineer suggested reversing the ventilation airflow. Although not ideal, this technique worked. In doing this, ventilation air was cooled slowly and the fog cleared significantly.

- Just-in-time procurement of parts works fine in an automobile assembly plant, saving on inventory and double-handling costs, but it often does not work well on a construction job, particularly when business is booming, the construction site is remote, and prices are rising quickly because of inflation. Sometimes, it pays to buy early and inventory it until it is needed. Sometimes, construction owners' representatives need direction in making these decisions.

To conclude this section, the construction period is when reducing time (to completion) is money. Analyses of details should not be allowed to paralyze the schedule.

Conclusion

Mines are expensive to build and payouts (generally) are slow compared to many businesses. If the company enjoys a successful startup, when product prices are rising, the team's work will make them heroes, despite the mistakes that might have been made. If, on the other hand, the mine's startup has been only average and its products hit a falling market, these same mistakes will be magnified 100 times, expensive corrections might be more difficult to fund, and the payout will be prolonged.

It's vital to remember that it will only be possible to know what the final capital costs, operating cost, recoveries, and achievable ore grades will actually be after the mine has operated successfully for a year or two. There are many unknowns at the start of any mining project, so it's important to give maximum attention to the details from the beginning and to hope that any mistakes/errors will be compensating errors, both good and bad.

MINING PROJECT MANAGEMENT: THE SUSTAINABILITY CHALLENGE

The management of a mining project actually starts even before exploration begins, when the company decides where to do its exploration. The initial choice of a particular area to explore should be based not only on its geological potentiality but also on its political, social, and economic background. The area selected might be a historic mining district or in a greenfield site with no previous mining record, but in any case, the project manager should be concerned with learning the peculiarities of the site regarding political, environmental, and community issues. When the exploration stages lead to a discovery that eventually turns into a mining project, a myriad of latent situations, new or unidentified impacts, and stakeholders can appear.

The sustainability challenge to managers of a mining project is present in almost all project areas and tasks. This challenge relates to management problems associated with sustainability, such as

- Permitting, mining regulations, and standards;
- Environmental risks;
- Social risks and benefits from mining;
- The issues of social license and corporate social responsibility; and
- Management of human resources, contractors, and other stakeholders.

Some of these management problems are discussed in this section.

Permitting, Legal Frameworks, and International Standards

Because the mining industry is global and many mining companies operate across the continents, it is crucial for the project manager to become familiar with the provincial and local regulations applicable to the permitting process of the project in question and the government agencies in charge of those regulations. These steps are necessary not only for carrying out project financing, but also for public communication and for reporting the progress of the project.

Permitting, of course, is a complex and time-consuming management process. Depending on how many different permits are required from how many agencies, permitting sometimes takes several years to complete. Generally, the permitting process is easier and shorter when the company gets a social license to operate from the hosting communities and other stakeholders. In this regard, an important factor for companies is to voluntarily adhere to the international legal framework and international mining standards. Some of these regulations and standards are shown in Figure 8.2.

Legal Frameworks	Corporate Social Responsibility Mining Standards
International • United Nations Human Rights • International Labour Organization Conventions • European Union Directives **National** • Mining Laws and Standards • Tax/Financial Framework • Environmental Laws and Regulations • Water Management Regulations • Labor Laws and Codes • Other (Archaeology, Defense, etc.) **Regional** • Regional Mining Standards • Regional Tax/Financial Framework • Regional Environmental Regulations • Land Management Regulations **Municipal** • Construction Permits • Municipal Taxation	**Sustainability Standards** • Sustainable Development Framework of ICMM • International Institute for Sustainable Development Standards • Prospectors and Developers Association of Canada Standards • United Nations Global Compact • Organisation for Economic Co-Operation and Development Standards **Project Financing Standards** • International Monetary Fund Standards • World Bank Standards • Equator Principles • Securities and Exchange Commission Standards (Public Companies) **Company Standards** • Governance Codes • Sustainability Framework • International Organization for Standardization 14001, 9001, etc.

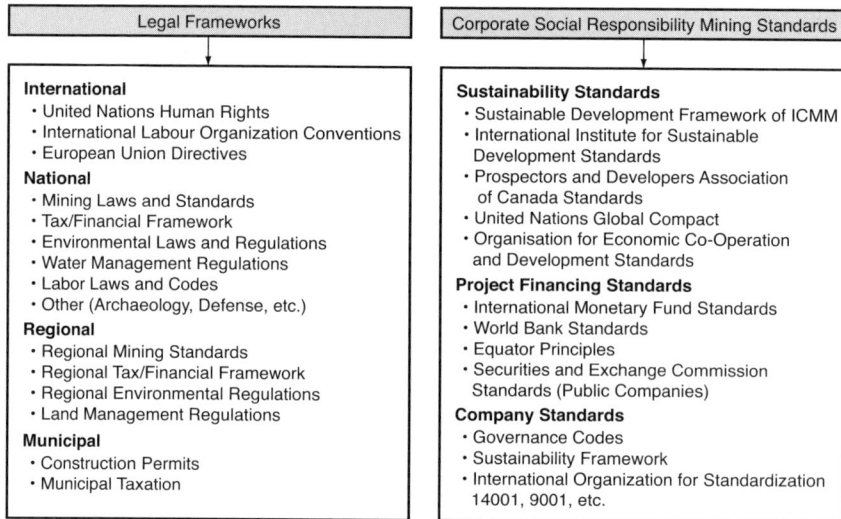

FIGURE 8.2   Typical legal and regulatory framework for mining projects

Management of Environmental Risks

Because environmental risks must be identified and quantified early, a preliminary Environmental Impact Statement should be part of the project scoping study performed during or on completion of the exploration stage.

Environmental Baseline Study

When the owner's team is on site and operational, an environmental baseline study (EBS) should be undertaken or, if already performed at the exploration stage, revised and detailed. The EBS is an investigation aiming to establish the baseline level of potential contaminants in soils, surface water, and groundwater, and to evaluate the initial status of other environmental risk factors, such as air quality, dust levels, gases, noise levels, ecosystems, and archaeological sites. The EBS should also plan for continual monitoring, through which changes in conditions can be documented.

Environmental Impact Study

Once the EBS has been completed, the environmental impact study (EIS) should be launched. The objective of an EIS is to evaluate the environmental impacts and risks associated with mining operations and determine the monitoring, control, and remediation actions that should be implemented in the project to minimize the environmental risks.

In addition to evaluating impacts associated with potential contaminants and environmental risks addressed in the baseline study, the EIS should evaluate other impacts derived from mining, such as water and energy use, cyanide, tailings, and waste dump management. Also required as part of the EIS or in separate documents are the decommissioning and reclamation plans and the postclosure monitoring and control requirements.

The EIS should include conservative estimates of all environmental costs, including those associated with regulatory oversight, reclamation, closure, and postclosure monitoring and maintenance. For important environmental risks, it is advisable to analyze best-case and worst-case scenarios and to develop appropriate response strategies in consultation with potentially affected communities.

Environmental Impact Assessment

Governmental regulations generally require mining companies to submit the EIS as part of the permitting process. The EIS is the basic document used by the U.S. Environmental Protection Agency or equivalent government agencies to issue the official Environmental Impact Assessment (EIA) for the project. However, regulations can vary from one country to another.

The EIS is the base used by the government to determine the financial-surety instruments to be required at permitting. Most governmental regulations consider the EIS a public document that regulates the publicity procedures and timeframe and ensures that all stakeholders can gain access to the information, so that participation in the EIA process is effective. Also, some companies provide technical and financial support to encourage stakeholders to participate in this process.

The EIS/EIA process is an important part of sustainable management and, as such, is subject to public debate regarding its efficiency as a sustainable development tool. As an example of this debate, the "Framework for Responsible Mining"[2] public debate platform of mining stakeholders on environmental and social issues associated with mining has developed "leading edge" recommendations to improve EIS/EIA standards:

- Minimizing water and energy usage and reducing greenhouse gas emissions should be a stated mine management goal.

- Companies should monitor and publicly report airborne hazardous emissions.

- Maximum noise level requirements should be implemented at the project boundary.

- Net acid-generating material should be segregated and/or isolated in waste facilities.

- Mine operators should adopt the International Cyanide Management Code and third-party certification should be utilized to ensure safe cyanide management.

- Reclamation plans should include plans for postclosure monitoring and maintenance of all mine facilities, including surface and underground mine workings, tailings, and waste disposal facilities. The plan should include a funding mechanism for these elements.

Management of Social Risks and Benefits of Mining

The management of social and environmental risks of a mining project is closely related to the company's achievement of a social license.

The concept of social license was first described by Pierre Lassonde as ". . . the acceptance and belief by society, and specifically our local communities, in the value creation of our activities, such as we are allowed to access and extract mineral resources . . ."[3] In practical terms, social license refers to the process of gaining the acceptance of the hosting community to conduct mining operations and to apply best practices standards, aiming to leave a benefit across social, economic, and environmental areas.

Social license is often won or lost through the initial interactions between local communities and the project management team. In this regard, the management objective should be to demonstrate that the project will achieve sustainable long-term benefits for the hosting communities. Furthermore, this objective should be pursued in consultation with stakeholders.

Good and fluid relations between the company and the community and other stakeholders should be based on trust as the only means to foster participation and commitment and gain social license to operate. To this end, most project managers seek consultation with local communities. However, in most cases, the consultation process is limited to informing the community about the expected benefits to the community and explaining how the community will be protected from any negative impacts, but no real negotiation takes place.

Clearly, building trust among stakeholders requires their effective participation in decision making regarding environmental and social risks and expected benefits, specifically

- Identification and evaluation of environmental and social risks associated with the project and agreement on the means of remediation and control, and

- Agreement on the benefits to be expected and the short- and long-term positive effects to be achieved from the project.

Therefore, a sustainable management approach to project permitting is to engage in a participatory process with the community and other stakeholders, fostering participation in decision making and agreement. This approach implies that the company should disclose all pertinent information regarding the mining project and should engage in consultation regarding the understanding of technical reports and facilitate independent audits as required by stakeholders.

One approach to a social license standard is the Seven Questions framework[4] for assessing sustainability. This framework was developed by the International Institute for Sustainable Development in 2002 as part of the Mining, Minerals and Sustainable Development (MMSD) project. Seven Questions is a standard procedure to determine if the net contributions to sustainability are positive over the longer term.

Another framework to guide interactions between mining companies and communities is the Community Development Toolkit[5] developed by the Energy Sector Management Assistance Program, the World Bank, and the International Council on Mining and Metals (ICMM) in 2005. It consists of seventeen tools to facilitate community development over the mining project life cycle, including exploration, feasibility, construction, operations, decommissioning and closure, and postclosure.

Both the Seven Questions and the Community Development Toolkit are described in detail in Chapter 3 of this book. In Chapter 6, the concept of social license is analyzed from the perspective of human resources management.

Social Risks

Many social risks are associated with mining operations located in remote, often underdeveloped, regions of the world. Others, like changes in land use, are inherent in mining, regardless of social environment. Some of the risks are

- Risks associated with long-term changes in land management and land uses;

- Risks to health and safety (e.g., dust and noise) inside and outside project boundaries;

- Risks related to short- and long-term impacts on surface and groundwater resources;

- Massive immigration associated with mining projects located in poor regions, with immigrants attracted by the job opportunities generated by mining;

- Risks related to resettlement and relocation of communities;

- Excessive economic dependency on mining; and

- Risks to livelihood of local population dedicated to artisanal mining.

Social Benefits

The benefits of mining are mainly associated with increased employment opportunities, increased financial capacity of community institutions, company sponsorship of sustainable community projects, and entrepreneurship aimed at sustainable economic growth and diversification.

Many companies are moving toward partnering with hosting communities and involving them directly in decision making. Partnering with communities and third parties for development is becoming instrumental in helping a company obtain a social license to operate.

Social License: A Business Perspective

Intangible assets, such as reputation, are widely viewed as comprising a substantial and growing portion of a company's value. This is especially true in the minerals industry, where reputation is a critical strategic asset.

Strategic Perspective: Access to Human and Mineral Resources

As it applies to corporate strategy, the concept of social license can be extended to a mining company's achievement of a positive public reputation on ethics and social responsibility issues. This business perspective of social license is rapidly gaining the attention of mining companies as a management approach to reduce the financial risk and uncertainty of new mining projects and also as a strategic advantage.

Mining companies perceive a positive reputation as an advantage when gaining access to mineral resources and also to human resources, two key strategic advantages in mining. For example, Rio Tinto states that it has been able to develop its Diavik mine in Canada and other projects that never would have come to fruition without its public commitment to corporate social responsibility. Rio Tinto also states, ". . . in the long run, the trust that we are creating by building sustainable relationships will enhance our ability to gain preferential access to the essential 'people resource.'"[6]

Many companies have begun to integrate social and environmental risk into their risk management system. As an example, ICMM's Sustainable Development Framework includes the following: "Consult with interested and affected parties in the identification, assessment and management of all significant social, health, safety, environmental and economic impacts associated with our activities" (Principle 4).[7] Failing to identify and manage social, health, safety, and environmental risks can cause significant economic losses.

Project Finance Perspective: Access to Financial Resources

Investors of all types are focusing more attention on environmental and social issues when investing in and insuring mining projects. The adoption of the Equator Principles,[8] a globally recognized benchmark for the financial industry for managing social and environmental issues, by several mining companies and many of the world's leading banks might be considered as a trend toward higher social and environmental commitments and social license by the minerals industry. Financial institutions adopting these principles are taking into account the positive financial consequences of social license in financing new mining projects. Similarly, mining companies aligned with the Equator Principles consider the benefits of an easier and more favorable access to credit when financing new mining projects.

The Equator Principles are a set of categorization, assessment, and management standards designed to identify and address any potential environmental and social risks that a proposed project might present. Equator banks have undertaken not to finance any project with a total capital cost of US$10 million or more unless the project can comply with those standards or there is satisfactory reason to deviate.

The International Finance Corporation (IFC) applies its Performance Standards to manage social and environmental risks and impacts and to enhance development opportunities in its private sector financing in its member countries eligible for financing. The eight Performance

Standards establish benchmarks that the client is to meet throughout the life of an investment by IFC or other relevant financial institution:[9]

1. Social and environmental assessment and management system

2. Labor and working conditions

3. Pollution prevention and abatement

4. Community health, safety, and security

5. Land acquisition and involuntary resettlement

6. Biodiversity conservation and sustainable natural resource management

7. Indigenous people

8. Cultural heritage

Standard 1 establishes the importance of (1) integrated assessment to identify the social and environmental impacts, risks, and opportunities of projects; (2) effective community engagement through disclosure of project-related information and consultation with local communities on matters that directly affect them; and (3) the continuous management of social and environmental performance throughout the life of the project. Standards 2 through 8 establish requirements to avoid, reduce, mitigate, or compensate for impacts on people and the environment, and to improve conditions where appropriate.

Though not a commonplace case, junior mining companies are becoming increasingly aware that social license can reduce social and environmental risks and so has a positive impact on their capacity to get funds in the stock markets and represents a sizeable added value when selling its projects. Insurers are increasingly addressing environmental and social risk issues. This has significant importance for project financing and capital cost. American International Group became the first major private insurer to adopt a policy to manage social risks. Other companies are expected to follow. Notably, political risk insurance will become more important as the need to access resources in environments with greater political complexity grows.

Other Key Project Stakeholders: Employees and Contractors

For a complete analysis of the management and interactions between a mining project and stakeholders, see the "Project Management and Stakeholders" section. The following paragraphs outline some important management aspects to be dealt with during the early stages of a mining project.

Human Resources

An important project risk is related to human resources and the human resources management organization and structure required after production startup. Therefore, investigating the labor market in the hosting and nearby communities is an important project task. Precise information and data on the availability and cost of skilled labor, occupational statistics, and unemployment rates will be required for the feasibility study. Early consultation with labor unions, human resources consultants, and employment agencies will help to obtain the information required and will foster the interest and the cooperation of these stakeholders during the feasibility, construction, and start-up stages of the project.

The important role of human resources in achieving and maintaining the social license is analyzed in more detail in Chapter 6 (Human Resources Management).

Contractors and Suppliers

Project contractors and suppliers are very important stakeholders during the project stages. They are a major source of information during the prefeasibility and feasibility stages and the leading force during construction. It is therefore important to investigate the local, national, and international markets for project contractors and suppliers and to manage creating a list of qualified contractors and suppliers for project use. Whenever possible, local contractors should be considered.

Local Industry Managers and Entrepreneurs

It is important to establish contact with local industry leaders and entrepreneurs, especially those in the minerals industry and related sectors in order to establish personal contact and search for shared professional and business interests. They can help the project and provide important management and cost information.

Landowners

Many projects require acquiring numerous small land lots from a large number of owners, which makes land acquisition a difficult, time-consuming, and challenging task—and a significant project risk. Landowners are important stakeholders, and early contact with them is the key to establishing positive personal relations between landowners and project management.

The management of land acquisition and the importance of a fluid relationship with landowners is analyzed in the "Project Management and Stakeholders" section of this chapter.

Other Interested Parties: The Project Summary Report

As soon as the project has been defined at the conceptual engineering level, a project summary report should be prepared for use as public information to all stakeholders and interested institutions. The project summary report should be widely distributed to internal and external stakeholders, landowners, local authorities, regional and national government agencies, and the media. It should be a top-quality 20- to 30-page booklet and should be presented as the initial step in a process of permanent dialogue with stakeholders.

Conclusion

An emerging standard of best practices in the mining industry aims to leave net positive benefits across social, economic, and environmental areas that are sustainable after closure. In the foreseeable future, there will be growing pressure to ensure that mining projects and operations are managed for sustainability and a legacy of economic, social, and environmental benefits long after the mine closes.

The achievement of social license for mining projects and creating a positive corporate reputation and image are perceived by mining companies as an important strategic advantage when gaining access to mineral resources, human resources, and financial resources.

Acknowledgments

The author of this section thanks Joaquin Duque of Tecnicas Reunidas and M. Norman Anderson of Norman Anderson & Associates for their contributions, ideas, and participation in the development of this section.

PROJECT MANAGEMENT AND STAKEHOLDERS*

Mining's interaction with the surrounding world has been a complex issue for centuries. Its major impacts, broadly negative in environmental terms, broadly positive in economic terms, and contradictory regarding labor, social, and local wealth effects, have led to agreements and clashes, clichés, and ignorance between mining and hosting communities. Mining is a unique world, frequently located in remote or uninhabited areas, mostly in rural settings, and in countries with varying degrees of political and socioeconomic development; mining has at times imposed its own rules, condoned to a greater or lesser extent in exchange for job creation, industrialization, and taxation benefits.

But times are changing for mining. A growing conscience has appeared about environmental impacts, public input, nongovernmental organizations (NGOs), union involvement, and indigenous rights, as well as internationalization of information and so on. In past years, this new scenario has opened up serious problems and increasing difficulties for mining companies already in operation and even more serious ones for new projects.

Even international best-sellers have focused on the environmentally negative legacy of mining, the scale of lasting residual impacts[10]—contaminated water from abandoned mines, sterilization of the land—and as a consequence of this legacy, one of the most recurrent requisites for a new mining project from its very outset is to address end-of-life issues, designing a sustainable (environmental and socioeconomic) and safe closure with an eye to the distant future. As recognized by an important workshop organized by the World Bank in 2002,[11] mine closure is one of the most important issues confronting the mining industry. The way responsible companies approach it will be seen as a very important step in promoting a change in the mining industry.

In the last decades, an attitude toward a two-way communication with the "outside" world has gradually been translated into corporate policies and transposed onto the day-to-day business of many other mining enterprises. Standalone sustainability corporate reports have now become a common business practice.

Project Background

How mining companies operate in their interaction with stakeholders depends on a number of factors. *Size* and *internationalization* of a company are among them. Some companies tend to be more sensible, experienced, and proactive in understanding and promoting such corporate policies. This approach is not philanthropic but based on mutual respect and equilibrium. The policy has to pervade the whole corporate organization in order to be really effective, credible, and perceived as such for most (if not all) of the stakeholders. Corporate structures (e.g., decision making, human resources, and compensation/rewards) must reflect this policy. It should aspire to be an example of how the company works and adapts to the realities of the host country. Some companies, usually local or the old ones, develop a strong cultural relationship with stakeholders over the years, rooted in a shared (although not necessarily easy) history with the host country, common development, and deep understanding of their values and needs. These companies can be valuable partners for other corporate enterprises seeking to enter into new mining scenarios.

Time is also a factor. Understanding and incorporating stakeholders' positions cannot be a matter of improvisation. Corporate policies have to be devised and agreed upon, then converted into actions, which in turn must lead to positive results. All this takes time, usually counted in years.

The so-called global village is essentially an urban (and intellectual) concept. Today's developed and underdeveloped worlds continue to present extreme differences, more significant in rural areas, where mining tends to be located. Agriculture and local industries as well as other

---

* This section was written by G. Ovejero Zappino.

local business initiatives develop naturally with local understanding of the country's peculiarities, whereas mining usually seems a strange culture, not only because of its technicalities and impacts, but also because of the international nature of the mining entrepreneurs. In addition to the purely technical aspects of mining, culture, language, work habits, and so forth also present challenges.

However, with the exception of political or legal issues, many feasibility studies or due diligence reviews still do not involve an in-depth study or simply fail to address or understand the cultural background and the stakeholders' positions where a project is located.

Whether located in a historic mining district or a greenfield site with no previous mining history, a new mining project requires a comprehensive, in-depth approach to the site's peculiarities and stakeholders' perspective. The same should also apply to existing operational mines, because they function in a continuously changing world.

In the case of projects discovered purely through exploration, this approach should be more manageable, because the initial decision to choose a particular corner of the planet is theoretically based not only in its geological potential but also on its political, social, and economic background. This means, in theory, that the main stakeholders influencing the development of the project will have already been taken into consideration. However, once the discovery becomes a feasible, exploitable project and, subsequently, a mine, a myriad of latent situations or previously unidentified or new impacts, and therefore stakeholders, can appear.

Information technology (Internet, e-mails, and cell phones) provides a tool for exploring and reporting between mining and stakeholders, but it also creates a parallel, virtual world that does not always coincide with reality. Anti-mining groups, for example, are much more active than mining companies, not only through the Web but on the street and in the media, and their views can lead to a distorted perception about how a mining project or operation is perceived by stakeholders.

Stakeholders

The stakeholder concept, developed and promoted in the 1980s by R. Edward Freeman,[12] refers essentially to a party that affects, or can be affected negatively or positively by, the company's actions.

The stakeholder concept is applicable to any business activity, but it has found an extensive application in the mining industry, due to the great diversity and magnitude of its potential (positive and negative) impacts and the controversial legacy of mining around the world throughout history.

Identifying and interacting with the relevant stakeholders is essential to producing good results. But much still needs to be learned about how to identify and involve stakeholders, as the World Bank[13] was to note, which suggests that a good way to identify appropriate stakeholders is to start by asking questions:

- Who are the "voiceless" for whom special efforts might have to be made?
- Who are the representatives of those likely to be affected?
- Who is responsible for what is intended?
- Who is likely to mobilize for or against what is intended?
- Who can make what is intended more effective through their participation or less effective by their nonparticipation or outright opposition?
- Who can contribute financial and technical resources?
- Whose behavior has to change for the effort to succeed?

TABLE 8.3   Mining project stakeholders

The Company's Directly Related People, Business, and Investor Institutions
Shareholders
Financial and lending institutions and financial analysts/the World Bank
Employees (and employees' families)
Contractors, suppliers, and customers

Government, from Municipal to National or International Levels
Legislators, regulatory bodies, members of the judiciary, security forces, and the executive/ministries
Ombudsmen

Intergovernmental Organizations
United Nations bodies (i.e., focused on corporate responsibility, human rights, environment, labor)
Regional bodies with authority for policy-making and enforcement (e.g., the European Union, Organisation for Economic Co-Operation and Development, Association of Southeast Asian Nations, Organization of American States)

Society
Directly affected parties: host and surrounding communities; landowners
Indirectly affected parties: Agriculture, private business, tourism, competitors, etc.
General public and broader society: local to global
NGOs: environment; development; humanitarianism, human rights; citizen platforms; etc.
Religious institutions
Religious protagonists with an effect independent of institutions (i.e., a local priest)
Media (press and opinion leaders)
Political groups
Labor unions
Professional associations and scientific/technological bodies and universities
Indigenous people and individuals
Minorities and other historically marginalized groups
The environment

A list of prominent stakeholders (not necessarily in order of importance) with direct or indirect interest in the mining project is shown in Table 8.3.

Normally, the groups with common interests are the shareholders, the employees, and the contractors, suppliers, and customers. But they operate in a very complex social, economic, regulatory, political, and even religious framework where other stakeholders must be considered.

High corporate standards, transparency ("far easier said than accomplished," Jackson),[14] clear targets and positions, and recognition of the perspectives of external stakeholders must, among others, be common factors when interacting with stakeholders.

The Company's Directly Related People, Business, and Investor Institutions

Interaction with these stakeholders is well regulated through official meetings and annual reports to shareholders. Lending institutions and their advisors, as well as financial analysts, have to be provided with clear information, beyond the project economics, related to sensitive factors potentially affecting financial exposure and stock value, such as environmental liabilities, permitting delays or no permitting, legal appeals from third parties, social opposition, and so forth. For financial analysts as well as for due diligence processes, social aspects are an important part of the list of items to deal with, such as the project's negative impact on third parties, anti-mining movements, or public resources depletion.

Employees, contractors, suppliers, and customers can be the project's strongest supporters. All of them, especially employees, will diffuse within the local communities the most accurate image of the company's behavior. Regular information should be provided on the project development through in-house newsletters, meetings with management, full site visits (including employees' families) and annual conventions.

Government

Government regulatory agencies are the administrative key to project permitting. Governments are the primary decision makers and implementers of policies and projects. One of the initial difficulties when permitting a project is insufficient or poor relationships with the numerous government agencies involved or even incompatible approaches and policies. Recommended proposals to the authorities to solve this issue (partially) can be (1) the allocation by each government agency of a responsible coordinating officer, and (2) to hold coordination meetings and workshops with presentations on the project, making the authorities aware of schedules and deadlines and, above all, agreeing on common permitting conditions and criteria.

It is essential to develop an impeccable, open, and frank relationship with the authorities in order to create credibility and common trust and to maintain their confidence by strictly complying with the legislation and permitting conditions. Demonstrating good business practices will secure project development and access to new projects.

One of the administration's main concerns and challenges at present is the aftermath of mining, confronting some companies' managers who might think this is an issue to be resolved by others in a distant future, when present players will not be personally involved. The better this issue is addressed, the greater the acceptance of the project.

Society

The general public constitutes the most complex group of stakeholders, not only because of its varied nature and the difficulty in assessing its requirements and concerns, but also because of potential conflicts of interest with project development. However, that said, beyond cultural differences, basic human needs and behavior are the same everywhere. Public hearings, direct consultation, and social baseline studies are the most straightforward ways of approaching the matter.

Public hearings, listening, and answering are necessary exercises to be carried out through formal and informal ways. Formal procedures are provided by regulated public access to the project documentation during the permitting process. This public enquiry usually leads to questions, recommendations, or allegations by public third parties, to which the mining developer has to reply satisfactorily in writing. Some can end in judicial appeals with different outcomes. In some countries, the legislation provides the interesting opportunity, before starting a full permitting process (which can take years with unknown results), to prepare a small project summary to be distributed by the government to a variety of government branches, NGOs, institutions, citizens, and so forth. The aim is to identify ideas, recommendations, potential conflicts of interest, and so forth to be gathered by the government and to be sent with comments to the mining entrepreneur. Although perhaps imperfect, it at least provides valuable insight into many government and interested/affected parties' views, plans, and positions and enables the permitting actions and documents to address the fundamental issues.

People sometimes complain that the usual 30 days formally allowed for public review (followed by more opportunities for closely affected parties) can pass unnoticed or are insufficient for a thorough review. The company should consider the advisability of reinforcing the information through parallel informal ways, such as public presentations, newsletters, or other means to encourage better understanding of a project and try to promote favorable opinion.

When interacting with directly affected parties, prearranged and ad hoc communication meetings to discuss particular issues are mandatory. The parties' requests have to be carefully evaluated and answered as promptly as is feasible. An adequate response must be given to prevent issues from growing with misunderstandings or wrong expectations. Project impacts have to be quantitatively (and qualitatively) assessed and addressed through constructive dialogue and actions. These assessments should be based, whenever possible, on previous baseline studies—indeed, it is most advisable to conduct preoperational baseline studies and community perception surveys.

Direct consultation and social baseline studies provide a better insight into stakeholders' issues, and are complementary to the formal procedures of public hearings. Founded in 1998, the Global Mining Initiative promoted the Mining, Minerals and Sustainable Development (MMSD) project[15] in 1999, based on an independent, multistakeholder-based review of how the minerals and metals industry can best contribute to sustainable mining development. The MMSD project ended at the Johannesburg World Summit (September 2002) on Sustainable Development. The final report, *Breaking New Ground*, included a number of basic recommendations with regard to stakeholder involvement:

- Greater cooperation among those stakeholders with similar interests and the importance of enhancing capacity for effective actions at all levels
- Constructive dialogue with key constituencies
- Effective community development management and tools

More recently, in November 2007, Anglo American used its Web site to make public the comprehensive, detailed Socio-Economic Assessment Toolbox,[16] which provides a way to develop a rigorous profile of the communities surrounding mining operations and a precursor to participating, practically and effectively, in community development priorities.

A commitment should be given to contribute to community welfare through existing development programs and by engaging local people, suppliers, and contractors with such programs measured to gauge their effectiveness. Third-party assessment helps to understand, plan, implement, and account for social and economic performance at a local operations level.

Above all, it is important, regarding social engagement and consultation, for mining companies to be aware of the potential for backlash if those consulted perceive their input is being disregarded by the project. Two key problem areas come to mind:

- Perception of insufficient benefits or inadequate development programs
- Perception that the project itself is being constructed/operated in ways that stakeholders consider irresponsible or disrespectful

This second point causes more animosity and is often overlooked by companies.

Interacting with NGOs requires specific approaches. Technically, an NGO is an officially recognized entity that is not affiliated with governments or companies, but common usage relates the term *NGO* to organizations geared toward a cause rather than toward profit. NGOs fit into a range of cross-cutting, nonexclusive categories: geographic (local, national, regional, and international) and subject area (development, humanitarianism, environment, human rights). Their positions on mining can range from pro-engagement to anti-mining: a number of NGOs participate in sustainable mining initiatives and projects, while others hold back purely because of mining companies' involvement. Some of these groups maintain a constant, close scrutiny (sometimes constructive and beneficial, sometimes neither fairly nor objectively based) of a company, its operation, and the government regulators, at times with a certain impunity in their declarations and activities.

Dealing with the media requires the company to provide information readily about project development (milestones, goals reached, environmental performance, technological innovations, ending with background corporate information) and respond to issues that arise (silence is not an option). Ideally, a personal relationship with media management and journalists should be maintained. When writing press releases or answering questions, short answers should be given ("less is more"), and one must stick to the facts, and use plain, understandable language. It is recommended to establish different levels for media interaction: (1) media specializing in business and economy, which is the natural medium for a mining project; (2) local or provincial information sections in the general media, used by the general public; (3) editors, managers, and owners of the chief media (press, television, radio, etc.); and (4) regional and local media present in the municipalities where the project or mine is sited.

It might also be wise to consider "the environment" as a distinct stakeholder. This is controversial because the term *stakeholder* typically refers to human individuals and entities. However, there is increasing support for the notion that the environment itself, and elements of it, have standing, regardless of how environmental change might impact human beings.

For practical reasons, some companies have edited specific and detailed guides for conduct on community relations, environmental and health, and safety items, covering from the very beginning when the first exploration team sets foot on the site up to the complex relationship developing over the various stages of a project.

As a final general note, there is no agreement on what constitutes a mining stakeholder, and not all mining projects affect or are affected by all these stakeholders.

Administrative Licenses

Any responsible mining activity has to be fully supported by administrative licenses based on the legal and regulatory framework of the host country, obtained throughout a permitting process. The permitting process has become an increasingly colossal, complex, tangled, and lengthy exercise requiring specific staff and highly professional legal advisors assigned to this task. The numbers of key and collateral permits, and the number of different regulatory authorities involved (municipal, regional/state, national/federal, even international), can reach surprisingly high figures. Permitting can easily take from 3 to 5 years, depending on a variety of factors, and it continues throughout the entire life of the project. Successful permitting requires the following factors at a minimum:

- Full knowledge of the existing legal framework and awareness of oncoming legislation. When needed, international or state-of-the-art regulations or recommendations, such as the ones from World Bank, provide useful references. Regarding future legislation, active participation in the open, public consultation process is highly recommended.
- Continuous and expert legal advice and support
- Conceptual permitting design and schedule, showing the critical permitting paths. Legal shortcuts should be avoided.
- Allocated staff and budget
- Preparation of high-quality mandatory projects (technical, environmental, social, feasibility studies). Identify/select the right consultants, using local ones whenever possible, especially in the environmental and sociocultural and socioeconomic areas.
- Active interaction with the administration/government bodies. If possible, a coordination committee involving the different administration players should be organized, extended if necessary to representative stakeholders.
- Continuous follow-up of permitting items and project development with the administration bodies, with a relationship based on common trust and ethical behavior.
- Persistence, tenacity
- Clear understanding of commitments assumed regarding the general and specific conditions of the awarded permits

- Strict compliance with the permit conditions
- Continuous dialogue with the government agencies that is regularly maintained and reinforced

Some of the main permitting areas required by a mining project are shown in Table 8.4.

Access to Land and Public Resources

Mining operations need infrastructure facilities, public resources, and services, which in many cases require the use of land and/or obtaining administrative licenses. Access to land requires negotiations with private, communal, and government owners, as well as other users of the land's natural resources (timber, fishing/hunting, tourism, etc.). Access to public resources (e.g., mineral rights, utilities, water) is usually acquired through concessions or contracts.

Land

In some areas of the world, two systems of land ownership coexist: formal/legal and de facto/informal/customary. Governments often view formal or legal ownership as the only legitimate system, whereas local communities often consider the second type to be more legitimate. The company has to navigate both systems simultaneously: to disregard or underestimate de facto or customary ownership can cause protests and jeopardize social license to operate; however, to treat both formal and informal systems as equal can irritate governments.

Understanding the structure of land property, including tenancy and its legal framework, is the first step. This will usually take place from the exploration stage, giving the company the opportunity to develop a fair, negotiated approach to landowners and laying the basis for future negotiations for land acquisition/rental should the project go ahead. This explains why, in some cases, the company staff members who deal directly with land acquisition are geologists or field exploration technicians, supported by legal backup.

TABLE 8.4 Main project areas requiring permits

Environment	Mining	Water
Environmental impact assessment	Exploration permit	Water concession
Effluent discharges	Mining concession/project	Surface and underground water management
Air emissions	Health and safety plan	Stream diversions
Solid residues disposal	Emergency plan	Water resevoirs, ponds
Noise	Explosives	Reutilization of discharge water
Management of hazardous materials	Yearly work plans	
Tree-cutting permits		

Infrastructures	Others	Municipal Permits
Power lines and substations	Demarcation and use of public domains (streams, roads, public paths...)	Town zoning plan permit
Ancillary installations		Activity, construction, and opening licenses
Mining camps/towns	Archaeological permits	
Access roads/road diversions	Construction permits (civil, industrial works)	
Railways	Aggregates, construction materials	
Pipelines		
Ports		
Crossings of existing infrastructures		
Telecommunications		

Many countries' mining laws grant the right of expropriation or compulsory purchase of the properties needed to develop a mining project (mine and treatment plant), including easement rights for off-site infrastructures such as pipelines, power lines, access roads, ports, or transportation railway lines. Even in the evaluation phase, the right to temporary occupation for evaluation work is contemplated. The juridical grounds are based on the declaration of public utility or social interest linked to the award of a concession to mine. The explanation for the public utility is that an ore deposit is a state-owned resource awarded as a concession to a public or privately owned company on the terms of a sound exploitation of a nonrenewable resource; the simple logic behind expropriation is the unmovable geographical and geological location of the ore deposit.

But beyond all these legal concepts and rules, expropriation is usually perceived as the intrusion of a private company that is conducting business to the detriment or destruction of other established (agricultural, forestry) businesses; this perception is more acute in the case of foreign companies. And, although perfectly legal, expropriation is, however, an act of force, and every reasonable effort has to be made to reach negotiated agreements with landowners to buy or rent the affected properties. The expropriation process starts once the initial attempt to negotiate has failed, and actually often helps to reach negotiated agreements. Agreements, instead of compulsory purchase, will always improve the relationship with the community and its municipal governments, will release pressure by the government agencies and officials responsible for the expropriation procedure, and will prevent anti-mining groups from finding another subject of complaint and certain communication media from exploiting an attractive social vein.

Frequently, expropriation can cause loss of agricultural, cattle, and/or fishing tenancies and jobs, negatively impacting the modus vivendi of the affected landowners. In that case, whenever possible, employment or other benefits created from the development of the mining project have to be seriously considered and implemented for the affected parties.

At the end of the mine's life, land should be returned, as much as possible, to a productive or recreational use.

Public Resources

Items or commodities such as ore and water follow complex permitting processes, sometimes in competition with other mining or nonmining users. Because of that, identifying the best technological methods for maximum ore yields from a mineral deposit is advantageous in creating higher revenues for the host country, among other benefits. In the case of water, mainly in arid or semiarid environments, the economic value and jobs generated per cubic meter of water consumed in the mine could be imperative for obtaining a water concession. Maximum water reuse and zero discharge, use of treated sewage water as primary water supply, recycling process water or reutilizing effluent for nonmining uses (e.g., agriculture) are examples of positive approaches to mine water management.

The potential impacts of mining on public resources such as superficial and underground water or soil must be conscientiously prevented because they can catalyze social movements that are difficult to counteract.

Social License

The social license concept[17] has become a frequent statement in corporate reports of mining companies. Companies today are fully aware, as demonstrated by examples around the world, that public, consistent anti-mining positions can prevent the development of projects despite obtaining all the administrative licenses.

At the feasibility and construction stages, gaining social license and credibility to operate implies a number of things:

- Assessment as necessary, to address stakeholders and community concerns and expectations. The most common items deal with the potential impacts of mining on public health, on water and air quality, destruction of unique ecosystems, and access to job opportunities. And, there are always the post-mine-closure issues, either in terms of liabilities involving long-term or indefinite commitments regarding safe mining residues (an aspect not yet unequivocally resolved by regulators) or the extinguishment of employment.

- Informing potentially affected parties of significant risks from operations, the measures to manage the potential risks effectively, and the cooperative emergency response procedures. The most frequent risks relate to the permanent chemical and physical stability of large tailings impoundments, as well as surface and underground water quality.

- Supporting the implementation of compensatory measures. This might be done through procedures integrating land use planning, biodiversity conservation, and others involving stakeholders, academia, and local communities.

The identification and appraisal (and ideally prediction) of social acceptance, needs and requirements, and the evaluation of significant social, economic, public health, safety, and environmental impacts can be achieved through several methods:

- Direct consultation with the regulatory entities and with stakeholders through opinion surveys, formal or informal public hearings, discussions at social gatherings, public meetings, and forums during the conceptual stages of the project. Depending on the host country, this can be a voluntary or compulsory process.

- Sociocultural and socioeconomic surveys/studies. These might be conducted by external local consultants and academics.

Several difficulties might be found when appraising sociocultural and socioeconomic items:

- Identification of the actual social representative forces
- Educational levels
- Demagogic messages
- Political interests, using the project as a tool against political adversaries
- Complex multifactorial, dimensional, and interactive issues

Role of the External Affairs Department

International mining companies work in many different environments and geographies, sometimes in uncharted territories. Understanding and appraising risks and opportunities is a difficult exercise, particularly for nonlocal companies, with frequent examples of misperception of multiple factors, sometimes with serious results.

The inclusion of an institutional or external affairs department, necessary from the initial feasibility and construction stages, has progressively become a frequent, standalone position in the corporate staff organization of mining companies. This department will specifically deal with permitting, land, legal, social, and communication issues; aspects of all of these are interconnected with a project's technical, economical, and environmental issues. One of the key purposes of this department is to help the company identify potentially divisive issues and address them effectively before conflicts with stakeholders arise or become difficult to manage.

The external affairs manager and his/her well-motivated, multidisciplinary team (sharing values and objectives) tend to be the familiar external face of an operation or company, acting frequently as the spokespeople dealing with authorities and stakeholders. Management of this department should be assigned to a member of the staff with some of the following ideal characteristics:

- Knowledge of the technical, environmental, safety, legal, social, and political aspects of the project, as well as its host background and the concerns of the public
- A professional profile from mining or other industries, government, or even NGOs. Sometimes, someone with only a technical, industry-specific background can be unable to (1) apply cross-industry learning, and (2) play the sensitive mediator between technicians and nontechnical outsiders that the role requires.
- Capacity for listening and for dialogue; a person trusted by and willing to spend time with stakeholders
- Communication skills that can be improved with exposure and training to deal with the public and the media
- A motivated, enthusiastic attitude and an inquiring mind

Assessing the complexity of factors, especially the cultural ones, requires help from sociological consultants and from local external public relations experts, and it takes time for the company, consultant, and stakeholders to understand each other, especially when a nonlocal mining company is involved. Some companies only realize the need for external help when a crisis situation has "suddenly" arisen.

The role played by a sociological consultant will be based on

- Causing no harm to communities or the company through its research and assessments;
- Carrying out an appropriate consultation program;
- Facilitating, guiding, and developing communication between communities and the company; and
- Helping to identify stakeholders' concerns and expectations.

The role played by a public relations consultant might overlap the role of a sociological consultant, but it should be seen as a complementary one. Giving priority to public relations for "selling" an image of the company is coming to be considered an outdated role, given that it can convey the message that the company sees its responsibilities to stakeholders as something within the realm of managing public information and perceptions. Far from this attitude, the public relations consultant will provide expertise in a variety of areas:

- Giving advice on local to national stakeholders, attitudes, and issues;
- Building understanding with media, including personal contacts with media directors, and business and environment correspondents;
- Identifying key opinion leaders; and
- Providing training for dealing with the media, for conflict and crisis situations.

Regular input both ways is critical; it is not possible for consultants to add value if they are working in a vacuum. Ideally, the progressive implementation of effectively integrated stakeholder engagement policies should simultaneously minimize the need for external permanent consultants.

Managing Communications and Reporting

Day-to-day business activities require honest, transparent communication and independently verified, audited reporting on environmental, health and safety, and socioeconomic performance indicators and on contributions to sustainable development concerning stakeholders. This can be made in writing (newsletters, press releases, official documents to regulators, sustainability and social corporate reports, etc.) through public forums or via multimedia (project presentations, media contacts, opinion surveys, corporate videos). The sustainability and social corporate reports have become a periodic, structured way of formally addressing project topics and issues. However, the emergence of many different approaches has created a problem of criteria uniqueness, leading to standardization initiatives, such as the Global Reporting Initiative, Towards Sustainable Mining, Dow Jones Sustainability World Index, FTSE4Good Index, and the Eco-Management and Audit Scheme. Even the European/International Organization for Standardization (EN/ISO) 1400, although primarily focused on environmental performance, promotes information, awareness, and commitment with the stakeholder community and the public at large.

In many cases, the road to success (or otherwise) lies in the direct relationships developed from the start, face to face, with landowners and immediate neighbors, as well as with other stakeholders (municipalities, regulators, and media). All public interactions, particularly at the beginning of a project, will persist in the minds of the community. The most difficult trial is construction, which involves the arrival of strangers, large equipment, and high but temporary employment and demand for local resources, which generate wealth but also reservations and confused expectations in local residents. Communication in this phase is important to keep communities well informed.

Communication has become a matter worthy of mining corporations' attention and a subject treated regularly in mining literature, demonstrating that poor public communication when launching projects or facing environmental/social crisis can provoke insurmountable local opposition to operations, difficulties in project financing, and even a drop in share price.[18]

A communication plan has to be prepared from day 1 and must be developed over the life of the project. Its ultimate aim is to obtain approval for the project. There are some key guidelines to good communication:

- Above all, the communication exercise is not a one-lane path but a two-lane street, so be sure that the genuine values, perspectives, expectations, and concerns of both companies and stakeholders are correctly identified and addressed. And, as Anderson and Lorber[19] write, listening is more than just waiting for your turn to talk.

- Communication has to be directed to the external but also to the internal (employees, contractors) stakeholders.

- The plan must be based on clear strategic objectives, with the approval of corporate management, but with local realities shaping implementation of corporate policies. Any statement that could make an impact in stock markets must be cleared beforehand with corporate headquarters.

- The plan should encourage knowledge about the project among all the target audiences, making it well known and giving it positive values; it should generate social confidence and acceptance.

- The plan must have a realistic budget and preferably be under the responsibility of an external affairs manager. One spokesperson is preferable to many; someone needs to be identified as the contact point who is always accessible and who responds promptly to stakeholders' questions and concerns.

- The plan should explain how critical issues will be managed responsibly and for the mutual benefit of company and stakeholders. The plan has to be reviewed and implemented periodically.

- The plan must incorporate government, local community, and media relations.
- The plan requires making a risk analysis and identification of issues of concern as well as of the stakeholders/allies/opponents, with the help of third-party research.
- An updated briefing summary on the project should be prepared and always available, ready for distribution to the media or other stakeholders. Its contents and delivery have to be corporately agreed upon and tightly controlled.
- It is important to develop key, clear, unequivocal, simple messages. Language cannot be a barrier. Simple diagrams and illustrations should be used, conveying an easy-to-remember message at a glance. Tiring the audience should be avoided. Meetings and presentations should be prepared carefully beforehand, making a predictive exercise of what is expected to be heard as well as potential responses. Spontaneity can be a real enemy of communication. Proposed questions and answers should be prepared in advance and one should always be on guard for the worst questions.
- The goal is to reinforce positive and neutralize negative facts or messages. News to the media should be provided (without saturation and avoiding trivialities) when reaching significant milestones: permits obtained, peak employment figures, H&S performance, environmental investments, and so forth.
- When talking of employment numbers and other benefits, time-scales, and so forth, it is important to be prudent. One should not speculate nor commit to vague, distant activities; spontaneous promises should not be made. This will prevent disappointments and enhance credibility.
- Stakeholders might want to see an operation firsthand. Stakeholders' visits to the operational site or to another company's operations should be planned; if needed, office space can be arranged with a small exhibit.

Conclusions

Project management starts before the project itself exists. At the initial decision stage, when selecting preliminary areas for grassroots exploration or entering into project acquisitions, a key point of departure, critical to the success of the venture, is to understand its environmental, political, sociocultural, economic, and legal framework. As the project develops, building an interactive, ethical, and transparent relationship with stakeholders is essential for becoming an accepted member of the host communities, and for demonstrating credentials to facilitate regulatory and social operating licenses. Project management will be more successful when guided by targeting the highest possible standards in technical, safety, and environmental issues and corporate responsibility towards social affairs and public health, and by openly sharing these core values with the wide variety of stakeholders (government, communities, NGOs, media, labor groups, etc.), throughout all project stages from exploration, conceptual and feasibility studies, construction, production, and closure/postclosure.

In summary, mutual respect, active partnership, and long-term commitment with stakeholders—to paraphrase Rio Tinto's[20] three principles—could synthesize perfectly the way to more successful project management.

PROJECT FEASIBILITY EVALUATION: NEW TRENDS*

The word *sustainability* can have many meanings and interpretations, and there can be considerable discord over how mining should be viewed within these various interpretations.[21,22] One view of sustainability, however, is incontrovertible, and that is the view that finite resources, if they are to be extracted, should be extracted efficiently.

Feasibility Decision and Sustainability

Economists define an efficient mining plan as one that extracts the maximum social value from the resource. Mine planning decisions that affect value include when to start mining, how to mine, and when to stop mining. Engineers might not recognize that each of these questions is addressed within a feasibility study, but they are there. Mine planning involves the use of optimization routines that specify pit limits, cut-off grades, mining rates, and even equipment selection with a view to maximizing the value of the asset. The feasibility study takes a broader view, assessing whether this maximized asset value is enough to warrant spending resources and tying up personnel to bring the asset into production.

Of the decisions as to when to start to mine, how to mine, and when to stop mining, perhaps the only decision that engineers do not consciously take into account in their feasibility studies is the timing of the project. Instead, once a project is identified as having a positive net present value (NPV), it is often developed as soon as possible. This, too, though, is an optimal decision rule when metals are in backwardation, meaning that prices are expected to fall. Production now is better than production later, at least for projects that are economic. It has been observed that negative NPV projects are held on the shelf until their economics improves, either through a lowering of costs, expansion of reserves, or increase in prices.[23] In fact, exploration managers and mining engineers have long spoken of the cyclicality of their industry, and make specific references to the ebb and flow of mining projects across this cycle.[24] The industry is in a commodity boom at the moment, where projects and reserves that have long been uneconomic are now being brought into operation.

To the extent that all of these management decisions aiming to achieve maximum value from the asset are done well, they will be in line with the sustainability objective. However, if this decision process is not done well or could be improved, sustainability can also be improved. This section discusses feasibility evaluation with particular reference to sustainability, and how a new evaluation technique called real options improves upon traditional NPV analysis as a means of managing mining assets efficiently.

The first step to efficient extraction of minerals is never to sink money into extracting a resource when the present value of expected revenues from sales of the extracted resource is less than the present value of expected costs. In other words, negative NPV projects should not be developed. Getting back to the idea of sustainability, it certainly is not sustainable to spend more money to extract a mineral resource than the mineral resource is worth to society. The requirement that one should be able to distinguish between positive and negative NPV projects might sound simple enough, but there are some complexities even here. When considering social welfare, the present value of costs must include all costs, direct and indirect.[25] Direct costs include everything normally thought of as being involved in mining. Indirect costs include what economists call opportunity costs. They are incurred whenever an input that is used up would have had value in some other use. Even though the mine might not have to pay for this input, if it is valuable, its use has a cost. Environmental costs are perhaps the best known of these indirect costs, where the use of the environment as a pollution sink might not show up as an explicit cost in the

---

* This section was written by G. A. Davis.

calculation of project NPV. Yet using up the environment is a cost, and it should be included when making the business case for sustainable extraction. In fact, it is the failure of mining companies to include the social and environmental costs of their activities in their feasibility studies that has upset so many environmental and sustainability advocates.

The second step is to compare these full costs properly against project revenues when computing a project NPV. There are two aspects to a mine's operation that make this difficult. The first is that the timing of the revenue stream does not usually match the timing of the cost stream. Costs are incurred up front, followed by a net revenue stream, followed by a remediation cost stream that could be perpetual. Given that a dollar today is worth more than a dollar tomorrow, or a lottery winning paid out today is a lot better than that lottery winning paid out in equal installments over 40 years, one cannot simply sum up revenues and compare them with summed costs. The cash flows must be first adjusted for the time value of money. The second aspect is even trickier. A safe dollar is worth more than a risky dollar. If a bird in the hand is worth two in the bush, then a dollar in the hand is probably worth two that might or might not materialize. If the cash flows associated with a mining project are not all equally risky, then they must also be adjusted for risk prior to being adjusted for time.

Standard engineering economics practice deals with these issues by multiplying each year of a project's cash flows by a single risk and time adjustment factor,

$$\frac{1}{(1+r)^t} \qquad \text{(EQ 8.1)}$$

where t is the year in which the cash flow occurs ($t = 0, 1, 2,..., T$) and r is the risk-adjusted discount rate. The risk-adjusted discount rate can be expressed as $r = r_f + \text{RP}$, where r_f is the risk-free rate or bond rate and RP is a risk premium appropriate for the nature of the project. Once the cash flows have been multiplied by this factor, they are deemed to have been adjusted for risk and time, and thus can be added together such that the revenues are compared with the cost to generate an NPV.

The problem with this approach, which incidentally has been in use for more than 150 years, is that it is in fact very difficult to determine whether or not a project has a negative NPV. There are many obvious reasons for this—the orebody is not completely defined at the stage of mine project evaluation, there is uncertainty as to future costs and revenues, some indirect costs can be difficult to price and are therefore excluded from the analysis—but there is one nonobvious reason that is more important than any of these. That reason is the absolute inadequacy of the factor in Equation 8.1 for adjusting cash flows for risk.[26] The essential flaw is that the factor presumes that the more distant a cash flow, the greater its risk. This might be fine for the evaluation of some types of projects, but it does not fit with the structure of uncertainty seen in mining projects.

Mine project risk is a combination of commercial risk and technical risk. Commercial risk is risk in interest rates, prices, and costs. These things do, in some cases, become more difficult to predict the further in the future the prediction is made, but after a point the uncertainty saturates. Metal prices, for instance, have equilibrium forces limiting the levels to which prices can rise or fall. A 20-year forecast of copper prices, for example, is likely to have the same error bounds on it as a 10-year forecast. The same is true for the main technical risk faced by a mine operator, ore quality. Again, grade uncertainty does not grow as the forecast horizon grows; ore grades 20 years out are not more uncertain than ore grades 10 years out. This means that beyond a certain time frame, say beyond the 10-year mine plan, uncertainty is approximately constant through time, much as the uncertainty in the weather exactly 100 days from now is the same as the uncertainty in the weather 200 days from now. Valuing 20-year cash flows at one-half or

TABLE 8.5 Impacts of uniform discount rate on project selection

Investment Decision Factors	Project A	Project B
1. Present value of cash flows @ 10%	$70 million	$70 million
2. Present value of cash flows using correct risk adjustment	$40 million	$90 million
3. Development cost	$50 million	$75 million
Measured NPV (1 – 3)	$30 million	–$5 million
Decision taken	Develop	Abandon
Actual NPV (2 – 3)	–$10 million	$15 million
Correct decision	Abandon	Develop

one-third the value of 10-year cash flows when the two are roughly equal in their uncertainty, which is exactly what Equation 8.1 does when discount rates of 7% and 12%, respectively, are used, leads to faulty evaluation.

Another problem with Equation 8.1 extends beyond its construction to its use. This can be called the uniform discount rate problem. Using copper as an example, some copper projects are riskier than others, if only because of decreased operating margins.[27] Yet Equation 8.1 is used without great variation in the values used for r, mainly because there is no known way of estimating r for each project. This means that risky projects are discounted at about the same rate as not-so-risky projects. The uniform discount rate approach causes low-risk projects to be undervalued, and high-risk projects to be overvalued. The result is a high probability of abandoning a low-risk project and instead developing a high-risk project.

Table 8.5 provides a numerical example. Of the two copper projects being evaluated, each has a net revenue payoff with a present value of $70 million if measured using a uniform discount rate. Project A is actually a very risky project, and proper risk evaluation would yield a present value payoff of $40 million. Given that development costs $50 million, this is a –$10 million NPV project. Yet it will be measured as being a +$20 million NPV project because of inattention to the specific nature of the project's risk. It will go ahead, even though, when measured correctly, it is a liability to society. Project B has less uncertainty around its cash flow expectations. Proper discounting for risk produces a cash flow present value of $90 million. The constant discount approach will value this project at –$5 million. It will not go ahead, even though it probably should.

What does all of this mean for sustainability? It means that the profession is likely making feasibility study errors that result in the development and extraction of mineral resources that should have been left in the ground as uneconomic for the time being, and the abandonment or deferral of mineral resource projects that should have been developed at the time. The evaluation tool is broken, in other words, and this is not good news for sustainability.

State of the Art in Feasibility Evaluation

In 1979, a Brazilian PhD student at the University of California, Berkeley, noted that the development of a natural gas field had similar characteristics to the exercise of a financial option on a stock.[28] The development cost of the natural gas field was similar to the exercise price on a stock option, the value of the natural gas in the field behaved as does the value of a financial stock of an energy producer, and the period over which the development decision could be made was similar to the time to expiry of a stock option. With this, real option applications—the application of financial option pricing techniques to real assets—were born. With the financial support of the Canadian federal taxation authorities, financial economists Michael Brennan and Eduardo Schwartz made the next major breakthrough, publishing a paper in the *Journal of Business* in 1985 that looked at the optimal development timing and management of a copper mine. That model extended Tourinho's gas field development model in two important ways. The first advance was a rigorous modeling of the commercial risk characteristics of mining. The second was to allow for operating options once the mine was developed; the mine owner had the option to close an open mine at any time for a fixed cost or open a closed mine at any time for a fixed cost. The closed mine incurred maintenance costs, and the mine could be abandoned at any time for a fixed abandonment cost. This brought the notion of the mine being a portfolio of sequential, compound options into the real options technique. The result was a model that gave fairly powerful insights into optimal mine management and the factors that influenced that management. Quoting from the paper's introduction:

> The general type of model presented here lends itself to use in . . . corporations considering when, whether, and how, to develop a given resource; to financial analysts concerned with the valuation of such corporations; and to policymakers concerned with the social costs of layoffs in cyclical industries and with policies to avert them. The model is well suited to analysis of the effects of alternative taxation, royalty, and subsidy policies on investment, employment, and unemployment in the natural resource sector.[29]

This work is not important for its sophistication. It is important because it identified that mining and oil and gas projects are options on commodities, and as such must be evaluated using option pricing ideas.

Real options are optimization at its best. Real options emerged as an area of study shortly after the development, in 1973, of a model for managing and valuing financial options. That model, by economists Robert Merton, Fischer Black, and Myron Scholes, resulted in what is known as the Black–Scholes or Black–Scholes–Merton option pricing formula. It revolutionized financial economics, and earned Scholes and Merton the Nobel Prize in economics in 1997. (Black died in 1995 of throat cancer, though had he been alive in 1997 he would undoubtedly have shared the prize with Scholes and Merton.)

A major insight that came out of the Merton, Black, and Scholes work was how to evaluate risk in complex assets such as mines. Another was the concept of optimal decision timing, in their case when to exercise the option (when to develop the mine). Waiting might or might not end up being the best action after the fact—the best that can be done at each point in time is to compute probabilistically whether to exercise the option immediately and reap a guaranteed cash flow or to wait an additional period and revisit this question as new economic information is revealed and possibly higher cash flows can be had. In other words, the option must be optimally managed using a forward-looking probabilistic model.

The value of the option at any point in time is based on the possible payoffs and when they might occur. Figure 8.3 is a schematic that depicts the management of a call option to buy a stock for a fixed exercise cost. This option would generate a positive cash flow to its owner were it exercised immediately (time 0). That value is the option's "intrinsic value." But the option might pay off even more if its exercise is postponed. If there is a substantial chance that the stock price will rise in the future, it will be optimal to hold the option and, as a result of this optimization decision, the option's current value will be greater than the payoff from immediate exercise. If the stock price is not likely to rise in the future, then the option is optimally exercised immediately, and its value will be its intrinsic value. The goal is to come up with an optimal exercise timing decision rule such that the current value of the option is maximized. Given possible future movements in the price of the underlying stock, the option in Figure 8.3 is computed to have substantial speculative value, and so it is not exercised at time 0. Price scenario 1 has the option being exercised at time T_1, and price scenario 2 has the option being exercised at time T_2. The bottom diagram shows that the option value at time 0 is the present value of the expected payoff from

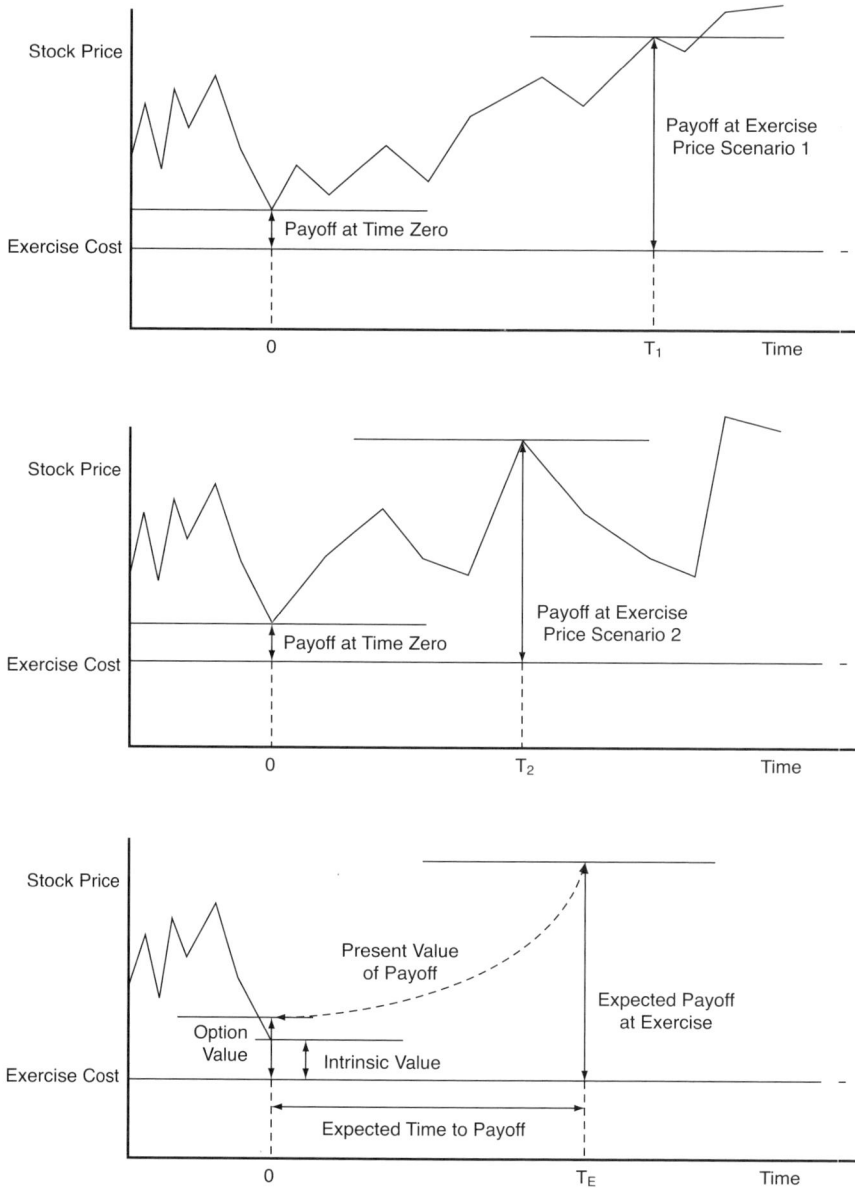

FIGURE 8.3   Call option payoffs under two stock price scenarios

holding the option, taking all possible payoff scenarios into account. The option value is greater for waiting, even though immediate development of the project would yield a positive result, which is called the option's intrinsic value. In a practical application, one can repeatedly harvest 1-year-old trees in a forest, but it is probably better to wait till they grow before harvesting them. The trees are options on lumber, and the decision is when to chop them down. In Figure 8.3, it is better to wait than to act now.

Prior to Black, Scholes, and Merton's work, there was no solution to this optimization problem. Traditional discounted cash flow was available, but it was not possible to use it to select the optimal time to exercise. The Black–Scholes formula made innovative use of differential equations to tell us how to time the option's exercise optimally and then value, in risk-adjusted present value terms, this payoff. The mechanics include correctly calculating the expected payoff, calculating its expected timing, and appropriately valuing in present value terms the deferred income (time value of money) and the riskiness of that deferred income (the risk premium). The Black–Scholes approach is a discounted cash flow technique, and as such is an extension of, rather than a break from, traditional NPV analysis.

Real options has also been called contingent claims analysis, modern asset pricing, market-based pricing, stochastic dynamic programming, and stochastic optimal control. Although there are subtle differences between these various techniques, they are various methods of conducting real options analysis. *Modern asset pricing* is perhaps the most appropriate terminology because that name reflects the fact that there has been a wide shift in finance theory since the 1970s, and that the theory is not only applicable to option pricing, but the pricing of all assets, financial and real. Real options has gained increasing acceptance in the mining world, and it is now the state-of-the-art method for asset management and valuation.

How Does Real Options Differ from Net Present Value Analysis?

The NPV technique involves forecasting the cash flows associated with an activity or decision, discounting those cash flows to the present using a risk-adjusted discount rate as in Equation 8.1, and then summing the resultant discounted values. Where there are a series of mutually exclusive alternatives, the NPVs are compared across the alternatives and the one that produces the highest NPV is selected. One alternative is to delay action. This is called the "now or later" alternative. Traditionally, NPV analysis does not include delay in its set of alternatives, and for this reason it had been deemed "now or never" analysis.

Although real options is also a discounted cash flow technique that compares alternative courses of action and selects that which produces the highest NPV, it differs from traditional NPV analysis in three important ways. First, real options analysis puts particular emphasis on dynamic decision making, the now-or-later alternative. This derives from real options' financial roots relating to optimal exercise timing (see Figure 8.4).[30] This results in, for example, mine plans that have an uncertain time horizon that varies as prices vary, derived from a dynamic "mine now or mine later" analysis. Traditional NPV analysis tends to assume a single price scenario and plans a fixed mine life according to that scenario. Second, real options uses financial theory to correctly price the risk associated with the future uncertain cash flows that are realized when the option is exercised. It is how the true present value of the cash flows in Table 8.5 are known. As is clear in that example, this pricing of risk is needed whether or not the option is exercised now or later. NPV analysis attempts to do this with a single risk-adjusted discount rate.

In real options, the risk discounting is done within the cash flow elements, rather than at the level of the net cash flows. The result is an "effective" time-varying net cash flow discount rate r_t that is an output of, rather than an input to, the valuation process. It is the exactly correct discount rate, that same rate that was impossible to know in Equation 8.1. What is even more astounding, this effective discount rate takes into account not only the inherent riskiness of the asset as it varies through time but also any modeled risk-mitigating actions, such as temporary closure or cut-off grade changes, that are likely to be undertaken by mine management.

Third, real options analysis correctly estimates future cash flows when there is uncertainty surrounding those cash flow estimates. It does this by taking into account the inherent nonlinearities

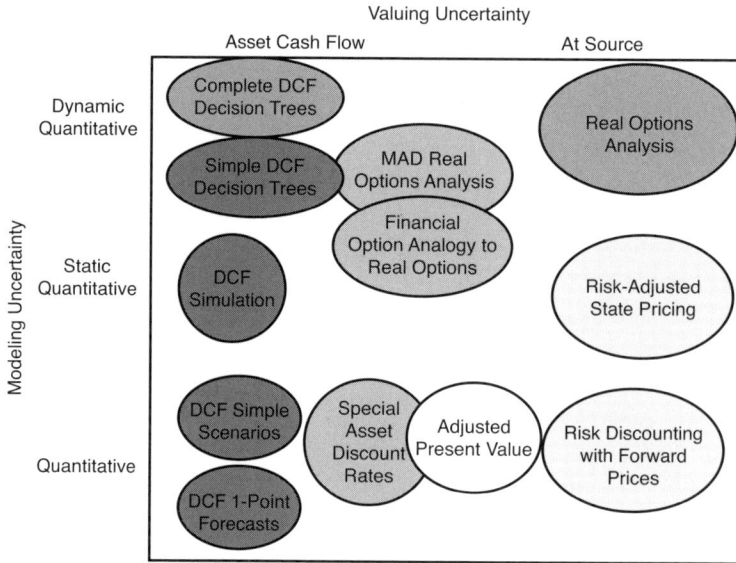

NOTES: DCF = discounted cash flow; MAD = market asset disclaimer.

Source: D.G. Laughton 2003

FIGURE 8.4   Banff taxonomy

in the cash flows and any optimal dynamic management of the asset and the inherent nonlinearities in the cash flows created by such dynamic management.[31] NPV analysis augmented by Monte Carlo simulation attempts to do the same thing but without the same rigorous definition of the uncertainties surrounding the cash flow estimates and without the ability to compute optimal dynamic managerial reactions to those uncertainties iteratively as new information is revealed to the manager. In essence, real options values the asset while at the same time anticipating how the asset will be optimally managed conditional on the information available and options available to the manager at each decision point.

These differences between traditional NPV analysis, NPV analysis augmented by simulation or decision trees, and NPV values calculated using real options have often been blurred. Laughton[32] prepared what he calls the Banff taxonomy to differentiate the analyses (Figure 8.4). That taxonomy has the treatment of uncertainty on the vertical axis and the pricing of uncertainty on the horizontal axis. Traditional NPV analysis is located in the southwest corner of the taxonomy. NPV analysis augmented by decision trees or simulation moves the analysis vertically but does not change the method of pricing risk. Only when full dynamic decision trees and simulation are combined with modern financial methods of pricing risk can the analysis be termed real options, located in the northeast corner of the taxonomy. The many other valuation approaches that have arisen in the past 30 years are also located on the taxonomy. None, however, match the uncertainty modeling and uncertainty pricing of real options analysis.

What Biases Arise with Traditional NPV Analysis?

Because a mine is a series of options to locate and extract ore, it can only be properly valued using real options theory. Through the application of real options analysis to mining problems during the past two decades, it has come to be learned that traditional NPV analysis tends to introduce five biases into the mine planning, development, and valuation exercise:

1. NPV analysis induces too much investment in capacity in mines where the commodity is subject to long-term equilibrium forces (i.e., all but the precious metals). This can also be seen as overvaluing investment in productive capacity and undervaluing long-term positive cash flows (long-life, low production rate mines).

2. NPV analysis under-represents the liability of long-term negative cash flows such as remediation costs.

3. NPV analysis underestimates the liabilities associated with contingent tax and royalty payments.

4. Where costs are less risky than revenues, NPV analysis undervalues investments that improve operating margin. Such investments might be investments to upgrade product quality or lower production costs.

5. Where NPV analysis is used in conjunction with decision trees, it undervalues the ability to manage downside risk and overvalues the ability to capitalize on upside potential. This means that abandonment options will be exercised later/at lower prices under NPV than under real options, because under NPV, abandonment (the avoidance of negative cash flows) is undervalued. Similarly, price-induced expansion options into low-grade material will be undertaken sooner/at lower prices under NPV than under real options. The net effect will be too little abandonment and too much capacity expansion.

One bias that is often alleged is that real options analysis produces a higher asset value than NPV. This is not true! Any NPV analysis with a low enough discount rate (inappropriately low) can produce a project value that is higher than the real options analysis. NPV analysis introduces valuation error, but that error is not necessarily always biased in the same direction (i.e., too low). Because of the biases mentioned previously, it is the author's experience that NPV analysis tends to undervalue world-class, high-grade deposits. It correspondingly overvalues small, low-grade deposits, many of which would have a positive NPV and a negative real options value.

This last point is very important, as there have long been complaints that there is too much long-term supply in the industry, forcing prices down. Real options is questioned in part because it is often seen as promoting the development of low-grade projects due to the flexibility to manage downside, whereas these projects would not be developed under traditional NPV analysis that ignores such conditional asset management. In fact, real options analysis would probably result in fewer marginal mines making it through the evaluation process. Moreover, in point 5 previously, it can be seen that NPV increases current supply from existing mines compared with supply that would be warranted under a real options analysis. NPV analysis is in many ways an aggressive approach to mine and project valuation, and real options analysis would curb that aggression, promoting resource conservation and efficient extraction, in line with the goals of sustainability.

What Can Real Options Analysis Be Used For?

Real options is a decision support tool that assists managers in making the optimal (value-maximizing) decision from a range of possible actions given an uncertain future state of the world. Therefore, it can be used any time there are alternative courses of action, including design decisions and timing decisions (when to undertake a given action) that require forward projections in the face of uncertainty. In an innovative application, real options has been used to compute the optimal time to get married and the optimal time to get divorced! There is no requirement that the risks being analyzed be transparently traded in the market, though the parameterization of the models is certainly easier when the risks are traded.

Real options is also a cash flow valuation tool and, in this sense, serves the same applications as traditional NPV analysis, where the optimal timing of a decision is "now or never" rather than "now or later."[33] Once again, the main difference between the two approaches is their calculation of the risk of a project, inclusive of any risk-mitigating actions that management will take once the project is underway.

What Are the Next Steps in Project Evaluation?

Real options analysis is now a pervasive component of academic work on mine design, management, and valuation. Because of the small numbers of experts in this area, the research program to date has been limited. Economists have focused on commercial risk, and there is the need to start to incorporate technical risks. Applications are slowly gaining complexity, but there is much work to be done before they can be reliably used for anything other than general inferences about project value and optimal project management. It is this author's opinion that the existing technology, traditional NPV analysis, is not reliable either, and in fact is less reliable than real options analysis even in its current state. Mine managers and owners do not use NPV analysis for anything other than broad qualitative guidance. In that respect, real options analysis, if the skill sets were available to implement it today, would be more informative (less "wrong").

Mining companies typically place mining engineers in mine design and valuation positions, and for this reason these analyses tend to be long on technical analysis and short on financial analysis. Surveys in the 1990s indicate that between 20% and 40% of North American mining companies still do not use NPV in their project evaluations. Only 10% of those surveyed use real options or Monte Carlo analysis. Of those firms who do use NPV analysis, 40% use a "subjective" risk-adjusted discount rate. To this author, an economist initially trained as an engineer, this is akin to using a subjective grade rather than calculating it from drilling data via a geostatistical package. Many companies use the same corporate-wide discount rate for all project evaluations, which seems like using the same average corporate-wide metal grade for all project valuations.

It took more than 50 years for NPV to replace payback period and accounting rate of return as the main project evaluation tools. Real options analysis is now in its third decade. Within 10 years the major mining companies could be devoting a specially trained staff to the use of real options on the most important 20% of their project decisions. These staff will mostly be trained in finance and economics departments, with master's degrees or doctorates in financial economics. This will give the major companies an additional advantage over the smaller companies; they will be able to manage assets optimally, and as a result they will derive more value from any mining asset than the small companies. They will be leaders in sustainability. Small companies will regress to the status of junior exploration firms, bringing prefeasibility projects to the large companies who can design and manage them efficiently. Mine planning software and much of the operations research applications now found in mining will come to incorporate real options ideas. Work on this has already begun at the Colorado School of Mines, McGill University in Montreal, and the Queensland University of Technology in Brisbane.

Courtesy of Cobre Las Cruces, S.A.

FIGURE 8.5   Las Cruces project location

CASE STUDY: LAS CRUCES AQUIFER PROTECTION SYSTEM*

Cobre Las Cruces S.A. (CLC) is developing a base metal sulfide orebody at the Las Cruces site (Figure 8.5), located about 15 km northwest of Seville, in Andalusia, Spain. The Las Cruces orebody is hosted by Upper Devonian–Lower Carboniferous shales and volcanics. The host rocks are unconformably overlain by Tertiary sandstones, which range from 0 to 30 m in thickness. The sandstones are in turn overlain by a massive marl unit that is 100–140 m thick over the orebody.

The orebody lies at a depth of 160–240 m and consists of chalcocite replacement in the primary massive pyrite body and in the associated stockwork. Current reserves are estimated at 17.5 Mt at 6.2% Cu.

The mining is by conventional open-pit mining methods, using a mining equipment fleet comprised of shovels and dump trucks. Mining of the overburden started in 2006 using a contractor and has now reached the –95 m level, a depth of approximately 130 m. Mining has commenced centered on the southwest corner of the deposit, and the pit will be expanded to the northeast in a series of phases. This approach will allow the southwest side of the pit to be backfilled and covered with marl material mined from the northeastern end of the pit. Ore mining will be completed after 15 years (2022) and partial backfilling of the open pit with marl will be completed 1 year later.

The ore produced will be processed on site using a ferric/acid leach system and a solvent extraction and electrowinning plant to produce copper cathodes. Production is scheduled to start in 2008 at a planned rate of 72,000 t of cathode per year.

After completion of mining and the partial backfilling of the open pit to seal off the aquifer, leaving an open pit with walls entirely of marl, the closure plan calls for the site to be converted into a disposal site for inert construction waste.

The Tertiary sandstones mentioned previously form a regional aquifer, which is considered an important supply or reserve of water. Protecting this aquifer during and after mining has been a major consideration throughout the project and has led to the design of the aquifer protection system known as the drainage reinjection system (DRS).

---

* This section was written by M. G. Doyle.

Drainage Reinjection System

In order to understand the purpose and the design of the DRS, it is important to understand how the regional aquifer functions, how the orebody and the area affected by mining interact with the aquifer, and how the mining project will alter this.

Niebla-Posadas Aquifer

The climate of the region is Mediterranean semiarid (i.e., hot and dry during the summer months of May to September, and warm and wet during the winter months of October to April). Average annual rainfall estimated for Las Cruces is 550 mm. Also characteristic of the climate are periodic and prolonged droughts; it is particularly because of these that the Niebla-Posadas (NP) aquifer is considered an important source of water. The aquifer is used both for irrigation and municipal water supplies. Important areas of groundwater use occur around the villages of Gerena and Guillena.

In hydrogeological terms, the Tertiary marl has a very low hydraulic conductivity and acts as a confining layer to the NP aquifer and the Paleozoic rocks. The properties of the NP aquifer vary significantly, but it is regionally extensive. It is generally 5–15 m in thickness but locally reaches 30 m in thickness, while in other areas (including directly above the deposit) it is absent. The aquifer is formed from thinly interbedded sands, sandstones, conglomerate, and vuggy, fossil-rich limestones.

The aquifer crops out approximately 4–5 km north of Las Cruces in an east-northeast trending zone. In the outcrop area, the aquifer is unconfined and recharge occurs through direct infiltration of incident rainfall and through seepage from local streams.

In the past, the main discharge point was the lowest point of outcrop of the aquifer, which is in the valley of the Ribera de Huelva to the northeast. The flow direction in the vicinity of the project was west to east.

Piezometric levels are usually within 10–20 m of the surface, though levels can vary seasonally by as much as 20 m and some areas are normally artesian. During a drought in the mid-1990s, the aquifer water levels dropped by up to 50 m but in the following 10 years recovered to previous levels.

The Palaeozoics normally have low permeability so that water movement regionally is principally along a thin, subhorizontal layer. However, in the area of the proposed open pit, the picture is far more complex. The orebody and the massive pyrite are both highly permeable because of extensive fracturing. The overlying gossan is also highly permeable, containing numerous fractures and vugs, and is also very porous. Also, in the area around the orebody, weathering and acid leaching associated with the formation of the gossan has locally increased the permeability. Pumping tests have shown that there is a strong hydraulic connection between the orebody/gossan/leached area and the NP aquifer.

The DRS was mainly designed to maintain the existing water levels. However, the existing water quality (as measured as salinity/electrical conductivity) being used by third-party wells varies from 400 µS to the northwest of the project (close to the recharge zone) to >4,000 µS to the east (due to a combination of regional flow patterns and pumping). In the area of the mine, the water quality changes from 600 µS on the northwest and north side to 2500 µS in the south and southeast.

Mining Aquifer Interaction

Dewatering of the Las Cruces orebody in advance of mining is required for several reasons:

- Dewatering the aquifer helps to underdrain the overlying marls, resulting in a decrease in the pore pressure levels in the marl. This has a significant effect on the slope stability.

- Part of the aquifer is formed of unconsolidated sands. Any water flowing out of the pit walls in this material will destabilize the sandy layers and then the overlying material.

- Any water not intercepted before the center of the pit will become contaminated because of the presence of large amounts of pyrite. This water would require expensive treatment before it could be released to the environment.

For these reasons, the aquifer piezometric level in the area of the open pit needs to be decreased by approximately 200 m.

To help with understanding the possible impacts of the dewatering required on the aquifer, a numerical model was constructed that represents all of the important properties of the aquifer system. The modeling package employed for this exercise was the U.S. Geological Survey three-dimensional groundwater flow code MODFLOW, and the system has been simulated using seven layers. All data available from the hydrogeological investigations has been employed to calibrate the parameters assigned to each layer, and the model has been used to predict potential impacts arising from the various mining scenarios.

From the model, the rates of groundwater extraction required to maintain the groundwater level below the pit floor were estimated at between 100 and 150 L/s. However, simply dewatering the area around the mining project would have two major problems:

- Modeling and test work shows that simply dewatering the aquifer would result in a very large depression cone (well over 10 km in radius), which would result in significant adverse effects on other water users in the area.

- The aquifer is classed as overexploited and, therefore, further permits for extraction are not normally granted. It is highly unlikely that permits would have been granted for the amount of extraction required.

Various options to deal with these problems were examined and the DRS was the option chosen to both resolve the mine requirements and prevent impacts on third parties.

DRS Design Basics

The DRS (Figure 8.6) is designed to lower the water levels to that of the open pit mine and to prevent significant negative impacts on surrounding users by reinjecting the extracted water. In order to minimize the amount of water lost, it is important to intercept the maximum amount of clean water before it reaches the areas of mining where sulfides are present.

The principal means of removing groundwater from the area of the open pit is the peripheral extraction wells, drilled from the ground surface and located outside the pit rim. The principal advantage of these wells is that groundwater flowing towards the piezometric depression caused by the pit is removed before it comes into contact with the sulfide-bearing rocks associated with the orebody. The potential for quality derogation of extracted groundwater is therefore minimized and the groundwater remains acceptable for injection. All of the extraction wells are screened within the Tertiary sandstone. Some of the extraction wells are also extended into the Paleozoic rocks. There are approximately 30 of these wells.

Because some of the water manages to bypass this ring, and to lower the water levels further, a second ring of wells exists within the pit area.

The water extracted is injected into a ring of approximately 30 injection wells placed 3–6 km from the edge of the open pit. The water is transported from the extraction wells to the injection wells by eight radial pipelines, known as sectors. The flow in each of the sectors is sealed off from

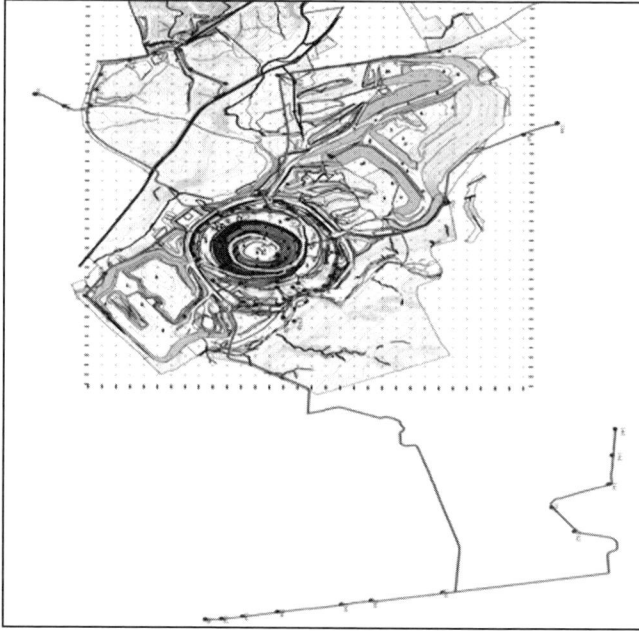

Courtesy of Cobre Las Cruces, S.A.

FIGURE 8.6   Layout of the drainage reinjection system

contact with the atmosphere or other material, minimizing the potential for changes in the chemistry of the water, which could result in clogging of the injection wells. The extraction, transport, and injection are all carried out by the submersible pumps in the extraction wells.

Part of the reason for the individual sectors is the change in groundwater quality across the project site mentioned previously. Having separate sectors avoids injecting poorer quality water from the southeast into the higher quality water areas to the northwest. The other reason for separate sectors is the presence of infrastructure, such as rivers and motorways.

Effect of DRS on Piezometric Level

While the DRS will produce a major drawdown cone at the project site, the presence of the injection wells, with no significant net extraction of water, means that the water levels in surrounding areas are only slightly affected. In general, the surrounding water wells will experience rises in the piezometric level of 1–20 m, depending on how close they are to the injection areas. Figure 8.7 shows the predicted piezometric levels for the end of 2008.

Effect of DRS on Quality

Figure 8.8 shows the existing distribution of chlorides in the aquifer water and Figure 8.9 shows the distribution at the end of mining.

The largest change in quality is to the east of the project, which was due to the distribution of the existing water quality and to the distribution of the extraction and injection wells. The DRS will be injecting water of approximately 2,000 µS in a zone currently running approximately 4,000 µS. This improvement will allow much greater use of the water for cultivation than presently exists.

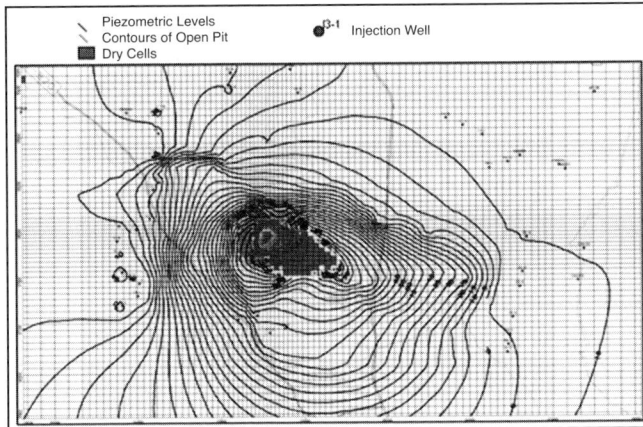

Courtesy of Cobre Las Cruces, S.A.

FIGURE 8.7   Predicted piezometric levels, end of 2008

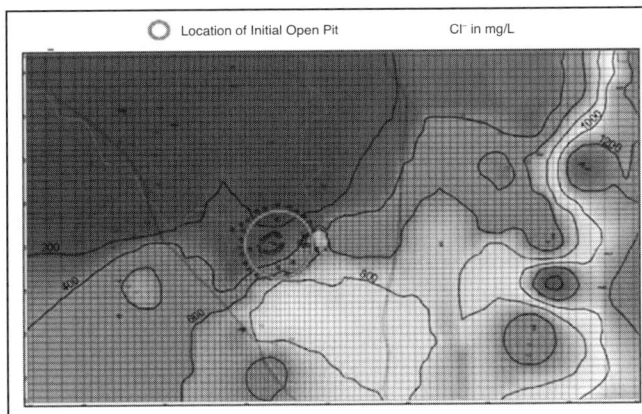

Courtesy of Cobre Las Cruces, S.A.

FIGURE 8.8   Initial chloride distribution in the aquifer

Consultation Process

The DRS forms part of a major mining project for which the field investigations, feasibility studies, and the permitting process took several years. During this process, there was extensive consultation with numerous stakeholder groups. The two major contact points were through the local municipal authorities and the regional authorities.

Local Authorities and Local Populace

The local village of Gerena uses the NP aquifer for the potable supply and numerous small landowners use it as a source of irrigation water. When exploration began in this area, southern Spain was in the middle of a major drought; therefore, the local population and the municipal council have been interested in CLC's activities and the impact that they would have on the aquifer from the time the very first exploration holes were drilled.

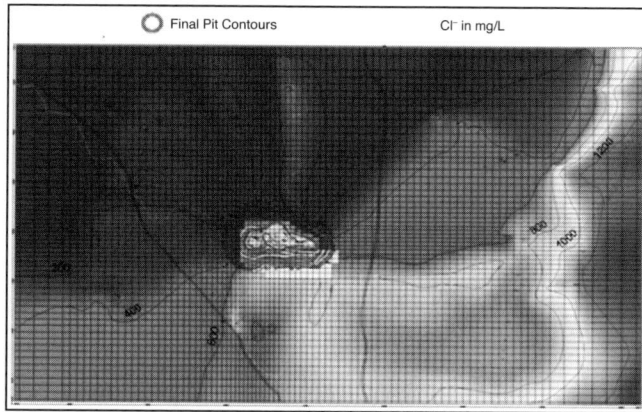

Courtesy of Cobre Las Cruces, S.A.

FIGURE 8.9   Predicted chloride distribution in the aquifer at closure

Over the years, CLC has held numerous meetings with the local populace and local authorities to explain the project. The protection of the aquifer has generally been one of the main points throughout these contacts. Except for a local activist group, the response has been very good, with the project being viewed positively by most of the local people and by the local authorities. It is believed that this positive relationship between CLC and the hosting communities is due to the following factors:

- The company has invested heavily in understanding the aquifer and other environmental issues. For example, dozens of wells and piezometers have been constructed and long-term pumping and injection tests carried out. The extent of the investigations over the years clearly shows how seriously the company has taken this and other environmental issues.

- The company has a fairly open policy and will explain the project to anybody with a legitimate interest.

- The company has held numerous meetings, conducted interviews with local radio and television programs, and released numerous newsletters.

Several other factors that have been vital to building local trust are somewhat uncommon on major projects but have been fundamental to improving community relations:

- The company has several employees from the local villages, one at the director level, and many have been employed by the company since the project started. Other members of the local population constantly approach these people when they want information about the project; these employees have been a (if not the) major interaction point with the community.

- The main company representative and contact point has been the same person since the start of the project. This has made it possible to build trust between the local authorities/population and the company.

- For much of the life of the project, the main company office has been in one of the local villages. This not only makes contact with the population/local officials easier but also contributes to the local economy and makes the project more "local" rather than something being done by a company in the capital city.

Central Authorities

The permits for the DRS are controlled by the Confederacion Hidrografica del Guadalquivir (CHG). Unlike most of the authorities with which the project deals, the CHG is a branch of the central government run directly from Madrid, rather than part of the Andalusian regional government. This is because most of the large river drainage basins cut across more than one of the semiautonomous regions.

A long consultation process took place with the water authority. This eventually led to the permits to operate the DRS being issued in 2003. The permit conditions under which the DRS must be operated are extremely tight, with little room to carry out operational adjustments to the system. Therefore, there is an ongoing process of negotiation and consultation with the water authority.

Performance Targets

No specific economic and environmental targets were formally established for the DRS construction or operation. However, the system was expected to comply with the strict environmental guidelines and policies of CLC. The economic target was simply to construct the DRS within budget.

Financial

During construction, several minor changes were made to the system, mainly to reduce construction costs or to speed up construction. The net result was that the system was constructed for approximately 10% less than the budgeted amount.

Operational costs are, however, slightly higher than expected because of the need for greater maintenance of the pumps than anticipated. Short circuits in the 30 submersible pumps or their cables require constant repair or replacement, with one team dedicating at least 50% of its time to this task.

Environmental

Operation is also subject to one of the conditions under which the permit was granted—that there be no net extraction from the aquifer. The system is being gradually modified to decrease water losses (from automatic filter cleaners and leaks).

Staff

The DRS would be progressively built and then operated for a period of more than 20 years. Because there were no similar facilities in operation in southern Spain, the decision was made to select staff who were predominantly local and then train them to operate the system.

Staff Selection

Three different types of staff members were required:

- Geologists/mining engineers: These staff members were chosen to supervise the construction of wells and then the operation of the system. Recently qualified people were selected and then trained.

- Mechanics/electricians: These employees were available locally; although it was unlikely they would have the exact kind of experience required, this was not considered a problem.

- Field technicians: Generally, lower-skilled staff members carry out routine maintenance and repairs and collect readings manually.

The geologists/mining engineers selected were from various areas of Spain because none were available from the local villages. All other personnel were found in the local villages.

It should be noted that the fact that most of the staff running the day-to-day operations of the system are local people is extremely important in giving the local communities a sense of confidence that nothing untoward is being done, that water is not being extracted from the aquifer, and that the aquifer is not being contaminated by the mining activities.

Staff Training

Very little training of staff was required directly as a result of constructing the DRS. The geologists were given basic training to cover the core logging required for drilling the wells; after construction, two geologists were given a course on groundwater hydrology. Fortunately, a good-quality postgraduate diploma course on groundwater exists in Spain, run in conjunction with the University of Catalunya. Apart from the exams, the course is presented online and provides most of the necessary theoretical grounding in hydrogeology. By being present during the construction and startup, staff also gradually built up a substantial amount of practical experience.

Significant training was required because the local people who would be operating and maintaining the system would be placed in the center of open-pit mining operations, with all the attendant dangers of this environment. The training necessary as a result of these hazards has varied from simply making workers aware of the risks posed by the large trucks and other machinery to specific courses on the risks of working on the open-pit benches.

A large amount of basic safety training is also being provided. The accident statistics for Spain (mining and construction industry) are very poor and there is a generally lax attitude about safety; for example, it is very common to see construction workers on roofs who aren't wearing safety harnesses or basic protective equipment such as helmets and boots. The general situation in Spain is improving rapidly. Because of the lack of a safe working culture, a significant effort is being made to improve worker health and safety awareness during operations.

Risk Management

During the project design, several third-party consultants undertook an evaluation of the technical risks, mainly focusing on possible operating problems and closure plans. In the following paragraphs, the main risk factors are analyzed.

Technical Risks

The evaluation focused mainly on possible problems with injecting the water due to well clogging and the possibility that the flows would be much larger than estimated.

To investigate the possibility of clogging, several tests were carried out. Actual extraction and injection wells that would form part of the final system were constructed and put into operation for tests lasting up to 4 months. Several unexpected problems were encountered, but no signs of any clogging occurred. The system has now been operating for almost 2 years and there are still no signs of clogging. This is mainly due to the simple chemistry of the water and the fact that the system is sealed.

If the flows were much larger, then more extraction wells and more injection wells would be required and possibly also larger pipelines. Adding on wells as required is simple as they are essentially individual units. Laying new pipelines, however, would be a major issue and cost. Therefore, the pipelines were designed to take approximately twice the estimated flow rate at 1 m/s velocity. This means that the existing pipelines could deal with significantly greater flows than they do now. In hindsight, it appears that one area of the risk analysis was poorly carried out.

Because of the reinjection of the extracted water, the piezometric level in the injection areas would rise and locally this could cause some irrigation wells to become artesian; in fact, some of the areas are commonly artesian. This was considered to be a positive impact because the farmers would have more water and would have to pump less, so little else was undertaken on this theme.

However, it turned out that, in some areas, the farmers do not consider an artesian well a plus. In two of the injection areas, the topography is very flat (old floodplain) and is drained by a series of ditches that connect to a nearby river. The ditches drain very slowly but are normally totally dry for at least 5 to 6 months of the year. With artesian flows of 1–2 L/s, two effects were found. First, the flow tends to pond up in certain areas and then spread out laterally through the soil. This resulted in the small areas (tens of square meters) being unworkable by farm machinery. The second, more important effect is that with water present all year round, the ditches rapidly started to generate thick vegetation. This vegetation worsened the already poor drainage of rainwater, affecting significant areas, and required cleaning out of the ditches every few months instead of every few years. This is obviously a cost the farmers do not want and the company is undertaking this work; nevertheless, it is still a disturbance to the landowners' normal operating procedures.

Although it would have been difficult to avoid the artesian flows, had the farmers been consulted about the "positive" impacts as well as the negative ones, more preparation could have been done to anticipate the negative responses these flows produced.

Closure

During the project design and permitting process, the closure plan for the general project was also developed. The initial plan, after consultation with stakeholders, was to convert the open pit into a large freshwater lake after backfilling to line the pit totally with marl. The surrounding 9 ha would be revegetated and converted into a nature reserve. The pit lake part of this plan was rejected because of the risk of eutrophication (increase in chemical nutrients) problems due to the nitrate/phosphate levels in the local rivers. The closure plan for the main project now calls for the pit to be used as a waste disposal site for inert construction waste, using this to backfill the entire pit. The remaining area will still be converted into a nature reserve.

In terms of the closure of the DRS, the aim is to allow the water levels to recover slowly by progressively decreasing the pumping rate, and to use a pump-and-treat method to deal with contaminated water. The only group that has been interested in the details of this plan is the regional authorities. Work and consultation with them is an ongoing process to define the precise details of the closure. Test work during the mining operation will also help to define these details.

Sustainability

For sustainability, there are two fundamental parameters: the volume of resource available from the aquifer and its quality.

Quantity

When the DRS was put into operation, the aquifer was artesian in some areas; in others, the water level was fairly close to the surface. The aquifer was basically full, having recovered from the strong drawdown that occurred during the last drought, which ended in the winter of 1995–1996.

It would be possible to inject some of the water into areas that are normally dry, but these are dry because any water there either leaves via springs or simply flows down to the areas that are currently artesian. It is unlikely that injection in these areas would achieve much in terms of increasing the volume of the reserves.

Quality

The water quality within the area of the DRS varies from near drinking quality in the northwest to saline in the northeast (500–1,000 mg/L Cl⁻) where the aquifer is artesian at this time. The DRS could be configured to flush some of the more saline water out of the system. Modeling indicates that, even as designed (without taking this opportunity into account), the DRS improves the water quality in the area where the water is of worst quality and where the water is only used for irrigation as a last resort. In theory, it would be possible to coordinate extraction from the poorest quality wells with injection to maximize the volume of poor quality water that was replaced with better quality. However, because of the legislation covering extraction of water, this would be very difficult if not impossible to achieve.

NOTES

1. D. J. Shields and S. V. Solar, *Sustainable Mineral Resource Management and Indicators: Case Study Slovenia* (Ljubljana: Geological Survey of Slovenia, 2004).

2. *Framework for Responsible Mining: A Guide to Evolving Standards*, 2005, www.frameworkforresponsiblemining.org (Accessed: February 2008).

3. Pierre Lassonde, "How to Earn Your Social License," *Mining Review* (Summer 2003): 7–13.

4. International Institute for Sustainable Development, *Seven Questions to Sustainability: How to Assess the Contributions of Mining and Mineral Activities* (Winnipeg: International Institute for Sustainable Development, 2002), www.iied.org/mmsd/mmsdpdfs/145_mmsdnamerica.pdf (accessed February 8, 2008).

5. ESMAP, the World Bank, and ICMM, *10 Principles*, www.icmm.com/our-work/sustainable-development-framework/10-principles, 2008 (Accessed: February 2008).

6. Rio Tinto, Sustainable Development/Social, www.riotinto.com/investors/7134_social.asp, 2008.

7. ESMAP, the World Bank, and ICMM, *10 Principles*.

8. The Equator Principles, *A Financial Industry Benchmark for Determining, Assessing And Managing Social and Environmental Risk in Project Financing*, www.equator-principles.com (Accessed: February 7, 2008).

9. International Finance Corporation, *Performance Standards on Social and Environmental Sustainability*, April 30, 2006, www.ifc.org/ifcext/sustainability.nsf/Content/PerformanceStandards (Accessed: November 15, 2008).

10. Jared Diamond, *Collapse* (London: Penguin Books Ltd., 2005).

11. Tracey Khanna, ed., *Mine Closure and Sustainable Development: Results of the Workshop Organised by the World Bank Group and Metal Mining Agency of Japan*, (London: Mining Journal Books Ltd., 2000).

12. R. Edward Freeman, *Strategic Management: A Stakeholder Approach* (New York: HarperCollins Canada, 1986).

13. World Bank Group, "Identifying Stakeholders," in *The World Bank Participation Sourcebook: Chapter III: Practice Pointers in Participatory Planning and Decisionmaking*, www.worldbank.org/wbi/sourcebook/sb0302t.htm.

14. R. T. Jackson, "Mine Closure: An Introduction," in *Mine Closure and Sustainable Development: Results of the Workshop Organised by the World Bank Group and Metal Mining Agency of Japan* (London: Mining Journal Books Ltd., 2000).

15. *Breaking New Ground: Mining and Minerals Sustainable Development* (Johannesburg World Summit: MMSD Report, 2002), www.iied.org/mmsd/.

16. Anglo American, *Socio-Economic Assessment Toolbox*, 2007, www.angloamerican.co.uk/aa/development/society/engagement/seat/.

17. Lassonde, "How to Earn your Social License."

18. J. Mawson, "Art of Communication," *Mining Journal* (April 2004): 18–20.

19. Gregory M. Anderson and Robert L. Lorber, Safety 24/7: *Building an Incident-Free Culture* (Results in Learning, 2006).

20. Rio Tinto, "The Way We Work: Our Statement of Business Practice" Rio Tinto plc (Rio Tinto Limited, 1998).

21. J. Cordes, "Normative and Philosophical Perspectives on Sustainable Development," in *Sustainable Development and the Future of Mineral Investment*, J. M. Otto and J. Cordes, eds. (Paris: United Nations Environment Programme, 2000): 1-1–1-53.

22. R. Eggert, "Sustainable Development and the Minerals Industry," in *Sustainable Development and the Future of Mineral Investment*, J. M. Otto and J. Cordes, eds. (Paris: United Nations Environment Programme, 2000): 2-1–2-15.

23. R. D. Cairns and G. A. Davis, "Strike When the Force Is with You: Optimal Stopping with Application to Resource Equilibria," *American Journal of Agricultural Economics* 89, no. 2 (2006): 461–472.

24. T. F. Torries, *Evaluating Mineral Projects: Applications and Misconceptions* (Littleton, CO: Society for Mining, Metallurgy, and Exploration Inc., 1998).

25. G. A. Davis, "Project Assessment Methodologies and Measures: The Contribution of Mining Projects to Sustainable Development," in *Sustainable Development and the Future of Mineral Investment*, J. M. Otto and J. Cordes, eds. (Paris: United Nations Environment Programme, 2000): 7-1–7-31.

26. M. R. Samis, G. A. Davis, D. G. Laughton, and R. Poulin, "Valuing Uncertain Asset Cash Flows When There Are No Options: A Real Options Approach," *Resources Policy* 30 (2006): 285–298.

27. Ibid.

28. O. A. Tourinho, "The Valuation of Reserves of Natural Resources: An Option Pricing Approach" (PhD dissertation, University of California, Berkeley, 1979).

29. M. J. Brennan and E. S. Schwartz, "Evaluating Natural Resource Investments," *Journal of Business* 58 (1985): 135–57.

30. D. G. Laughton, *The Banff Taxonomy of Asset Valuation Methods: Lessons from Financial Markets for Real Asset Valuation in the Upstream Petroleum Industry* (A paper presented to the SPE [Society of Petroleum Engineers] workshop, September 15–17, 2003, Banff, Canada) www.davidlaughtonconsulting.ca/docs/banff_taxonomy.pdf (Accessed: August 2, 2007).

31. M. R Samis, G. A. Davis, and D. G. Laughton, "Using Stochastic Discounted Cash Flow and Real Option Monte Carlo Simulation to Analyze the Impacts of Contingent Taxes on Mining Projects," in *Proceedings, Project Evaluation Conference* (Melbourne, Australia: Australasian Institute of Mining and Metallurgy, June 19–20, 2007).

32. Laughton, *The Banff Taxonomy.*

33. Samis, "Using Stochastic Discounted Cash Flow."

CHAPTER NINE

Mine Planning and Production Management

N. Mojtabai

INTRODUCTION

This chapter gives an overview of units of operations in mining that play an important role in the success of the project and, therefore, in meeting the production requirements at an optimum point, while minimizing the undesirable side effects of mining operations on the environment and the communities. The sustainability of the operations depends on the economic success and protection of the benefits of the company, the communities, and stakeholders. Although the topics that are covered in this chapter are technical, there is no focus on the design or technical aspects of these units of operations. The intended goal is to draw managers' attention to how these operational components could result in serious problems such as environmental impacts, damage to property, harm to the health and safety of both employees and residents of nearby communities, and economic outcomes that can result in early or premature closure of the mining operation.

Mining is an interdisciplinary operation that consists of many different units of operation. These units of operation are all integrated into a single system that runs the mine. Some of these units or sections of the mine might be working together, either in parallel or in series. Although each unit operates as a single entity, it is interconnected with the rest of the operation. All units of operations rely on each other's performance and any lack of efficiency or productivity will carry across the entire operation. It is the management's responsibility to make certain all components at the mine operate in harmony at their peak productivity.

This chapter has 12 sections.

"Mine Planning and Grade Control" starts with the discussion of long-term and short-term planning, and ore grade control. A brief introduction to tools and techniques that are used in planning and grade control follows.

"Rock Fragmentation by Blasting" concentrates on rock fragmentation and drilling and blasting operations. The emphasis is on safety and control of adverse effects of blasting on the surrounding area and neighboring residences. Important parameters that control the outcome of a successful blasting operation are discussed. Specific issues and parameters related to safety and adverse effects are summarized as well.

"Loading and Hauling" provides an overview of important aspects and parameters that must be taken into consideration when designing and planning the material-handling systems. The focus is on how the loading and hauling equipment should be integrated into the mining operation. There are no discussions or details on specific types of equipment. It is assumed that the management and those in charge are familiar with all material-handling systems.

"Ground Control" focuses on the importance of ground control as part of a successful and safe mining operation. Issues related to ground control are summarized. Again, no theoretical background on this subject is given. The goal is to bring the importance of ground control to management's attention.

"Mineral Processing" gives an overview of mineral processing and all the steps involved in the complete processing cycle. Various techniques are introduced with a limited amount of technical and theoretical discussion. The environmental issues related to mineral processing are discussed.

"Leaching" covers issues that have been of major concern with leaching operations. Leaching is becoming a very popular and economically attractive alternative to milling and processing. However, it does present some environmental challenges, which are discussed in this section.

"Mined Rock and Tailings Management" concentrates on the problems associated with the large amounts of rock and tailings produced by mining operations. These products of the mining operation have no value and must be handled and placed on surface. The volume of tailings and mined rock has been consistently increasing as mining operations become larger. Managing these materials is very challenging and mistakes can have catastrophic outcomes.

"Reclamation and Closure" discusses the reclamation and closure processes, important aspects of mining operations that require special attention, planning, and managerial control. The importance of reclamation and closure related to public health and safety is discussed.

"Maintenance Management" focuses on the importance of maintenance to the success and continuity of the operations. Safety and environmental issues related to responsible maintenance are discussed.

"Case Study: Grade Control Systems at El Valle-Boinás Mine" illustrates the importance of ore grade control as part of planning and managing a mining operation.

"Case Study: Reliability Assessment of a Conveying System at Atlantic Copper " is related to processing and maintenance. This section describes a methodology for improving maintenance practices based on the application of reliability-centered maintenance and mathematical modeling for the conveyor belt system at Atlantic Copper.

"Case Study: Overview of the Aznalcóllar Tailings Dam Failure" is related to a classic problem with mine tailings dam failures. It reviews the lessons that can be learned from such events.

MINE PLANNING AND GRADE CONTROL*

Short- and long-term production planning is a primary influencing factor for sustainability in mining operations. A sustainable mining plan should minimize the effects to the mining business process caused by variability in commodity pricing, inflation of mine costs, declining ore grade, and declining mining conditions such as increased depth of cover. Design factors with the capacity to increase sustainability include layout, application of new technology, mine infrastructure, proactive communications, and interfacing with land, water, socioeconomic forces, and postmine reclamation. These planning considerations will be examined in the following paragraphs for surface and underground mines.

The main function of the mine design and planning team is to provide the necessary technical support to ensure the optimum operational efficiency and sustainability of mining operations. Although the long-term mine planning process determines the production and economic objectives, the short-term planning and control processes are key to the efficient use of resources, performance evaluation and control, and the timely implementation of corrective actions.

This section aims to describe the main mine planning functions and processes and highlights the management aspects that are key to sustainability.

---

* This section was written by C. Castañon and J. A. Botin.

Long-Term Planning

Long-term production planning and scheduling is concerned with the development of an optimum mining sequence for ore and waste required to sustain production. The production schedule takes into account operating constraints, blending requirements, economic considerations, reclamation, and other operational and sustainability constraints.

Long-term mine planning includes a variety of specific functions:

- Detailed definition of minable ore reserves in sufficient quantity to sustain production
- Detailed engineering and layout of mining blocks/stopes, optimum mining sequences, and production schedules to serve as the basis for 1-year and 5-year operating plans
- Evaluation of performance, productivity, and costs required for the preparation of the 1-year plan
- Evaluation and revision of dilution and cut-off grades used in long-term planning
- Advice and data management role for the preparation and control of the annual operations plan and budget
- Preparation and annual revision of 2-year and 5-year plans
- Update of planning, design, and scheduling of waste dumps and tailings dam
- Preparation, updating, and control of environmental management and mine closure plans
- Ongoing equipment replacement analysis and economic evaluation work
- Reporting functions on all the preceding functions
- Maintenance of the database management systems, the general mining packages, and other mine planning tools used in the planning process

Short-Term Planning

Short-term planning is concerned with schedules on a daily, weekly, or monthly basis, as well as grade control and mine geology functions. The goal is to furnish the requirements of the operating plant with ore of uniform quality to ensure its operating efficiency. The short-term plan has to comply with the long-term plan, take into account equipment availability, and accommodate blending requirements.

Daily, weekly, and monthly short-term plans are required at a mining operation for sustainability. These plans are derived from the yearly plan. A robust short-term planning process with commensurate follow-up provides the feedback that planners need to correct long-range plans. Short-term planning includes a number of main functions:

- Shift/day plans: First-line supervisors plan and mine at specific locations that maintain continuity, and identify and mitigate on-shift risk.
- Weekly plans: General supervisors plan down shifts, construction, and spoil placement, as well as design updates and forecast weekly production.
- Monthly plans and weekly updates: Superintendent- or manager-level personnel update monthly forecasts to production and costs.
- Quarterly plans and monthly updates: A rolling plan adds a new monthly plan to replace the month just completed.
- One-year plan and quarterly updates: The 1-year plan is the basis for the annual operations plan and budget. The plan is normally updated quarterly.

- Ore grade control and mine geology functions: This is a shift-by-shift definition of ore–waste boundaries, ore grade forecast, loading faces, waste dumping points, mill through-put forecast, and so forth.

- Mine surveying (mine, stockpiles, dam, etc.): This is the daily staking of ore–waste boundaries and blastholes, ongoing toe-crest surveying, and so forth.

- Blast engineering and design: This includes blast engineering and design, fragmentation studies, blasting performance studies, vibration control, and so forth.

- Ground control

- Ore stockpile control

- Mine–mill grade/tonnage reconciliation

- Performance and time studies

- Reporting functions on all of the preceding functions

Ore Grade Control Function

Grade control refers to a decision-making process in which the classification of a block as ore or waste is revised on a day-to-day basis. This is done by ore graders and mine geologists through blasthole data processing, visual observation of faces, and operational considerations.

Concept of Grade Control

The grade control process comprises a number of data processing, studies, and management activities (supervision, meetings, and coordination) aiming to take the final decision on the destination of production units (e.g., a haulage truck, scooptram), to the plant or to a specific mine dump. It also involves a number of control functions related to the analysis of deviations between the planned and actual values of grades, dilution, tonnages, and metallurgical recovery, all of which are critical for sustainable management.

Scope of Ore Grade Control Function

The scope of the grade control function is different for each mine and depends on the mining method (open-pit/underground), the type of resource (precious metals, coal, etc.), the size and shape of the orebody, and other factors. Furthermore, for a specific mine operation, the grade control function is a continuing improvement process that evolves with time.

Depending on operational requirements, the grade control function can range between a simple visual delimitation of the ore–waste boundaries to a highly sophisticated process usually required at economically marginal precious metal operations.

Ore grade control focuses on the following issues:

- Recalculation of minable reserves. Often, the tonnage, grades, and spatial distribution of the ore reserves reported at the feasibility stage lack the precision and detail required for medium- and short-term planning.

- Definition of ore–waste boundaries at mining faces. This is important and not always easy.

- Day-to-day definition of the minimum size of unit mining blocks. This depends on bench height and the type of ore loading equipment used.

- Day-to-day control of moisture. This is especially difficult in soft or oxidized ore.

- Control of dilution

- Monthly and year-to-date reconciliation of mine, mill, and concentrate sales statistics

- Design of the size and geometry of production blasts, aiming to optimize dilution by selective blasting of materials with different blasting behavior
- Management of mining ore–waste faces and broken ore inventories, aiming to ensure that enough loading places are available and blasthole assays and other relevant information is available on time
- Follow-up and control of bench toe and crest elevation
- Bench height optimization studies. The higher the bench height, the higher the mining productivity but also the higher the dilution.
- Mine planning and design changes in relation to changes in annual production plans. When production objectives increase, it might be necessary to review the design and operating parameters.

In conclusion, it might not be easy to quantify the economic and sustainability returns from grade control, but it can be said that money invested in improving grade control systems will improve the bottom-line results of the company.

Definition of Cut-off Grade

Break-even cut-off grade is defined as the lowest grade of a mining block that can be mined and processed, considering all applicable costs, without incurring a loss or gaining a profit. In calculating the cut-off grade, sustainable management requires that all costs incurred during the entire mining life cycle, including costs associated with environmental and social responsibility actions, be taken into account.

In a specific mining situation, the term *cut-off grade* applies to the grade value used by the mine as the decision criteria for the classification of a mining block as ore or as waste. This value could be equal to or greater than the break-even cut-off grade.

The choice of the cut-off grade influences the profitability and the life of the mine, two key aspects of sustainability. Depending on the planning objectives, various cut-off grade concepts are applicable:

- A long-term planning design cut-off is required for the design of the ultimate pit outline. In this case, the cut-off grade is defined as the lowest grade of a mining block that can pay for its mining and processing, the mining of the incremental waste blocks associated with it, and all other applicable costs.
- An operational—short-term—cut-off grade is used to decide whether a mining block at the face should be sent to the mill or to the mine dumps. In this case, the cut-off grade is defined as the lowest block grade that can pay for the difference in cost between each alternative.

When the cut-off grade is used for preliminary evaluations during the exploration and feasibility stages, where reserves have not been proven in significant quantity, approximate cut-off grades are determined from cost information obtained from similar projects.

Equivalent Grade for Multielement Ores

When the orebody contains more than one element of economic value, the value of the block can be referred to an equivalent grade of the principal element (the element bearing the maximum value).

The concept of *equivalent grade* is frequently applied to resource calculations in polymetallic orebodies, where several elements are present, some payable (Cu, Zn, Pb, etc.) and others penalizable (As, Sb, etc.), where one of the elements bears most of the block value (principal element) and the rest are by-products.

Equivalent grade is then defined as the grade of the principal element that alone would account for the total net value of the block. By using this concept, the economic value of the block is represented by a single grade value.

The equivalent grade is calculated as the sum product of the grades of the elements and the relative unit value factors. Relative unit value factors for each element are estimated as the ratio of the unit value of the element (positive or negative) and the unit value of the principal element. Obviously, the unit value factor of the principal element is the unity.

$$E_1^{eq} = E_1 + E_2\frac{u_2}{u_1} + E_3\frac{u_3}{u_1} + \dots \qquad \text{(EQ 9.1)}$$

The calculation shown in Equation 9.1 should take into account the effects of the metallurgical recovery and other factors that might affect the net unit value of the elements. In fact, any aspects that can have a differential effect on each element must be considered, such as

- The mill feed mix that might be required to optimize the metallurgical process—metal recovery and concentrate grade—of each element;

- The effects on plant performance related to the variation of plant feed grades and different elements;

- The variations in element recovery as a function of its grade in the mill feed ore;

- The proportions in which each element reports to the different concentrate products obtained; and

- The concentrate sales contract of each concentrate product and the penalties applied.

Another approach is to reduce grades of by-products in proportion to the actual payable metal content (payable recovery) after all metal losses, smelter charges, and penalties, thus calculating the metal content equivalent to the income received.

Use of Information Systems and Mine Planning Software

Integrated mine planning software can be used to process and evaluate iterations of geologic interpretation and geostatistical routines efficiently, optimize mine layouts to identify major expansions and mid-life development (such as moving to a new mining district, a major pit pushback, new refuse dumps, and new shafts or access), and essentially evaluate all alternatives and identify the most cost-effective mine plan.

Specific optimization tasks include interrogating the geologic data set with all modules of geostatistics, searching for the best estimating method for ore grade block values. Iterations of the mine layout are done to minimize development time to reach the production phase of mining. Such planning can sustain mining when the mineral price increases, thus lowering cut-off grade. This phenomenon occurs frequently in tabular and lenticular reserves such as coal, trona, potash, and limestone. The planning process can identify the cost of leaving conveyors and infrastructure in place, and when economics would dictate that the equipment be removed. The goal is for the mine planners to become intimately familiar with the deposit—to mine the reserve on paper. Such hands-on familiarity will always lower the mining cost per unit during execution.

Most commercial mine planning packages include mine design modules for open-pit and underground mining. Open-pit design modules use dynamic programming algorithms such as those of Lerchs and Grossman for the optimum design of ultimate pit limits and pit pushback stages, pit smoothing, and road design. Underground mine planning modules use solid modeling as a planning tool, where drifts, raises, ramps, shafts, pillars, and stopes are represented as "solids" that constitute the unit planning blocks.

Integrated software packages are powerful tools for efficient, interactive evaluation of different production plans and schedules and mining scenarios. Though the plans being compared might not necessarily be optimal, the capacity for easy and fast generation and evaluation of many different mining scenarios by experienced planning engineers yields optimum results.

Sustainable Management Issues

Control is an essential management function. Sustainable management should focus on the control of those operational aspects that are most critical for sustainability. Some of these key aspects are discussed in the following paragraphs.

Grade Control Quality Indicators

An overall quality indicator is obtained by comparing tonnage and grade estimates and actual mill feed values. If the difference on a monthly basis is less than 10%, grade control quality might be considered acceptable. Here are some other quality assurance indicators:

- Control of the grade control sample preparation system
- Control of possible systematic errors of the sample assay laboratory by repeating 5% to 10% of the sample assays using an external laboratory and performing statistical analysis of variance and trends on a monthly basis
- Monthly reports on reconciliation of grade control, ore stockpiles, and mill feed

Reconciliation of Mine and Plant Production Statistics

Reconciling mine and plant production records is an important quality assurance process, but in some cases, attempting it causes confrontations between the mine and plant superintendents and produces no practical results. This always happens when management lacks a formal reconciliation standard defining the procedure and the steps to be followed.

The mine–plant reconciliation process aims to reconcile the following three production statistics:

- The year-to-date tonnage and grade of the run-of-mine (ROM) ore estimated by the annual mining plan;
- The actual tonnage and grade fed to the plant as calculated by the grade control process; and
- The year-to-date metallurgical balance, which gives the actual tons and grades treated by the plant, as measured at the mill feed weight meter and online sampling.

The analysis of deviations among the three data sets is an important part of the continuing improvement process. The following is suggested as a general approach to this process:

1. Estimate the volumes of ore mined precisely.
2. Estimate dilution for each bench or mining zone separately.
3. Develop a ROM ore sampling method allowing for a good estimate of ROM tonnage and grade.
4. Install and maintain a reliable online sampling system for the mill, with automatic samplers for mill feed and tailings and, possibly, at the head and tail of each plant process line.
5. Issue a monthly mine–plant reconciliation report, showing deviations and trends.

The most important aspects of the reconciliation process are ore grades and metal content in the mine, the mill, and the concentrates and products sold. In this regard, it is important to consult with experts in the areas of sample preparation and quality control.

Management of Ore Stockpiles

Mining operations use ore stockpiles for flexibility and better control of the process. Here are some examples of ore stock:

- Stocks of ore by grade: Used to optimize mill feed grade by mixing ore from different stockpiles.

- Stocks of ore by grindability: Seldom used but would optimize grinding capacity and increase recovery when grindability of ore is very variable.

- Stocks of ore causing process problems or smelter penalties: Used to blend this ore with normal ore in a controlled proportion, so that the process is not affected negatively, or to keep the undesirable element below the penalty grade.

- Marginal ore dumps: For temporary stock of ore below the operational cut-off, but having the potential to become economic in the foreseeable future.

Here are some important considerations on stock management:

- Ore stockpiles use large land surfaces and might contain reactive elements, which could cause environmental risks. Therefore, the stockpile should be properly designed and engineered to protect the environment from any potential risk. Measures such as impermeable liners and drainage systems are often necessary.

- When reconciling tonnages in large stockpiles, it is recommended to implement the means (topographic surveys and dump statistics) to measure the in-situ density (swell factor) so that precise volume and tonnage calculations can be performed at any time in the future.

- Also consider the progressive oxidation of ore, which will affect metallurgical recovery when the ore is processed.

A more detailed description of the grade control functions and processes is presented in the "Case Study: Grade Control Systems at El Valle-Boinás Mine" section.

Information Management Systems

Sustainable mine planning requires using the capabilities of modern integrated information management systems and mine planning software to optimize the mine plan. This software gives planning engineers the capacity to manage and process information, analyze different mining scenarios, and react to operational changes and take advantage of all possible opportunities. Most commercial mine planning software includes a drill-hole database, block modeling, pit optimization algorithms, long-term planning applications, production scheduling, and grade control.

Orebody block modeling and associated databases are powerful tools for sustainable management because they organize a variety of important types of data:

- Geological data: Stratigraphy and petrology data is necessary for the characterization of different types of waste low-grade material and its allocation to mine waste dumps. It is also important for metallurgical studies, mill management, and so forth.

- Geotechnical data: This includes characterization of the strength of the block rock, rock quality designation, joint type and direction, and other data required for mine design, grade control, ground control, blast design, and vibration control.

- Groundwater parameters: This information includes the water table, permeability rates, and water quality and its evolution during the mine life cycle.

- Land management: Terrain models covering the mine site and hosting communities can be used as a tool for sustainable management throughout the entire mine life cycle. They include land uses, surface water quality control, soil quality control, land reclamation, and so forth.

- Sales contract management: Block databases including the multielement assays of the ore for contaminants, penalty/bonus elements (e.g., As, Sb, Hg, Bi, Se), are useful in the sales planning and control processes for contract management.

- Environmental risk assessment: Block model data is also useful in the advance assessment of potential environmental risks associated with certain contaminants.

- Operational performance data: ROM ore grades and dilution, metallurgical recovery, operating cost and performance, and other operational variables can be correlated to ore type, rock types, geology, structure, and so forth. The storage of these variables in the block model is important for continuous improvement of operations.

Regarding ore grade control, the use of information systems is essential. Production must comply with plant feed requirements, and the grade control function must be performed as quickly and efficiently as required. Therefore, the grade control function must be equipped with the data processing equipment and software systems required for timely definition of ore–waste boundaries.

Acknowledgments

The authors gratefully acknowledge the contributions of Alberto Lavandeira and Andrew Schissler, whose ideas and information assisted us in preparing this section.

ROCK FRAGMENTATION BY BLASTING*

Rock fragmentation is the first stage of production of mining operations. All construction and development stages such as removal of overburden, shaft sinking, excavations of accesses to the orebody (adits, inclines, raises, etc.) require breakage of the rock material into sizes that the next unit of operation can handle. The efficiency and success of other units of operations (e.g., loading and hauling, crushing, and grinding) following fragmentation depend on the effectiveness of rock fragmentation. Fragmentation can be achieved by means of mechanical breakage or the use of explosives.

Rock blasting is the most efficient technique for fragmentation, specifically when large volumes of rock must be fragmented over a short period of time. However, the process of rock fragmentation has several undesirable side effects that require major attention. These issues are discussed in the following sections.

Economics of Rock Fragmentation

Economical and effective rock blasting requires knowledge and experience in terms of design of a blast round. However, strict compliance with safety and security, environmental standards, effects on nearby structures and residences, and induced damage to the remaining land boundaries of the excavation (tunnel perimeters, pit slopes, benches, etc.) add additional complexities to the blasting operations. The blasting engineer and the mine manager must be fully aware of

---

* This section was written by B. Cebrian.

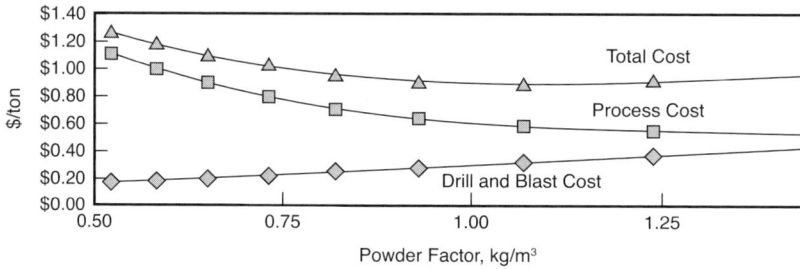

Source: Eloranta & Associates 1999

FIGURE 9.1   Mine-to-mill approach for optimum mining–milling comminution cost

the consequences of these side effects in order to prevent authorities, associations, or groups of individuals from inquiring and attempting to shut down these operations. In other words, blasting operations could become a social problem and will affect and reduce the sustainability of mining operations.

Mechanical excavation has been used extensively in tunneling (tunnel boring machines, soft ground tunneling machines, etc.), shaft sinking, raise boring, and production of soft ores such as coal, potash, salt, and so forth. These excavation tools are very effective and generate a minimum of undesirable side effects. However, rock fragmentation using explosives is still the most effective and economic choice in large operations where mechanical excavation is no longer feasible.

Rock Breakage with Explosives

Drilling and blasting is a mechanical, repetitive operation. Regardless of this, it requires taking a series of key aspects into account, because there can be dramatic short- and long-term effects on safety, economy, production rates, mine stability, and ore recovery as a result of improper blasting techniques and management.

First, blasting must not be considered as an isolated operation in the chain of production in a mine. Because blasting is the first step of excavation, right before the loading-hauling and comminution processes, proper blasting can benefit these units of operation by fragmenting rock efficiently and uniformly within an appropriate size distribution suited to the material-handling equipment. Therefore, an optimum fragment size range must be obtained in order to minimize the handling and crushing costs. The consideration of the incidence of proper blasting in the overall process is known as the mine-to-mill approach and is currently at the leading edge of hard-rock mining management concepts. Figure 9.1 illustrates the unit operating cost of drilling and blasting as well as material handling as a function of powder factor (i.e., explosive weight or energy per unit weight of blasted rock). It must be noted that as the powder factor increases, the resultant fragmentation will increase. This in turn will reduce the cost of loading, hauling, and crushing of the ore, mainly at the primary crusher. However, at the same time, the cost of drilling and blasting at a higher powder factor or degree of fragmentation will increase. As Figure 9.1 shows, there is an optimum powder factor or fragment size where the total cost is at minimum.[1]

Achieving this optimum fragment size requires significant knowledge of many factors and parameters that control the outcome of a blast design. These factors can be divided into three categories:

- Geology: Geological conditions of the rock mass to be blasted play an important role in the behavior of the rock when blasted. Important parameters are hardness of the intact rock and natural fractures and weakness planes.
- Explosives: Explosive properties must be taken into consideration very carefully. The type of explosive(s) must be suited to and compatible with the site conditions.
- Blast design parameters: These parameters are blasthole diameter and length, spacing and the pattern of blastholes, amount of charge per blasthole, and timing and sequence of initiation of each hole.

All of these parameters are interrelated and must be taken into consideration when designing a blasting operation. The operators must also be aware that the results of a blast are site specific, so close observation and some testing might be required. Details of the blast design are not within the goals and the objectives of this book and will not be discussed here. However, management must be sure that the blasting crew and the design team understand the connection between the blasting operation and the downstream operation. The manager is responsible for interaction and coordination between drilling and blasting and mine planning, production, and processing departments. Planning and scheduling of blasting must be set in such a way that there is always sufficient blasted rock available to maintain the peak design production without any delays.

Blasting quality depends on many factors (e.g., drilling, explosive type and quality, firing sequence), but proper drilling and explosive quality are probably the two most critical for good blasting results. If drilling is not accurate, well designed, and properly implemented, blast efficiency and costs will be greatly affected. This is why drilling and blasting are usually organized under the same supervisor.

A drilling and blasting engineer needs to coordinate with the planning department and the load–haul staff on a short- and medium-term basis. This allows for a well-scheduled drilling (drill rig movement along the mine, maintenance, meters drilled rate predictions) and blasting (explosives provisions, blasting crew, and explosives-loading trucks) process.

Mechanical Excavation

Because of the lower cost and higher efficiency of using explosives to break the rock material, drilling and blasting seems to be a method that will be widely used well into the future. It will continue be so as long as no new digging machinery technologies are developed that are capable of high-rate, low-cost production in excavating intact hard-rock masses. Mechanical excavation is more effective and economical than blasting when conditions are suitable. This method of rock breakage can be applied in most civil projects and some development sections in underground mining. Furthermore, mechanical excavation techniques do not cause the social and environmental problems associated with blasting, which can create major issues when blasting operations are conducted near populated areas. These issues are discussed in later sections. However, when large volumes of rock need to be broken, such as in mining operations, blasting is the most feasible and effective fragmentation method.

Sustainable Management Issues

The importance of achieving an optimum fragmentation was discussed in the preceding paragraphs, but the side effects of blasting must not be overlooked. Blasting can have a severe impact on the surroundings if the designers concentrate only on the fragmentation and primary cost of blasting. With the increase in population, residential communities are expanding into rural areas and encroaching on mining sites. As the vicinity of mining areas becomes populated, the problems associated with blasting will become more and more severe. The mine operators must pay close attention to the impacts of blasting on these residential communities. These impacts could range anywhere from structural damage to health problems and annoyance among residents. Litigation and paying for the damages can become very costly, and it can result in the complete shutdown of the operations. Controlling these problems will add challenges and complexity to the design of a blast round. Various impacts of blasting operation and problems that must be taken into serious consideration are summarized in the following paragraphs.

Ground Vibrations

Blast-induced ground vibration is the main problem when there are residential and other structures in the vicinity of the blasting area. These problems can range anywhere from major structural damage for extreme cases down to annoyance and disturbance of humans and animals where the vibration levels are within their perceptions. If these issues are not properly handled, the operators might have to deal with major complaints and costly litigation, and the outcome might be a complete shutdown of operations.

The level of ground vibration at any point depends on the amount of charge per delay, distance, sequential timing of each blasthole, the frequency of ground vibration, confinement of the charge, type of explosive, and geology of the site. All these parameters, as well as the safe limits set by the regulations (if any) or the level that is acceptable to the neighbors, must be taken into consideration when designing a blast round. A survey of nearby homes and structures that might be susceptible to damage must be performed prior to the start of blasting operations. The purpose of this survey is to establish the condition of the structures and pre-existing cracks and damages. A third-party consultant should be used to perform the preblast survey. Monitoring of ground vibration at the most sensitive structures must be performed as well. It is also recommended that additional measurements at various distances to be made in order to establish the site-specific ground vibration and wave transmission. This will allow the designers and operators to make predictions of what vibration levels can be expected for future designs. Again, it is a good practice to use a third-party consultant to perform the monitoring.

Timing of the blast is another important factor that affects the response of people to ground vibration. Blasting should be done during the time of day when people are most active and the ambient noise is high. This will reduce the level of perception and annoyance.

Keeping records of all blasting, such as design parameters (charge diameter, length of blasthole, weight of charge in each hole, initiation system used, spacing of blastholes, delay timing, and pattern) with a drawing, time of blast, and all the measured vibration levels is very critical and important.

Airblast

An airblast is the result of overpressure, low-frequency compressional waves traveling through the air. Airblasts are generated in addition to ground vibrations when blasting. The main causes of airblast, as with ground vibration, are improper blast design and blasting practices. Airblasts are generated when explosive gases are suddenly released into the atmosphere. Another main cause is the high-velocity movement of the rock face at the instant of detonation of the charges. Factors that affect the airblast include maximum charge per delay, delay timing and direction, depth of charge, exposed charges, temperature gradients, wind speed and direction, topography, and atmospheric conditions.

Damage from airblasts from blasting is very unusual and rare. The most notable damage is broken windows. However, airblast could become the main cause for complaints if the levels of air overpressure are high. Studies have shown that there are no health or psychological hazards to humans due to airblast from blasting.

Airblast could become a problem in underground mining as well. Excessive overpressure could be hazardous and could cause damage to underground structures and ventilation control devices, such as doors and regulators.

Flyrock

Flyrock is the main cause of injury in blasting. Flyrock is classified as uncontrolled fragments of rock reaching great distances and outside the perimeter of the blasted area or the mine. It can consist of different sizes, from a few centimeters to sizes larger than boulders. Flyrock can also damage homes, structures, and equipment.

The main causes of flyrock are excessive amounts of explosive, too much or too little burden, an insufficient amount of or ineffective stemming, the presence of voids and weak structures or formations in the close vicinity of blastholes, and poor delay patterns and timing.

Fumes

Detonation of explosives generates large quantities of gases. These gases play an important role in fragmentation and placement of the blasted rock. Under ideal conditions, gases primarily consist of carbon dioxide, nitrogen, and water. However, other products can be generated, including poisonous gases such as carbon monoxide and nitrogen oxides (NOx), which could cause health concerns. These fumes are not usually a major concern in surface blasting if they can be dispersed by wind or movement of air. If those in charge of detonations anticipate that excessive fumes will be generated, the direction of the wind must be taken into consideration if it is likely to blow toward residential areas and communities. In underground mining or blasting in general, fumes can be a serious problem, and proper ventilation and suppression of these gases is required. Explosives that generate excessive fumes must be avoided.

There are a variety of causes for the formation of excessive poisonous gases: poor explosive mixture or formulation (non-oxygen-balanced), contact with water when non-water-resistant blasting agents are used, poor initiation, and anything that can cause poor and incomplete reaction of the explosive. It must be noted that formation of these gases as the result of poor or incomplete detonation of explosive is accompanied by low energy and therefore ineffective breakage.

Dust

Dust is classified as fine particles that can remain suspended in the air for a long period of time. Blasting can generate large quantities of dust. Dust can become a health issue, particularly if it consists of material that can be toxic. The generation of dust depends on the material being blasted and the blast design parameters (too much explosive, poor timing, not enough stemming or confinement, etc.). Just as is the case with fumes, wind direction at the time of blasting must be taken into consideration if excessive dust is expected and there are nearby residential or other populated areas.

Safety in Blasting Operations

As the result of extensive research and development in explosive material and devices, the safety records in blasting have improved significantly over the years. However, a small accident or incident in blasting can and will result in a catastrophe or loss of life or limb. Such accidents could very easily result in loss of production and possible shutdown of the operation. It is the responsibility of the management to make sure that safe and approved blasting practices are followed. The main health-related risks associated with blasting operations are described briefly in the following paragraphs.

Flyrock.   Flyrock is the main cause of injuries in blasting. The causes of flyrock were discussed in previous paragraphs. A minimum safe distance of 800 m should be maintained when large-diameter blastholes are used. This minimum distance refers to a flat, even surface and it should be considered as an initial cautionary measure. In this sense, any incident of flyrock falling near control points or outside of the estimated safety area is a warning sign and an indication of the need to review procedures and safety distances.

Misfires.   A charge or portion of a charge that has not detonated is classified as a *misfire*. A misfire is the most dangerous situation in blasting and must be treated with the maximum level of care and caution. Misfires can be the result of *cut-offs* in the initiation lines (electric wires, detonating cords, nonelectric lines, etc.), poor connections or discontinuous lines between charges, defective initiators or explosives, or insufficient current when electric blasting is used. Causes of cut-offs include the explosion from the previous hole breaking into the next hole, shifting of ground and cutting the lines, and the impact of flyrock on surface initiation lines. An unexploded and unidentified charge can be impacted by the loading equipment or any other type of equipment, which can very easily detonate the charge. The explosive charge with the primer can be carried to the processing plant and cause a serious incident at the plant.

Misfires must be avoided at all costs. The continuity of connecting lines must be checked prior to blasting. Usage of proper timing and initiation systems can reduce the risk of misfires. Old and deteriorated explosive material should not be used. After each blast, the blasted area must be inspected for misfires. Only authorized and trained personnel must enter the area, and, in the event of a misfire, they are the only individuals who should handle the problem.

Several techniques and procedures can be used to remove and eliminate the dangers of misfires. If there are regulations regarding the handling of misfires, they must be followed. A sufficient time interval must be allowed before taking any action to remove the misfire. The unexploded charge or charges must be identified first. Any exposed remnants of unexploded charge can be removed very carefully. In the case of ammonium nitrate/fuel oil, it can be simply washed out with a water hose. Drilling and firing a hole charged with small amount of explosive near the unexploded hole can cause either sympathetic detonation or expose the unexploded charge.

It is the responsibility of the management to make sure that well-trained and qualified individuals are in charge of the blasting operation and handling misfires. The managers must also make sure that all the regulations and required procedures are followed.

Improper handling, storage, and transportation of explosives.   Improper handling, storage, and transportation of explosives and initiation devices can lead to catastrophic accidents. All laws and regulations regarding handling, storage, and transporting explosives and initiation devices must be understood and followed.

Storms, static electricity, or mechanical stress on explosives/detonators.   Electric storms, static electricity, and mechanical stress such as impact can result in premature and uncontrolled detonation of explosives. These problems can be eliminated if all the procedures and regulations for handling and usage of explosives are followed. Therefore, only qualified and trained personnel must be in charge of handling and transporting the explosives and other explosive devices.

In general, the occurrence of accidents is usually low, but consequences are often fatal (risk = probability × consequence). Only trained and supervised personnel should be involved with the handling of explosives. Procedures should include external auditing of safe practices, because routine sometimes hides real danger or threats. Incorrect, unsafe practices can commonly become routine in blasting without crews being aware of the specific dangers into which they are falling. Communication between operators and mine access control points/crew should be clear.

All procedures must be commonly and explicitly understood, with no dead points at which no communications are possible.

Blasting in Populated Areas: A Case Study of La Araña Cement Quarry

The Financiera y Minera quarry provides limestone to the Italcementi Group cement plant in Malaga (Spain). This operation is located in an urban environment that has experienced great expansion due to its coastal location. Decades ago, when the cement plant and quarry were established, that was not the situation, but a highway and several communities and beach resorts currently surround the site.

Years ago, the blasting operations used electrical detonators along with detonating cord on the surface. As the mining operations became larger and got closer to the surrounding communities, concerns grew regarding the impact of blasting operations. The existing blasting practice was causing airblast and ground vibrations at levels that became a major concern. The vibrations due to operations in the vicinity of the residential structures were reaching a level that needed attention. Although the majority of vibration levels were within the legal limits, they were high enough to be perceptible and very noticeable by people. As better and safer initiation system technology (nonelectric) appeared on the market, management at the plant decided to study new ways to approach rock blasting. The first measure taken was to contract with a responsible and experienced blasting team that would take major steps to protect the communities surrounding the cement plant. The new blasting contractors started a completely new blasting procedure, consisting of accurate drilling, a nonelectric detonating system (eliminating the detonating cord), proper stemming, and continuous recording of vibration levels.

More detailed site-related studies were done to understand vibrations at the Financiera y Minera quarry. In June 2007, an on-site vibration analysis was performed to establish the specific characteristics of the rock mass in different directions toward the sensitive areas. Also, a complex shooting sequence plan was studied to analyze possible destructive interference of that rock mass, which could help reduce overlapping of ground waves. It has to be noted that, although PPV (peak particle velocity) values are well below the legal limits, human sensitivity to vibration and preventing complaints were reasons enough for the mining company's management to implement a continuous monitoring and vibration reduction program, along with explaining and communicating with community representatives and the most affected individuals.

In a further step, to combine vibration level reduction with acceptable fragmentation on the rock pile, a different set of drilling grids and use of gas bags (to reduce the amount of explosives used) were tested in a series of blasts. Although these tests proved useful in fragmentation analysis terms, those blast designs showed no clear benefit from the vibration perspective.

Finally, results from double-decking (dividing the explosive charge in the holes) blasting showed potential for keeping vibrations low and around the 1-mm/s level (Figure 9.2), which is below the level of human sensitivity (down to 2 mm/s). Also, electronic detonators provided useful information on tailored sequencing of the blast at different areas of the quarry. Figure 9.2 shows the progressive changes in vibration levels as different designs were used. It is clear that the vibration levels are reduced when double-decking loading is used. This procedure can increase the cost of blasting, but the benefits will recover these costs. The benefits include better fragmentation and therefore a lower cost of material handling and crushing. Other major benefits include reduction of complaints and legal battles, as well as harmony between the operation and the community, which is a very important factor in sustainability.

As shown in Figure 9.2, the average values of PPV in millimeters per second at the quarry increase over time. Values greater than 2 mm/s are darkly shaded (1% chance of complaints

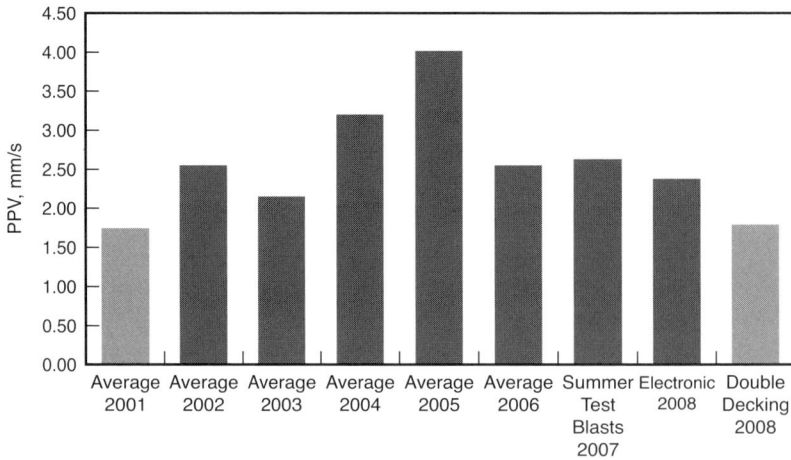

FIGURE 9.2 Evolution of average annual PPV values at Financiera y Minera quarry

according to John Floyd[2]) and values less than that level lightly shaded. In the summer of 2007, specific designs reflect four test blasts for both an increase in fragmentation and a decrease of vibration. Two electronically sequenced blasts have been recorded and processed as this book was written, with an average of 2.37 mm/s. Two double-deck blasts have been fired, but only one offered a valid value of 1.77 mm/s at the same control point; there was no reading on the seismograph in the other blast. More test blasts are being performed as this book is being written in order to obtain a valid average value for both systems.

Acknowledgments

This section's author expresses sincere and deep gratitude to the following individuals and companies for their contributions to this section: Ignacio Navarro, Jesus A. Pascual, Nando Nasca, Manuel Lopez Cano, and Jose Maria Fuentes, all of Maxam-UEE Explosives Company, and Stephen Jeric of Dyno Nobel-DNX.

LOADING AND HAULING

Loading and hauling represents a major part of material handling in the mining industry. After processing, loading and hauling is the most energy-intensive process and therefore is accountable for the second highest operating cost. The success and sustainability of a mining operation is heavily dependent on the overall efficiency and productivity of the loading and hauling system. It is the responsibility of the designers and planners to make sure the loading and hauling operation meets the production requirements and goals safely, economically, and effectively. It is then the responsibility of the managers and operators to perform the tasks according to the designs and implement all the required steps in order to meet the production goals. The hauling and loading system or equipment is integrated into all other units of operation, as well as the design of various components of the mine. Therefore, the system must be compatible with all other units of operation in the mine.

Equipment Selection Procedure

The equipment selection process is a difficult and complex task. There are a large number of suppliers and equipment types, and the technology changes rapidly. The equipment to be used will be dictated primarily by the mining plan and production goals.

An evaluation must be made of all of the heavy equipment related to price, availability, product support, local vendor capabilities, and so forth. Product support is very important in the selection of any equipment. This will dictate the maintenance staffing requirements, as well as capital expenditures on such things as shops and tools. Some local vendors supply very little support; others provide very significant support in both parts and supply availability, as well as supplying contract maintenance personnel and off-site component rebuild. Availability of parts and supplies in remote areas is a major concern. An evaluation must be made regarding delivery of parts and supplies to the mining operation, which might include use of the product transportation systems such as railroads, trucking, and so forth.

One of the main supply items of any mining operation is the source of fuel and lubricants. Care must be taken to evaluate all possible suppliers, looking at price and supply capabilities. There are a variety of sources of information for equipment selection:

- Personal experience
- Trade magazines
- Technical literature
- Trade expositions and conferences
- Talking to other users
- Visiting similar operations
- Visiting and contacting other operations in the area

Several alternatives are possible when selecting a piece of equipment:

- Make it: It is possible to manufacture some pieces of equipment in-house. The advantage of this is that the costs could be lower and it might be the only alternative when certain equipment is not otherwise available. The problem with this alternative is that it requires manufacturing facilities and skilled machinists and workers. Some replacement parts can be built in-house, which could reduce costs without major requirements in terms of facilities.

- Buy it new: New individual pieces of equipment or a fleet of equipment can cost much more than used equipment. However, new equipment can be expected to have much higher reliability, efficiency, productivity, and lower operating costs than used equipment. A simple economic analysis over the life of the operation or the equipment can determine the economic advantage of new versus used equipment.

- Buy it used: Buying used equipment can be a viable alternative to new equipment. Used equipment might cost less to purchase, but it will not have the same productivity as new equipment and it will have higher operating and maintenance costs. Again, a simple economic analysis can determine the advantage of used versus new.

- Lease it: Leasing equipment can be considered as an alternative to purchasing. Generally, large machinery with a long expected life and use will not be economical to lease. Leasing can be advantageous for fleets of small vehicles for general-purpose uses.

Important Factors That Affect the Productivity of Hauling and Loading Equipment

The selection and operation of mining equipment or a fleet of equipment depends on production scheduling and mining plans. It is useful to understand the definitions of some terms and factors that play an important role in performance and therefore in the selection and design of a hauling and loading system. Kennedy[3] has defined these factors, which are summarized in the following paragraphs.

Time Frame

During the feasibility study and mine evaluation, all data and estimates are based on a 1-year time frame. Economic analyses are always on an annual basis. Whether the analyses are for short-term or long-term planning, the costs and production rates are projected over a number of years or for the entire life of the project. The life of a mine depends on the reserves and the level of daily production designed, which must be determined very carefully.

Operations Scheduling

Operations scheduling relates to the number of annual hours of scheduled operation. The importance of establishing this at the early stages is that it controls the output of the material-handling equipment over 1 year. Most very large mines are scheduled to operate 365 days a year, three shifts per day. Not all sections or operations are scheduled to work 7 days per week and three shifts per day. For example, sections such as the administrative, engineering, and technical group might work only 40 hours per week on day shifts only. Drilling and blasting operations might be scheduled differently as well. It is not common to drill and blast at times other than the day shift, and blasting during the weekends is limited as well. Most heavy maintenance functions are performed on day shifts. Some mines can plan a total shutdown for legal holidays and so forth.

Overall Job Efficiency Factor

Overall job efficiency refers to the percentage of time that a machine or a system will operate at its peak output while in service in the mine. This can be evaluated as an hourly factor and it is an average number of minutes per hour. The loss of efficiency is due to the delays that occur during the scheduled operating hour. Causes of delays or downtime during the hour include the following: fueling and servicing of equipment; recess and lunch time, if lunch must come out of an 8-hour shift; poor coordination of different components of a system (e.g., shovels and trucks); and crowding at the loading, dumping, transferring, or exchange points. This factor is mainly controlled by the planning and management of the operation and the coordination of various groups or sets of unit of operations that work in series or parallel.

Mechanical Availability Factor

Much like the job efficiency factor, mechanical availability is the percentage of time that a piece of equipment is ready and available to perform work. Loss of mechanical availability refers to time when the machine is completely out of operation as a result of breakdown during the period when it is scheduled to work. This downtime should not be confused with the loss of time due to lack of job efficiency. A machine could be 100% available but not scheduled to work. This available time should not be included as part of the mechanical availability. The equipment manufacturers tend to overestimate the mechanical availability of their products. It is a good practice to determine the actual availability factor for each machine during the operations. Equipment downtime could be due to a variety of factors: mechanical conditions, operating conditions, the skill of the operator, the quality of the equipment, the complexity of the machine or the system, and most importantly, the maintenance program at the operation. The mechanical availability factor is used to determine the number of back-up machines needed to maintain the designed production rates. For example, if 10 trucks are needed to perform a task (with the efficiency factor taken into consideration), a mechanical availability of 80% requires a total of 13 trucks in the fleet (10/0.80).

This is where the maintenance program plays an important role in the productivity of the operation (see the "Maintenance Management" section). A good maintenance program will increase the availability of the equipment. It must be noted that the cost of increased maintenance must be taken into consideration when evaluating the economics. The additional cost of maintenance could exceed the savings from a higher availability factor. In the example outlined in the preceding paragraph, if the availability factor is increased to 90%, the total number of trucks needed in the fleet will be 11. This could be a major savings in the capital costs for the fleet. Although equipment manufacturers tend to overestimate this number, the overall number for all machinery in surface mining is about 85%.

When calculating the total operating cost, the time that the machine is down as a result of mechanical failure should not be included. This is not part of the costing hour where the machine is on the job. The lost time due to lack of 100% efficiency should be included as part of the costing hour because the machine is at the site operating but not producing.

Annual Outage Factor

The annual outage factor refers to the total time, on average, that is lost as a result of complete loss of production at the mine in a year. This total stoppage could be the result of (1) bad weather conditions, such as a heavy snowfall or flash floods that cause water to build up in the pit and block roads; (2) loss of electric power due electrical storms and snowstorms knocking out transmission lines and substations; (3) moving large units of equipment; (4) slides in the mine; and (5) external causes, such as breakdowns in the transportation systems, strikes in some other segments of the industry, and local labor disturbances. This factor will have an effect on the entire operation, not just the loading and hauling equipment. With no available data, a factor of 95% can be used, meaning that the operations will be delayed 5% of the scheduled time. A higher value can be used in areas with mild climates where natural effects are small and limited.

Production Utilization

Production utilization can be considered as the product of all the other factors discussed previously (equal to job efficiency × mechanical availability × annual outage factor).

It must be noted that in all operations, achieving 100% efficiency from a machine or a system is not possible. A machine might be operating at 100% efficiency for a short time but, on average, over an hour, shift, day, or month, the amount or percentage of useful work from the machine could be much lower. All the factors described previously must be applied to the peak output of the machine in order to determine the actual production rate of the equipment. During a feasibility study, these factors are not known because the study is concerned with an operation that does not yet exist. Therefore, no productivity data is available. The engineer must make assumptions and use judgment and experience with similar operations and equipment to determine these factors. However, these factors must be reevaluated during the operation so that a more accurate estimate is available for future equipment purchases and the updating of mine planning.

During both planning and operation, there are a number of goals related to equipment:

- To keep as many units working efficiently as possible;
- To have minimum equipment doing the job in an assigned period of time; and

- To achieve the first two goals with high probability, at a constant production flow within the scheduled time.

Equipment Selection Criteria

Before the start of the equipment selection process, a set of criteria must be developed. These criteria might be unique to the operation and can vary from operation to operation. The following general questions should be considered during equipment selection:

- Does it fit into the entire operation and other handling systems?
- Does it help to optimize material flow with maximum output and minimum cost?
- Can it operate continuously with minimum interruptions?
- Is it as simple as is practical? It's advisable to avoid complex systems.
- Is it capable of utilizing its maximum capacity?
- Does it use a minimum of operator time or labor?
- Does it utilize gravity wherever possible, depending on the mining system?
- Does it require a minimum of space?
- Does it handle as large a load as is practical?
- Does it operate safely?
- Does it use the maximum level of mechanization and automation?
- Is it flexible and adaptable?
- Does it have a low deadweight-to-payload ratio?
- Does it require a minimum of loading and unloading time?
- Does it need little or no rehandling?
- Does it require as little maintenance, repair, power, and fuel as possible?
- Is it reliable and does it offer maximum availability?
- Will it have a long and useful life?
- Does it perform the handling operations efficiently and economically?

Degree of mechanization is very important in equipment selection. As the degree of mechanization increases, the unit operating cost decreases. However, an increase in mechanization results in an increase in capital cost and maintenance due to the complexity of the system. Therefore, the optimum level of mechanization must be used. An "ideal" piece of equipment is always available and working.

Equipment Selection Procedure

As mentioned previously, the loading and hauling equipment must be compatible with the mining system and meet the production goals. All the important factors and criteria, such as those in the preceding list, must be established and defined prior to the process of equipment selection. In general, the following procedures should be followed:

1. Relate all factors pertinent to the problem.
2. Determine the appropriate degree of mechanization.
3. Make a tentative selection of equipment type.
4. Narrow the choices.

5. Rank and evaluate the alternatives.

6. Check the selection for compatibility with the rest of the system.

7. Select the specific type of equipment.

8. Prepare specifications.

9. Procure the equipment.

Multicriteria/multifactor analysis can be used when several alternatives with a large set of factors and criteria are considered. This technique allows for a quantitative comparison by assigning a weighting to each criterion. A scoring scheme can be established so that each factor or criterion will receive a score. Then each score is multiplied by the weighting factor and the final total score is compared. Generally, the main controlling factors for alternatives are cost, standardization, and reliability. The use of computer programs (artificial intelligence) is very beneficial.

Basic Steps in Selection Process

The process of selecting a loading and hauling system involves the following steps:

1. Determine required production.

2. Determine reach or haul path.

3. Calculate cycle time.

4. Calculate capacity.

5. Repeat to improve productivity.

6. Calculate fleet size.

7. Repeat to reduce ownership and operating costs.

Sustainable Management Considerations

The sustainable management of the loading and hauling requires a management focus on efficiency and safety, with some environmental issues (e.g., dust control and air pollution) also requiring special attention.

Safe Working Conditions

A large loading and hauling unit can create a very dangerous working environment. However, through proper training and enforcing all the safety rules, it's easy to create safe working conditions. Attention must be paid to the design and placement of haulage roads. The minimum width and sight distance must be carefully designed and implemented. Keeping all material-handling units isolated as much as possible will significantly reduce workers' exposure to unsafe conditions. There are federal, state, and local laws and regulations, as well as those set by the companies, related to safe equipment operation. These laws and regulations are based on many years of experience, observations, studies, and experiments. Operators and managers must follow and enforce these laws and regulations. Accidents will result in loss of time, production, equipment, and, worst of all, loss of life and limb.

Environmental Issues

The main environmental concerns are those associated with equipment that uses internal combustion engines. This problem is mainly a concern in underground mines. Through proper ventilation and maintenance of the equipment, the problem can be eliminated or reduced. Use of electric power instead of internal combustion engines can completely eliminate this problem.

Dust generation from the haul roads can become a major health problem, as well as a safety problem, because excessive dust will reduce visibility. Well-maintained roads and continuous application of dust-suppressant reagents will minimize this problem.

Maintaining High Productivity

It is the responsibility of the chief operators and managers to ensure that the material-handling system can operate at its peak efficiency and utilization. State-of-the-art technology must be used in selecting and operating a fleet of equipment. However, the best available equipment might not work at its peak efficiency if the units are not compatible and scheduling of assignments is not done properly.

The overall efficiency of the material-handling system can be improved by implementing a good training program for the operators and proper scheduling and synchronization of all the units. The effect of skilled operators on the safety and performance of any equipment is obvious. Delays and bunching times at the loading or dumping sites, as well as crossing points, will increase the cycle time of the haul units and cause major reduction in overall job efficiency. Computer-controlled dispatch systems will solve this problem in a complex system. In many cases, mining operations have managed to increase their efficiency and productivity by more than 50% by introducing these dispatch systems. Automation can significantly improve the productivity because the human factors are either eliminated or reduced. However, such systems can become too complex and expensive.

Equipment that operates continuously, like conveyor belts, has higher productivity. This type of equipment must be considered and used where the conditions make it suitable. However, the complexity and the presence of large numbers of moving parts will reduce the reliability.

If several different material-handling systems are working continuously, the overall utilization of the whole system is the product of the utilization of each individual unit. In this case, if there are delays or a failure in one unit, the entire system will be delayed or stopped. Placement of storage, either in the form of stockpiles or bins, at the transfer or contact points will reduce or eliminate this problem.

GROUND CONTROL

Ground control is a major component of the design and operations of both surface and underground mines. The integrity and stability of all openings and excavations such as shafts, drifts, stopes, haulage tunnels, inclines and declines, benches, and slopes are extremely important in all mining operations. The costs associated with support requirements in underground mining could be extensive. It is always a major challenge for the designers to protect the openings while keeping the costs low. In surface mines, steepening the slopes can save a large amount of money by reducing the amount of waste to be mined. On the other hand, steeper slopes will increase the chance of a failure and could result in loss of life and serious damage to the property and the operation.

Ground control should be performed in parallel with the mining operation. It is an integral part of the planning and production. It is the responsibility of the management to understand the requirements, processes, and interaction of the ground control systems, as well as the basic theory. The management must ensure that all the ground control systems are implemented and installed according to the plans and designs with the highest level of quality control. Ground control can become very costly. With unlimited resources and funds, it is easy to provide a support system that never fails. However, in mining operations, the profit margins are very narrow and funds are limited. This makes the job of the engineer in charge very challenging. The engineer is forced to push all ground control systems and devices to their limits. The engineer in charge as well as the management should not allow safety to be compromised for profits.

Role of Ground Control in Mining

The role of ground control in mining is to establish a safe and stable working condition. The responsible geotechnical engineer must understand all aspects of the mining operation, such as production requirements, type of material-handling system used, mining method, ore control, drilling and blasting or any other means of excavation, associated costs, economics, and so forth. The engineer must also have a clear knowledge and understanding of the geology of the mine, as well as the shape and extent of the orebody. He/she must be well aware of the ground conditions as well as plans for future mine expansion. Ground control must work in parallel with operations and be able to interact without hindering or slowing down other mining activities. The engineer must also understand how the ground reacts as mining operations progress, and how the mining activities affect the implemented support systems. The main challenge is that each rock mass has unique properties. There is no single or standard solution to ground control problems guaranteed to produce certain and correct answers consistently. The engineer has to come up with a practical solution from the basic and limited geologic data available as he/she progresses through the design and planning process. The engineer has to know how to use and combine all the available tools to solve the problems associated with ground conditions that change constantly.

The role of ground control in mining is very different from the role it plays in civil engineering projects. Generally speaking, the safety factor used in the design of structures in a mine is much lower than what is used in civil-engineering-type structures. However, this is not because safety is less important in mining. This is where the ground control becomes challenging in mining. As mentioned previously, the profit margins are narrow in mining and the total mining costs must be kept to a minimum. The major difference between mining and civil structures is the useful and expected life of the structures and facilities. Basically, the life of a civil engineering project is very long—more than 10 years. In mining, only certain parts such as shafts, main haulage drifts, underground shops, and storage areas are considered to have long lives. These components are designed with a higher safety factor. However, structures like stopes and temporary accesses have short lives and they are not designed with a high safety factor. Again, the role of the ground control program is to keep the opening stable and safe for the duration of its use and life. One important aspect of a ground control system that designers must understand is that the role of the support system is to make the surrounding rock or material self-supporting, and it should utilize the rock as the principal supporting structure. Again, this must be achieved with as little disturbance as possible during the excavation and the usage of the structure.

The designers face the challenge of achieving an optimum design without compromising what is considered economically acceptable and safe.

Economic Consequences of Instability

Any type of collapse or failure in mining is costly in one form or another. Collapse of a section of a mine or the entire mine has severe consequences that are not acceptable. The monetary losses come from closure of access, loss of equipment, covering the exposed orebody with waste in case of slope failures, loss of haul roads, and so forth. For example, it will cost the company a lot of money to replace the lost equipment, to remove extra waste in order to expose the ore, to rebuild and reroute the haul road and accesses, and, in general, to employ all the remedial measures to recover the losses and fix the damage. In case of loss of life or limb, it is very challenging if not impossible to determine the cost of recovery in terms of dollars. Putting a dollar value on the loss of life or limb is a very difficult ethical issue with which to deal.

Implementation of ground control devices and systems can be costly and requires special attention. All the outcomes of the collapse must be assessed and the costs of both recovery and prevention must be evaluated. In many types of mining systems, the roof or the surrounding rock is allowed to collapse, or controlled and induced cave-ins are part of the mining method, such as in longwall mining, block caving, and sublevel caving. The effect of these closures and cave-ins must be assessed very carefully on the nearby structures and surface. Subsidence is a major issue when these techniques are used.

Failures and accidents could also have other negative effects that cannot be assigned a dollar value directly. Whenever there is an accident in which fatalities are involved, the negative publicity about mining operations will have an unconstructive and downbeat effect on the public. This can severely hurt the sustainability of the mining project.

The benefits of a good and successful ground control program are unlimited. Investing more money in the ground control program early can produce major cost savings in the future. Safe and stable openings also increase the recovery and therefore the income. For example, steepening the pit slopes on a large mine by a few degrees can save millions of dollars in terms of the cost of overburden removal or the same amount of income as a result of access to more ore. The cost of supporting and stabilizing underground openings and slopes depends on the level of safety that is required. What is considered to be safe depends on the application and usage of the structure. For example, Barton et al.[4] suggest the following categories of underground excavations based on the support requirements:

    a. Temporary mine openings;

    b. Vertical shafts;

    c. Permanent mine openings, water tunnels for hydroelectric projects (excluding high-pressure penstocks), pilot tunnels, drifts, and headings for large excavations;

    d. Storage rooms, water treatment plants, minor road and railway tunnels, surge chambers, and access tunnels in hydroelectric projects;

    e. Underground power stations caverns, major road and railway tunnels, civil defense chambers, tunnel portals, and intersections;

    f. Underground nuclear power stations, railway stations, sport and public facilities, underground factories.

As the categories move from a to f, the requirements for a high level of safety are increased. Note that the application, duration of usage, and importance of the categories increases from a to f. Consequently, the cost of support requirements and construction increases as well. For example, in case of category a (temporary mine openings), the outcome of a failure is not severe and the risks are acceptable. The probability of failure might be high, but the severity of a failure is low. Furthermore, in this particular situation, only trained and experienced individuals like miners are exposed and they are exposed for a limited time. On the other hand, this is not the case if you look at excavations that fall in category e or f. Again, in many mining situations, a safety factor of one or less has been used. A common case is in large open-pit mines where the slopes are designed and allowed to move. This is done by monitoring the slopes very carefully.

Modes of Failure

As in all engineering works, understanding and predicting the mode of failure is very crucial. Mode of failure dictates what type of support and remediation tool should be used.

Underground Mining

Underground openings are confined structures that are excavated through a rock mass. The behavior and response of a rock mass to any excavation or disturbance must be well understood and must be predictable. Hoek and Brown[5] categorize the modes of failure of underground structures as follows:

- Instability due to adverse structural geology: This is a potential mode of failure when the rock mass is faulted and contains a major set or sets of joints. Discrete blocks of rock form when these joints intersect. These blocks are potentially unstable when the excavation exposes them. They can either slide or fall freely into the opening. Sliding along a single dominant joint set is possible as well. The orientation of the dominant discontinuities with respect to the excavation plays an important role in instability of these blocks. The most effective support system for this condition is installation of rock bolts, resin dowels, and cables. If reorienting the excavation is possible, the stability can significantly be improved.

- Instability due to excessively high rock stress: Stress-induced instability occurs in the case when the rock is very weak or the ground pressure with respect to the strength of the rock mass is very high. This is the situation at great depths or at a shallow depth if large excavations are made. The result is formation of an overstressed zone around the opening. In this overstressed zone, the rock has failed and severe stability problems exist. Formation of the overstressed zone depends on the in-situ state of stress, the shape of the excavation, and the material properties of the rock mass. The size of the excavation controls the extent of the overstressed zone. Changing the shape of the opening can reduce or redistribute the stresses around the opening to a more favorable condition. Reorienting the opening might help if that is possible. Installation of supports can control the instability if stresses are not too high and severe.

- Instability due to weathering and/or swelling: Some rocks contain minerals with properties and behaviors that change as soon as they are exposed to air or water. They will undergo a severe weathering process and they will become very weak and flaky. In cases where the rock contains clay minerals such as montmorillonite, swelling will take place. Swelling is associated with expansion as a result of increase in volume, and will apply excessive pressure on the support and liners in the tunnel. The best method for remediation and control of this behavior is to apply shotcrete immediately after excavation.

- Instability due to excessive groundwater pressure or flow: This problem can be encountered under any conditions. It can be severe if the excavation is below the groundwater table and the water head is very high. High flow or high hydrostatic pressure will be a major concern if any of the conditions mentioned here are present. Dewatering and grouting are the main remedial measures that can be applied to control excessive water.

Open-Pit and Other Surface Mining

Slope failures are very common in surface mines. Unlike underground mines, rock masses are not confined and are completely exposed to open faces. Instability is mainly the result of exposure of dominant weakness planes. These weakness planes consist of faults, joints, bedding planes, contacts, or any form of a plane that can act as a sliding surface. In very weak rocks and formations, failure through intact rock is possible but very rare. These are the main types of slope failures:

- Circular failure: This type of failure is typical in overburden soil material. It is very rare in hard rocks. However, very weak and heavily fractured rock could potentially exhibit circular failure. In mining, this type of failure should be evaluated when overburden rock piles and tailings are being evaluated for stability.

- Plane failure: This type of failure is common in slopes where two sides are open and exposed and a single dominant joint or joint set is intersecting the slope. A single block or multiple large blocks can potentially slide along this sliding plane.

- Wedge failure: This is the most common mode of failure in rock slopes. When two discontinuities or joint sets intersect, they form a wedge. This wedge is potentially unstable if exposed in the slope face.

- Toppling: Toppling is not a common type of slope failure. However, when steeply dipping and closely spaced discontinuities are exposed at the face and dipping into the face, toppling can occur. The discontinuities form columns of rock and, under favorable conditions, these columns can turn and topple over.

Information on discontinuities is the most important data of this type, as will be discussed later. Water by itself does not pose any hazards in terms of instability. However, when discontinuities are present, then joint water pressure will be the main factor in instability.

Important Factors in the Design of Ground Control Systems

Several factors and parameters must be well understood and considered in the design and planning of the ground control program for various mine sections. These parameters must be determined as accurately and as early as possible in order for the design engineers to plan the ground control program. These factors are summarized in the following paragraphs.

Geology and Ground Conditions

The geology and the rock mass conditions are the most important factors in ground control planning. The stability of the openings strongly depends on the geologic parameters and conditions. These parameters must be determined at early stages of planning and design of mine structures and sections. The data collection, testing, monitoring, and observation must continue throughout the operations and construction in order to update the information and implement new designs. There are a number of important geologic parameters:

- Rock type: Whether the rock is sedimentary, volcanic, or metamorphic determines some of the general behavior of the rock mass.

- Rock engineering properties: Important engineering properties of the rock are those that determine the response of the rock mass when it is disturbed or stressed. These properties are strength parameters (cohesion and internal friction angle, for example), Young's bulk moduli, and Poisson's ratio. These parameters are very important in underground structures and are used to determine the reaction of the ground and whether the rock mass is overstressed. The behavior of the rock mass is also used to select the best support system if needed. These are important input parameters when failure is due to severe stresses, as discussed previously. These properties are determined through lab and in-situ testing.

- Structure of the rock mass: As mentioned in the discussion of modes of failure, discontinuities play the main role in formation of potentially unstable blocks in both underground and surface operations. Important properties of weakness planes are the orientation (strike and dip) and shear strength parameters (cohesion and friction angle). Other useful information is length of the discontinuities, spacing or frequency of joints, and condition of joints (roughness, alteration, fillings, and separation). Strike and dip are measured in the field at the exposed faces. The friction and angle can be determined through direct shear testing. The cohesion is very difficult to measure, but back analysis can be done to estimate it.

State of Stress

The state of applied stress is an important parameter in the design of underground structures and ground control programs. Stress levels can be estimated using the overburden pressure. In-situ stress measurements are very reliable and useful. However, they are expensive and require access and special tools and skills. During the preliminary design stage, stresses must be estimated using some assumptions. Computer models and numerical analyses (such as finite element, boundary element, and discrete element analyses) are very useful and popular tools. Scenarios can be investigated using various stress conditions.

Hydrologic Conditions

Both surface and groundwater hydrology conditions should be studied. Surface hydrology is important in designing flood control and diversion channels, as well as erosion control structures. Surface water can cause some stability concerns in slope stability, particularly in overburden soil piles and tailings. If the surface run-off water filters in the slopes, tension cracks and discontinuities will cause a severe problem in rock slope stability. Excessive water pressure from the groundwater can become the main controlling factor in underground excavations, as discussed previously.

The location of the water table, quantities and direction of water flow, as well the quality of the groundwater must be evaluated. The quality of the groundwater, such as whether it is clean, brackish, acidic, alkali, or any other condition, is important to the operations. It is important to know whether the water can be used as needed for the operations, or if it is harmful or corrosive to the equipment and support systems. Knowledge about the quality of the water also determines whether it needs to be treated before releasing it back into the environment.

Shape and Size of the Excavation

The shape and size of the excavation do control the stability condition of the openings. The shape and size are determined by the application or use of the opening (i.e., shaft, adit, incline, decline, raise, main haulage access, etc.), the size of the equipment, the mining method, and the overall dimensions of the orebody and therefore the stope. The larger the excavation, the more instability problems that might be encountered. For example, when a block caving operation switched from the conventional gravity method or grizzly system to an LHD (load, haul, and dump) system, the ground control became a much more challenging issue. The difference between the two methods is that the openings are significantly larger when an LHD system is used. However, with the grizzly system, the size of the openings is smaller while the number of openings is significantly larger.

Production Rate

Production rate affects the ground control program indirectly. The designed production rate controls the expansion rate of the mine openings, which is a factor in design and implementation of the ground control program. It also determines the size and number of pieces of equipment, size of the facilities (both surface and underground), and so forth.

Economics

Like any other unit of operation in mining, the cost and benefit ratios play an important role in the design and selection of ground control methods. The economics of ground control systems were briefly discussed in the previous section. The engineer in charge must look at a set of support alternatives and select the most economic option.

Planning the Ground Control Program

The ground control program starts with the review of the mine plans, mine system, production requirements, type of material-handling equipment or system, and all other pertinent parameters associated with the mining operation. The planning then goes through the following stages of data collection and analysis:

- Stage 1: This stage involves the preliminary collection and interpretation of geological and regional data from existing historical documentation and the review of literature, geological maps, surface mapping, and core logs. This data, together with relevant mine operation information, should be used to assess the ground conditions that can be expected and to identify possible ground condition scenarios to be evaluated (e.g., favorable, most likely, and unfavorable).

- Stage 2: This includes preliminary analysis of all the data to establish any patterns. Those areas and conditions that require additional data and information are identified. Additional data will be collected if possible and the analysis repeated.

- Stage 3: The next step is to divide the proposed mine area into design sectors based on the geologic conditions and the mining activity. Each design sector has a unique property and might have different construction and support requirements. For each design sector, the expected mode of failure is predicted and ground control method(s) is recommended.

- Stage 4: Next, the support requirements are implemented as the mining operation commences and continues. Data collection, sampling, testing, and analyzing continue as operations go on. The process is involved with a continuous learning process through close observation and monitoring as mining is carried out. The ground control systems should be redesigned and updated as needed.

Monitoring

Monitoring should be proposed as a major part of the ground control program. Monitoring provides constant data to update and improve the tools and techniques that are used as part of the stability program. Monitoring also provides an early warning system to predict a potential failure and identify the location and source of instability. This increases the safety level and provides safe working conditions so the workers and operators can work with high confidence and comfort. The mine operators and managers are strongly encouraged to implement an extensive monitoring program as part of their operation. There are a number of common monitoring systems:

- Surveying points with a robotic totaling station: This tool provides a continuous record of movements. This has been used in open-pit mines extensively.

- Radar: This also provides a continuous readout of the slope movements.

- Optical surveying: This has applications in both surface and underground mines where access is available.

- Convergence measuring tools: This is normally carried out by means of a rod extensometer between the walls and roof of an excavation.

- Borehole extensometer: This measures the displacements in the rock mass surrounding an excavation. The data is very useful for understanding the behavior of the rock.

More and more advanced techniques are coming out, and operators should always consider using the most advanced techniques available.

A classical and most spectacular example of a successful monitoring program is the case of the slope failure at Chuquicamata mine in Chile in 1969. The details of this event can be found in an article by Kennedy et al.[6]

In June 1968, after signs of instability were noticed, a major monitoring program was implemented. An attempt was made to mine the unstable slope as much as possible to reduce the load. However, by late 1968, it was evident that a major slope failure was inevitable. It was not possible to stabilize the slope. Steps were taken to reroute the main haul road and stockpile material for the mill.

The monitoring program continued and on January 13, 1969, the plotted displacement data was projected to predict the date of failure. Based on the data from the fastest moving target, engineers predicted that the slope would fail on February 18, 1969. The sloped failed at 6:58 PM on February 18, involving approximately 12 Mt of material.

Full production resumed on February 19 after a shutdown of the pit for 65 hours. The mill continued working throughout this time from the stockpiled material.

What is spectacular about this case is not the accuracy of the prediction. It could have been off by several days or even a week. What is worthy of attention is that all the negative consequences of such a major failure were avoided and there was no loss of life, equipment, or time at the mill.

Ground Control and Sustainability

The longevity and safety of all mine structures and components rely on a successful ground control program. Any uncontrolled and unpredictable failure can cause major losses and even total shutdown of the mine. In this section, the importance of the ground control was reviewed. No data or information regarding the technical and theoretical steps in design of mine openings was given because it is not the responsibility of the mine managers to design the structures and the support systems. However, it is their responsibility to make sure that the mine components are designed to be safe and their integrity is preserved. However, management must have sufficient knowledge and background in geomechanics and support systems.

The mine can remain sustainable if it is safe and economical, and the ground control team is responsible for achieving that goal.

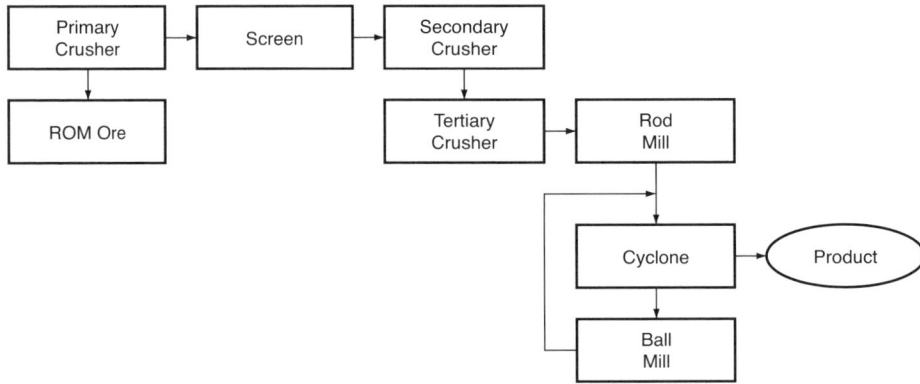

FIGURE 9.3 Conventional comminution process with three stages of crushing

MINERAL PROCESSING*

Mineral Processing Principles

Mineral processing is the process of extraction and concentration of economic minerals contained in ore.[7,8] It can be described by a multistage process:

1. Comminution (blasting, crushing, and grinding);

2. Separation by size and classification;

3. Separation by physical/chemical properties (flotation, cyanidation, etc.); and

4. Grade concentration.

Despite the fact that blasting is not considered part of mineral processing, the economics and efficiency parameters of blasting and plant processing must be studied as a whole. In this regard, blasting with explosives should be considered as the first stage of comminution.

In a mineral processing plant or "mill," size reduction takes place as a sequence of crushing and grinding processes (Figure 9.3). Crushing reduces the size of ROM ore to such a level that grinding can be carried out (usually from 12 to 19 mm top size). Grinding continues until mineral and gangue are substantially in the form of discrete particles.

Because most minerals are finely disseminated and intimately associated with gangue, they must be initially "unlocked" or "liberated" before separation can be undertaken. This is achieved by size reduction or comminution, in which the size of the ore is progressively reduced until a sufficient fraction of free mineral particles are available for separation.

Crushing takes place by compressing the ore against rigid surfaces or by impacting the material against surfaces in a constrained path. This is contrasted with grinding, which is accomplished by abrasion and impact of the ore by the free motion of a grinding media such as steel rods or balls, or ore pebbles.

Crushing is usually a dry process, although more modern techniques are beginning to apply waterflush technology in cone crushers. Typical crushing plants operate in stages, with reduction ratios from 3 to 6 at each stage. Conventional crushing processes are carried out in two or three size stages (primary, secondary, and tertiary crushing)

---

* This section was written by A. S. Rodriguez-Avello.

Grinding reduces the ore size from 19 or 12 mm down to 1.5 mm for iron ores and to –65 mesh for most base metal ores. In some industrial plants, the product is ground to the fineness of talcum powder (on the order of 5 μm or less). Wet grinding is generally the preferred method, but dry grinding is used in some special cases.

Crushing

The devices most frequently used for primary crushing are gyratory or jaw crushers. Secondary and tertiary crushing uses cone crushers or hammer mills.[9]

Jaw crushers are devices in which two opposing nonparallel plates trap rock in an alternately expanding and shrinking aperture. Rock pieces are broken into progressively finer sizes until they drop through the gap at the bottom of these plates. Jaw crushers accept feed sizes up to 1,200 mm and produce nominal product sizes as small as 19 mm. Product size is determined by the distance between the lower ends of the jaws. This gap dimension can be adjusted by shims behind the stationary jaw assembly. The jaw crushers can be choke fed from hoppers or conveyors.

Gyratory crushers consist of two cones, one being a large truncated cone with the apex down; the other smaller one, with the apex up, is mounted inside the large one and driven eccentrically. This structure provides a progressive breaking action similar to that in a jaw crusher. However, unlike jaw crushers where production is cyclic, gyratories are capable of continuous output.

Cone crushers have a fixed compression surface with the shape of an apex-up truncated cone; the moving surface is another apex-up cone inside the larger one. The eccentric motion of the inner cone (or mantle) produces a reciprocating motion in the same pattern as in jaw crushers, and with similar effects. However, as with the gyratory crusher, there is no cyclic production. The large discharge opening of cone crushers makes their use practical for fine crushing. The Symons crusher is the most widely used cone crusher. It is manufactured in two forms: *standard* for normal secondary crushing and *shorthead* for fine or tertiary duty. Standard crushers can handle feed material from 250 to 100 mm top size and generate a product from 50 to 30 mm in size. Shortheads are designed to produce a top size from 19 to 6 mm.

Secondary crushing plants treat ore of 250 to 100 mm top sizes and produce a product with a top size of 19 to 12 mm. Shorthead crushers often work in closed circuit with screens and, in some cases, the feed to standard crushers is prescreened to remove the fines, thus increasing crusher capacity.

Hammer mills (also impact crushers) are made up of a set of cast iron hammers to crush the ore with repeated impacts. These units are designed to crush fragile or fibrous materials such as asbestos, fluorite, limestone, or coal. The hammers are made of Mn-steel or nodular cast iron with chromium carbide for abrasion resistance. To increase hammer life, each hammer pivots as it contacts material, although fixed hammers are used for coarse product discharge.

Grinding

Grinding[10] can be performed in two stages using rod mills and ball mills or in a single stage using autogenous grinding (AG) mills or semiautogenous grinding (SAG) mills.

Rod mills are tubular mills in which milling action is achieved with steel rods. They are capable of processing feed as large as 50 mm and discharging a product as fine as 300 μm; reduction ratios are normally in the range of 15–20:1. Normal top size of rod mill feed, however, is 19 mm.

Rod mills are considered for coarse grinding applications or, operating in two stages, discharging into a ball mill. In a rod mill, smaller particles slip through the spaces between the rods and are discharged without appreciable reduction. As a result of this action, the particle size distribution of a rod mill is much tighter than that of an equivalent ball mill.

Ball mills are tubular mills that use steel balls as the grinding medium. They are better suited for fine grinding. Circuits employing a ball mill are always operated in closed circuit with a mechanical or hydraulic classifier such as a cyclone or a classifier (i.e., rake, bowl, or spiral).

Pebble mills are tubular mills with a single compartment and charged with hard, screened ore particles as the grinding medium. Because the weight of pebbles per unit volume is 35%–55% that of steel balls, the power input and the capacity of pebble mills are correspondingly lower. This is compensated for somewhat by operating such mills at high critical speed relative to conventional rod or ball milling. Thus, for a given ore at a particular feed rate, a pebble mill would be larger than a ball mill, with correspondingly higher capital cost.

AG mills, especially with ROM or primary-crushed ore, have become increasingly important in recent years. With suitable ores, these mills reduce grinding media costs and can produce lower percentages of slimes or fines than do conventional rod and ball mills. Instead of steel media, the AG mill uses the action of large pieces of ore on smaller ones to produce size reduction. SAG refers to grinding methods using a combination of ore lumps and a reduced load of steel rods or balls as the media.

AG mills can be operated wet or dry. Dry mills cause environmental and occupational-health problems, do not handle clay content well, and are more difficult to control than wet mills. In certain applications involving grinding of minerals such as asbestos, talc, and mica, dry semiautogenous milling is used exclusively.

Separation by Size and Classification

There are two types of particle sizing separators: screens and classifiers. Screens are used normally for coarse size separations down to 6 mm; classifiers are used for fine separations as low as 200 mesh. Wet screening applications are beginning to compete with classifiers in some cases where close size separations are needed.

Classification[11] is a method of separating mixtures of minerals into two or more products on the basis of the velocity with which the grains fall through a fluid medium. In mineral processing, the fluid is usually water, and wet classification is generally applied to mineral particles that are considered too fine to be sorted efficiently by screening.

Many types of classifiers are on the market. They are designed to separate one or more ore fractions using gravity, water flow, heavy media, and magnetic or electric fields. Some of them are hydrocyclones, spiral classifiers, hydraulic classifiers, heavy-media separators, and magnetic separators.

Flotation

Froth flotation[12] is a versatile process that is used for many ores whenever the liberation size is too fine for gravity separation, and magnetic or electrostatic methods are not applicable. It separates ore particles by exploiting the differences in their surface behavior in the presence of air bubbles and water. To produce separation, particle surfaces are made water-repellent so that they readily attach to an air bubble or are kept water-wetted (hydrophilic) so they sink. In practice, this involves adding chemical reagents to an ore pulp to create conditions that favor the attachment of gas bubbles to one class of mineral particles. A conventional flotation circuit in a mineral processing plant is shown in Figure 9.4.

Several types of flotation cells are available on the market, each having its own unique characteristics. The principal difference between self-aspirated and forced-air designs is the position of the rotor inside the cell. Self-aspirated flotation cells (with elevated rotor) are expected to recover more ore particles with weak particle–bubble bonds or easily dislodged coarse grains.

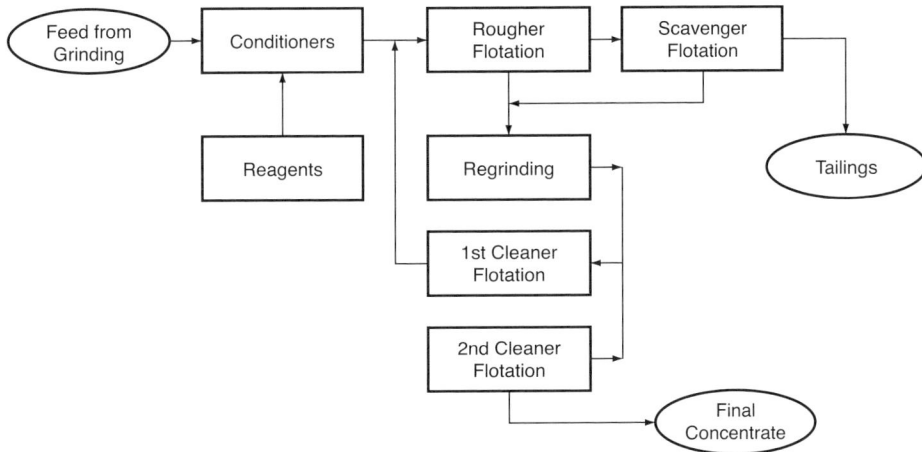

FIGURE 9.4   Flotation circuit

The bottom-based rotor cells have better selectivity when recovering fine particles, as a deeper draining froth can be used to allow entrained material to fall back into the cell.

In recent years, an increasing number of mills are installing column flotation cells to supplement or replace conventional flotation machines. Column cells consist of a cylindrical or square vessel, from 150 mm to 3.6 m across and up to 10 m high. Pulp is introduced about halfway down the vessel, and tailings pulp is extracted out the bottom through a valve controlling pulp level. The pulp is kept above the point of feed entry, and a very high froth column (up to 1.2 m) exists above the pulp. To stabilize this froth and provide a washing action within the froth, water is gently rained down from above the froth.

Cyanidation

Cyanidation is a simple process that exploits the solubility of gold in dilute cyanide solutions and its ease of recovery by cementation with zinc dust or by adsorption to active carbon. The major features of the process are the formation of a gold–cyanide complex ion, the requirement for air, and the sensitivity to pH.

Cyanide can also form complex ions with numerous other heavy metals such as copper, iron, mercury, and zinc. Thus, the presence of other ions derived from sulfide or oxide minerals in the ore can present difficulties in processing gold ores because the presence of these ions (referred to as cyanicides) can greatly increase the cyanide consumption, thus making the process economically unfeasible.

Dewatering

With few exceptions, most mineral separation processes involve the use of substantial quantities of water and the final concentrate has to be separated from a pulp in which the liquid/solid ratio can be high. Dewatering, or liquid/solid separation, produces a relatively dry concentrate for shipment.

Dewatering methods can be broadly classified into three groups: (1) sedimentation (thickeners), (2) filtration (filters), and (3) thermal drying (dryers). Dewatering in mineral processing is normally a combination of these methods. The bulk of the water is first removed by sedimentation or thickening, which produces a thickened pulp of perhaps 55%–65% solids by weight. Up to 80% of the water can be separated at this stage.

Filtration of the thick pulp then produces a moist filter cake of 80%–90% solids, which might require thermal drying to produce a final product with about 95% solids by weight.

Sustainable Management Issues

The sustainability of a mineral processing operation requires a management focus on operational efficiency (e.g., metallurgical recovery, water and energy consumption), the environmental issues (e.g., tailings disposal, cyanide use, and air pollution), and safety and health issues. In the following paragraphs, the main sustainable management issues are discussed.

Energy Consumption

Power consumption is the largest cost item in most mineral processing operations; therefore, optimizing energy consumption should be a major management objective. The crushing and grinding processes are normally the largest energy consumers in the mineral processing plant, but large energy savings are achievable through design optimization and efficient operation of other areas of the facility.

Mine–mill energy balance. Optimizing the mine–mill interface to increase efficiency and reduce energy consumption is a major challenge in mineral processing. This can be achieved reliably and inexpensively by measuring the performance of blasting operations as it relates to the subsequent improvement of crushing and grinding. Systematic estimation of the size distributions of blasted ore is critical to establishing a baseline for a predictive model to be used for the evaluation of process performance. Although conventional sampling techniques are not adequate for the large fragment sizes in muck piles, trucks, or crusher feed, video sampling systems and analysis of the raw images by specific software are one example of useful tools that could be used.

Mill motor efficiency. About 50% of the total energy used in mining operations is consumed in the mineral processing, mainly grinding. Therefore, improving mill motor efficiency is one way of achieving a significant amount of energy savings in the plant. A recent U.S. Department of Energy study determined that 44% of industry motors operate consistently at less than 40% of full load. Energy efficiency technologies for electric motors are a practical solution for variable load equipment such as crushing and grinding mills. A motor operates most efficiently when the load is 70%–95% of the rated load. There are a number of ways to accomplish this: shutting off an idling motor, right-sizing the motor, using frequency drives when it is essential to the process to change the speed, or installing reduced-voltage motor controllers.

Ball mill discharge: Overflow versus diaphragm. Overflow discharge ball mills are used when a product with high specific surface is more important than the particle size distribution curve. Diaphragm discharge-type mills are used to maximize power application to the coarse particles without production of excessive slimes. Twenty percent more energy is used than in an overflow mill of similar dimensions.

Dry grinding versus wet grinding. As a guideline, wet grinding consumes significantly less power than equivalent dry grinding, requires less physical space and less support structures, and reduces or eliminates required emissions (dust) control equipment. Dry grinding increases capital and energy consumption when processing moist or wet materials (requires predrying to less than 1% moisture), consumes less media and liner material, and eliminates the need for filtering and final drying later in the process in an open-circuit configuration. In areas where water is not available in sufficient quantities or the use of water is not allowable (e.g., cement, talc, potash), dry grinding is a viable option.

Energy-efficient screens. For larger feed between 40 and 150 mm, circular or linear motion inclined screens are often used. The latest advances in screen design are wider screens and more efficient, higher tonnage capacities. A recent development in high-frequency, low-deck-angle screens offers interesting new features in linear screen technology. It incorporates electromagnetic motors and specially designed resonators to amplify motion and cause only the screen panel assembly to vibrate at resonance, with low energy consumption.

Use of high-pressure grinding rolls. High-pressure grinding rolls (HPGRs) are now installed in crushing circuits as a replacement for conventional tertiary and quaternary crushers, increasing capacity and reducing power consumption. An additional advance in this technology is the use of HPGRs in parallel with cones in so-called reverse closed circuit with prescreening.

Operational Efficiency

Operational efficiency may be optimized at the design and the operational stages. An efficient plant design should achieve a proper balance between capital and operating cost. During operations, maximum efficiency is achieved through a management focus on maintenance, plant availability, and optimum metallurgical recovery.

Balancing capital and operating costs. Size reduction costs are important factors, not just in capital and operating items, but also in expected production, availability, and total service life. Investment paybacks are now in a time frame of 3 to 5 years. Effectively balancing capital and operating costs requires an experienced and practical understanding of how to use the energy efficiently and configure the crushing and grinding circuit for the best possible wear material service life.

Mill liners. The performance and life of mill wear parts must be carefully considered. There is great variation in the design, life, and cost of mill liners. In fact, liner life and change time are critical factors for mill availability and, therefore, plant performance. Mill relining machines can now place one 5,000-kg liner with the same speed as one 1,000-kg liner before, thus relining can be reduced from 160 hours to just 60 hours. The future could include robotic solutions, but the question is the capital cost, which is three times higher than with manually operated machines.

Screening panels. Screening media (panels) are made of polyurethane or rubber, although the sand and gravel industry still uses wire cloth. Although the polyurethane and rubber panels are much more expensive, they are also much harder wearing, and modular types using these materials have become increasingly popular.

Most of the major improvements that have been made are in wear protection systems using polyurethane and special high-durability coatings. Recent innovations include wear parts that last longer, extended periods for maintenance, and reliable monitoring systems, together with preventive (predictive) maintenance strategies that ensure better availability of the equipment, with less downtime and improved profits.

Plant maintenance. Good preventive maintenance programs can result in lower repair costs and greater equipment availability and, therefore, greater productivity. Newmont reports that after a comprehensive and effective training program, scheduled major maintenance outages that once consumed 30 days of production per year have been reduced to just 23 days; unscheduled maintenance stops have also been reduced.

Preclassification of ROM ore. The ROM ore delivered to the crushing plant contains particles of sizes less than the crusher setting. If this undersize material is fed to the crushing process, it will reduce crushing efficiency and produce higher wear, dust, and hold-up problems (if the ore is of a sticky nature). Because of it, this undersize material should be removed first or bypassed so that the available force will be applied to the coarser material.

Cylindrical flotation cells. Round tanks have proven to be more efficient and easier to operate than the rectangular cells because there are no corners, and mixing is more uniform, with minimal sanding effect. Also the froth bed is steadier across its entire surface area. The use of cylindrical cells yields triple the volume from a doubled surface area. In large cylindrical cells, the bubbles travel a longer path and can therefore collect more valuable minerals, thereby improving selectivity. However, there is a limit to the mass of mineral that froth can carry out of the cell, and a warning is required with higher-grade ores, which need more surface area to remove sufficient froth to maximize recovery.

Flotation software and control systems.   Economies of scale must also be considered, not just the relative capital cost of the flotation machines but also the installation footprint, installed power, and air requirements. Operational costs include mainly the energy used for agitation and air generation per volume of cell; these costs can also be affected by the number of wear parts, lubricants, belts, and so forth. Computational fluid dynamics modeling is a powerful tool for cell hydrodynamic optimization and determination of the optimum setup for each process. Other flotation software available provides simplified flotation models that rely on assumptions of probabilities for collision and bubble–particle attachment and detachment to describe flotation kinetics. In any case, mathematical modeling must be validated at the laboratory and at full scale.

Controlling reagent addition, air intake, and overflow rate of froth efficiently is critical to get the best results. To achieve this, most flotation plants are equipped with on-stream X-ray analysis of pulps and vision control technologies.

Design improvement trends.   Today, more grinding circuits are using combinations of SAG/AG for primary grinding, followed by ball mills; these combinations tend to dominate over traditional three-stage grinding in cones followed by rod/ball mill circuits. The addition of a cone crusher to downsize the recirculation pebbles from the SAG/AG mill is normal practice today.

One alternative to a rod mill in a three-stage crushing circuit is the use of vertical shaft impactors or waterflush cone crushers.

Environmental Issues

Clearly, the main environmental issue in mineral processing is the sustainable management of the very large volumes of water and tailings that must be handled in the process. For example, processing 0.5% grade copper ore to produce 1 t of copper would generate 245 t of mill tailings and 600 m^3 of process water.

Tailings disposal.   Disposing of tailings is a major environmental concern, becoming more important as we continue to mine ever-reduced grades of deposits. Apart from the visual impact on the landscape, the major problem is usually pollution arising from the discharge of water contaminated with solids, heavy metals, reagents, sulfur compounds, and so forth.

Tailing dams are designed to retain the solids safely and control water flows for recirculation or discharge. Seepage must also be avoided using polyethylene liners (very expensive) or clay layers impermeable to water flow.

Tailings water.   Total recycling of water is the best method to prevent environmental problems with tailings water, and it might also be the most economic solution, considering that wastewater treatment for discharge can be very expensive. Large holding tanks are needed to promote precipitation reactions and settling. Pumping costs are high and must be designed to cope with spring runoff conditions. Chemicals such as alum and/or ferric chloride are added, increasing the costs further.

However, total water recycling can cause problems in the flotation circuits and reduce recovery. Furthermore, tailings water invariably contains significant amounts of sulfate ions, which when fed to the warm, alkaline conditions of primary grinding can precipitate gypsum and plug launders and lines. Water treatment to render tailings water suitable for recycle is usually cheaper than treatment for discharge to the environment and should be considered to increase the use of recycling.

Use of environmental management system.   The company must have implemented a suitable management system to ensure compliance with permits and environmental laws and to interface with local authorities and the community to establish and maintain workable relationships over the life of the plant. To this end, many companies incorporate risk management systems, such as ISO 14001, to integrate a range of environmental health and safety issues in their operations. Indeed, certification to ISO 14001 standards is today becoming a contractual requirement for companies operating in the United States and the European Union.

Safety and Health

Sustainable management of a mineral processing plant requires safety precautions to protect personnel who may be working in dusty environments, handling toxic reagents, and operating mill equipment.

Safety training.   Ongoing safety training of plant personnel is imperative and is considered one of the most vital and monitored features at most mineral treatment operations. Another fact is that dry crushing plants are subject to bearings wear if dusty conditions prevail. Both circumstances encourage management to strive to avoid the problem by using air-swept sealed circuits, including bag filters, scrubbers, cartridge collectors, surfactants, water sprays, sonic fog, and so forth.

A training manual covering operating procedures, safety and health, emergencies, and other activities should be developed as part of the engineering and design process for use by plant supervisors and operators. After startup, the knowledge and skills can be honed by providing continual access to education and training.

Safety equipment.   The responsibility for providing a safe working place and to enforce safe job practices rests by law with the employer rather than the employee. Vendors can also assist in the testing, selection, and purchase of all safety supplies and personal protective equipment such as hard hats, face shields and visors, safe suits and gloves, belts and lanyards or harnesses for fall protection, and so forth.

Industrial hygiene.   One important aspect of a complete safety and health program is industrial hygiene, which relates to the effect on employees of physical and chemical agents within the flotation plant environment. Literature on the reagents used must include material safety data sheets showing the chemical product and supplier identification, composition, hazards and potential health effects, first aid measures, recommended handling and storage, personal protection, physical and chemical properties, stability and reactivity, toxicological and ecological information, and disposal considerations.

Dust control.   Dust control is very important from an occupational health and safety viewpoint. Dust is generally silica-bearing, which presents obvious silicosis dangers for plant personnel. Most crushing plants employ simple classifying devices such as dust-settling chambers, cyclones, vortex tubes, baffle dust collectors, and water aspirators that slurry the dust into the main separation plant. Bag filters and electrostatic units are required with hazardous dust such as asbestos fibers.

Air pollution.   Air pollution can be an additional environmental concern in mineral processing. Dust generated from crushing or grinding and combustion gases from concentrate drying and roasting can have a significant environmental impact and must be cleaned to comply with air emission control regulations.

Also, the mill reagent preparation plants can generate toxic gases such as hydrogen sulfide, sulfur dioxide, hydrogen cyanide, and ammonia, so they should be subject to control and inspections.

The Cyanide Management Code.   The "International Cyanide Management Code for the Manufacture, Transport and Use of Cyanide in the Production of Gold" (Cyanide Code) is a voluntary industry program for the gold mining industry to promote responsible management of cyanide used in gold mining, enhance the protection of human health, and reduce the potential for environmental impacts.

Companies that become signatories to the code must have their operations audited by an independent third party to demonstrate their compliance with the code, and audit results are made public on the International Cyanide Management Code Web site (www.cyanidecode.org).

Maintenance and housekeeping.   Designing for easy cleaning and maintenance reduces labor force requirements and improves safety. Adequate guards must be provided for all exposed moving machinery. Plant structures should be provided with adequate designated walkways, elevated platforms and catwalks, toe boards, guard rails, and building exits. Standard building codes applicable to the area should be checked and followed carefully. Judicious use of colors in painting can provide a positive effect on safety. Safety color codes established by the National Safety Council provide visual identification of hazardous machinery, guards, piping, railings, and structures.

Safety and health regulations.   It is easy to predict that ever-tightening occupational health and safety legislation, plus the labor availability and costs considerations, will push new developments into flotation equipment. Noise regulations and restrictions on working in confined spaces will set new demands for safety. As there will be fewer people running and maintaining the plant, it is evident that an approach based more on a preventive maintenance and sophistication in predicting wear and maintenance intervals, as well as online condition monitoring, will be making inroads in flotation equipment.

LEACHING*

Mining industry designers and operators have major responsibilities to develop sustainable management strategies for handling the large piles of spent ore produced by heap leaching technologies, taking economic, environment, community, and government demands into consideration.

Considerations/Characteristics

The widespread application of heap leaching technology in gold/silver and copper operations since the early 1980s is a clear testament to its economic advantages. Capital and operating costs of heap leach projects are lower than projects requiring milling and processing. Therefore, heap leach technology is well-suited for metal recovery from lower-grade orebodies.

Large volumes of low-grade ore are stacked on liner systems to recover the metals through dissolution by the lixiviant that is applied to the surface. These solutions flow through the heaps under unsaturated conditions. Large areas are taken up by the heap leach facilities and, in many cases, they are constructed to greater heights, more than 200 m in the case of expanding and some valley leach pads. The disposal areas containing spent ore from dynamic leach pads can also occupy large areas. The potential impact on the landscape posed by these large structures containing high volumes of material and covering large areas of land must be considered in the site selection and design of heap leach facilities.

A dilute cyanide solution applied at pH > 10.5 is used for leaching gold and silver ores. Dilute sulfuric acid is the lixiviant for copper ores. In general, the operational and long-term environmental impacts of sulfuric acid as a lixiviant can be more pronounced than that of cyanide. However, the emotional issues related to cyanide are often more contentious as a result of

---

* This section was written by D. Van Zyl.

the legacy attached to its past misuses. In the author's experience, even scientists and engineers unfamiliar with the potential environmental impacts of these two lixiviants can be more concerned about environmental impacts of the use of cyanide. The mine closure issues associated with each of these lixiviants will be discussed later in this chapter.

The important management concerns are emotional issues related to the use of cyanide because they can make it difficult to obtain a social license to operate in countries and regions with little familiarity with the safe operational handling of cyanide, as well as postoperational conditions. Inviting experts to provide their views of the environmental and human health risks to the communities might not do any good because they will not address the emotional issues and therefore will not contain the "outrage" factor Sandman[13] describes. He defines risk with respect to stakeholder acceptance and tolerance as follows:

$$risk = hazard + outrage$$

Sandman[14] then continues to identify reasons for and ways to address the outrage. Stakeholder engagement about cyanide usage at a site and, therefore, the related transportation, handling, health, and environmental concerns must be carefully planned. A negative outcome to these considerations can contribute to opposition to the development of projects. An example of this problem is the Esquel gold mining project in Argentina,[15] where the strong opposition of the hosting community of Esquel (expressed in a referendum) forced the mining company to halt the project.

A heap leach project has the potential for a number of major environmental impacts, including the visual impact associated with the large volumes of material involved and the potential impact to groundwater and surface water quality should a leak or spill occur. Project design and operational goals and procedures must be developed to prevent the occurrence of leaks and spills.

Sustainable Management Issues

This section has described some of the specific sustainable development issues that are addressed in the development and operations of heap leach projects. The author recommends reviewing the seven themes to sustainability on a regular basis for all projects (see Chapter 5). The seven themes can be used to establish the project goals and criteria, as well as audit and review the project with respect to its contributions to sustainability.

All economic, environmental, and social considerations must he evaluated on a project life-cycle basis. Short-term economic benefits must be weighed against longer-term impacts. Economic analyses should include life-cycle costing, considering all the aspects of project development, operations, and closure.

Water management is by far the largest issue facing all heap leach projects. Water supply on projects undertaken in dry climates is an important consideration and must be weighed against other potential uses of the water, as well as other potential technologies for providing sufficient operational water. For example, a number of mines located relatively near the ocean (present applications are less than 200 km from the coast) in Chile and Namibia are developing plants to desalinate ocean water for use in their processing. In wet climates, it may be necessary for mines to treat and discharge excess water.

The technology available for the containment of the leach solutions on well-designed and constructed pad and pond liner systems is mature. A full commitment on all projects to apply the best engineering approaches to create high-quality liner systems is essential. Saving money at the expense of lower design and construction standards is a foolish way to approach this essential component of all heap leach projects.

Cyanide leaching of oxide ores is economically feasible and very attractive. However, at a few projects, the incorrect or incomplete characterization of the ore resulted in the placement of sulfide materials on the heaps. While the projects suffered economically toward the end of the leach cycles as a result of higher reagent consumption and lower recovery, significant impacts followed closure. Acid drainage is generated at these projects and the pad effluents must be treated in perpetuity. Complete material characterization of the ore with respect to acid drainage is an essential part of all successful heap leach operations. Acid drainage is the biggest environmental problem the mining industry faces, and it will be addressed in further detail later in this chapter.

MINED ROCK AND TAILINGS MANAGEMENT*

This section begins with a discussion on site selection because it is an important tool in limiting the operational and postclosure impacts of mined rock and tailings management facilities. Other options for management of these materials include mine backfilling and below-water placement; these topics are touched upon briefly. The term *mined rock management facility* (rather than waste rock dump, rock pile, overburden storage, etc.) is used in this section to reflect the importance of terminology and its implications. While waste rock dump or other terms are now more widely used in the industry, it is essential that the terminology used reflect the approach and intent of the level of care. In this sense, mined rock management facility reflects the present level of care given to the design and development of these facilities.

Site Selection

A large number of factors determine the location of mined rock and tailings management facilities (MRTMFs) with respect to the orebody. After the mining and processing methods have been established, selection of MRTMFs and other facilities provides an opportunity for the long-term mitigation of environmental and social impacts that must be addressed. This is an important tool for the mine developer that must be used in the prefeasibility, feasibility, and final design phases, making sure that the mine is designed for postclosure.

Rigorous site selection methods are available for the location of MRTMFs. It is essential that a multi-disciplinary team consisting of mine developer, environmental professionals, engineers, social scientists, and others work together in selecting a site. The different perspectives involved in site selection are important.

The location of a mine will determine the potential options for mined and processed materials. It might be decided to consider land disposal, underwater disposal in lakes, humanmade impoundments, or oceans, pit, or underground disposal in mined-out areas, and so forth. Ideally, a comprehensive siting study should be done without any preconceived ideas about a potential preferred option. Site selection should not be done in isolation from the selection of the most appropriate control and closure measures that will be applied at each site. Sites unsuitable for use when applying one type of control and management technology (i.e., slurry tailings) might be eminently suitable when another technology (such as stacking of thickened or paste tailings) is applied.

Many trade-offs are made in selecting the location of a site for a specific facility. For example, the preferred site for a tailings management facility might not be the first choice of any stakeholder group (engineers, local communities, accountants, etc.) but a compromise of all views. At the extreme, there can be cases where the choice is between no mine and a mine with a tailings management facility in a location where there will be some impacts. This is not a threat but the

---

*   This section was written by D. Van Zyl.

reality of it, and making that choice should be the job of all concerned, not least the local communities.

Here are the basic physical steps in site selection for land disposal of mined and milled materials, taking into account the climate of the area and other appropriate factors:

1. Identify the area of interest for locating the facilities. An example is: a radius of 10 km around the orebody.

2. Identify the site and project-specific factors that must be considered in the site selection (e.g., excluding sites in national parks, areas of mineralization, consideration of wetlands and other sensitive environmental areas, the relative priority of economics and other factors, how local communities will be treated). It is clear that the multi-disciplinary project team must accomplish this task with input from regulatory personnel and stakeholders (including the communities that might be affected). Establishing clear siting criteria will make the rest of the process easier.

3. Eliminate zones in the area of interest from further consideration based on the site and project-specific siting criteria; this is also referred to as a fatal flaw screening.

4. Perform a screening of the remaining area to identify possible locations of facility sites; this might result in 20 or more possible sites. List the characteristics of the remaining sites: physical characteristics including capacity, environmental issues and risks, social impacts (including the possibility of resettlement), relative capital and operating costs, and so forth.

5. Develop conceptual plans for the intended use of the site using alternative development, control, and closure technologies. Select the most appropriate technology for each site.

6. Perform an alternatives analysis of the sites, with the associated best technology, using qualitative or quantitative methods.

7. Investigate the remaining sites in more detail (i.e., field reconnaissance, mapping, drilling). Expand the conceptual plans for site development using the most appropriate control and closure technologies and do the final alternatives analysis and selection.

There is a greater possibility that the optimum sites will be selected when a rigorous approach is used. Another significant advantage in using a rigorous, well-documented approach is that it provides a basis for review of the methodology and results. It is therefore a transparent approach to use during consultation with regulators, surrounding communities, and other stakeholders, which should be ongoing throughout the process.

A variety of specific factors play an important role in the selection of sites:

• Economics: There are economic advantages in locating the mined rock management facilities near the orebody because it will reduce the operating costs. It is common to transport tailings as slurry, and locating tailings management facilities near the orebody is less important from an operating cost perspective. Although there are operating cost advantages to locating heap leach facilities near the orebody, capital cost considerations (such as extensive earthworks) might dictate that the facilities be located away from the orebody.

• Climate: Mines are located in all climatic regions of the world, from the Atacama Desert in Chile to high-rainfall tropical areas of Indonesia to polar regions of Canada and Russia. Climate is a major factor determining the environmental performance of a mine and has a large impact on the site selection of facilities. Intercepting and storing large volumes of runoff can also affect the stability of facilities.

- Site seismicity: Seismic activity at a site is determined by its location. Much is known about the impacts of seismicity and dynamic loading on the performance of earthen structures. Specific attention must be paid to location of structures sensitive to seismic loading (such as tailings management facilities); their location and design must consider the potential risks posed downstream. The same is also true for pipelines.

- Topography and hydrology: Steep topography in the immediate vicinity of a mine often makes it very difficult to locate sites for the various facilities. While mined rock management facilities can be located in steep terrain (with special attention to design and operating conditions), it is very difficult to locate tailings management facilities in steep terrain when the tailings are transported as slurry. The size of an embankment in steep terrain will require a large volume of structural fill and can result in a very small volume of the site remaining for tailings storage. This makes the storage capacity of the site inefficient. It is not uncommon to pump the tailings a long distance to a suitable site in flatter terrain. For example, in Chile, one of the tailings storage facilities of El Teniente is 75 km from the mine. Runoff volumes at a site are determined by the precipitation upstream from the site, as well as the area of the drainage basin. Placing the MRTMFs near the upper reaches of the drainage area will reduce the amount of runoff at a site that must be diverted or stored.

- Surface geology of site (foundation conditions for facilities): Some sites might be ideal for storage of mined and processed materials from a physical perspective but completely unsuitable based on the surface geology. Thick layers of foundation materials with low strength and high compressibility might make it impossible to locate such facilities, especially if the site is located in a high seismic zone.

- Local communities and land use: Not all orebodies are located in remote areas with low population density. Very often, an orebody is located in an area where there are settlements and where the land use has benefit to these communities, such as agriculture. This presents an opportunity and an obligation to involve the communities in selecting the locations of the mine facilities and to get input in the design and final closure. For example, assume that the area near the mine does not have much flat area for cultivation; the mined rock management can be constructed in such a way to increase the flat area for cultivation. It might also be necessary to reprofile these facilities to allow future cultivation. Aesthetic values and impacts must be considered for placement of these facilities. Mining changes the land use in an area and provides special opportunities for new thinking on postmining land use. Using premining land use as a basis for long-term land use planning might not be the best approach, although it is widely used. Local and regional land use might change during the mine life as a result of population influx, regulatory changes, and so forth. Flexibility in the regulatory framework and planning at the mine site must be maintained throughout the mine life to make adjustments so that postmining land use can be productive.

- Other environmental issues: Careful consideration must be given to all environmental issues, including protected areas and biodiversity. Using multi-disciplinary teams for the site selection process will increase the awareness of these issues.

Backfilling of mined and processed materials into underground workings or open pits has certain advantages and disadvantages. A major advantage is that the materials will be placed below the ground surface and therefore will not take up further space on the land. Mined and processed materials used to backfill underground mines also improve the stability of the underground workings and minimize postoperational subsidence. As a result of the increase in volume of the mined rock when excavated, as well as the requirement to leave some remaining openings underground to provide access to the ore, it is not possible to backfill all the materials removed. Up to about 60% of the materials can be replaced underground; the rest is usually placed in surface management facilities.

Backfilling of mined pits is only possible in some cases where there are separate pits or an elongated pit or open cast. In a number of mines, pits have been partially or completely filled with mined rock. In some instances, this approach might result in greater environmental impact than leaving the pit open, such as when the mined rock and pit walls are highly acid generating and a steady water level cannot be maintained in the pit, or where the regional groundwater is recharged from the pit.

There are also concerns about covering up potential future resources because low-grade mineralization at the bottom of the pit would be effectively removed from future exploitation. Double handling of materials for disposal can be a very important economic issue; therefore, it is not cost effective to backfill a pit at the end of operations unless there are very specific environmental and other advantages. Placing acid generating waste beneath a stable water table in the mined out pit can stop acid generation and provide for a long-term solution to chemical instability.

Marine disposal of mining waste is used at a number of mining and mineral processing operations around the world. Mined rock can be discharged at the shoreline or to deeper water from a barge. Tailings and other processed materials can be conveyed through a pipeline as slurry and discharged at beach level or at depth via a submerged pipeline.

Marine disposal is sometimes considered as a disposal option at coastal and island sites, where viable land disposal options are limited or virtually nonexistent. What is known as submarine tailings placement or deep-sea tailings placement involves a different set of criteria and potential impacts than shoreline and very shallow water marine disposal. The environmental impact of marine placement and its acceptability to local communities, governments, and civil society vary according to many factors.

The alternative to land-based management that deep-sea placement represents, and the associated risks, are not widely agreed upon between the industry, academics, and civil society. Perceptions can vary considerably and points of divergence need to be identified to assess how this mined rock and processed materials management contributes to the overall discussion of mining at a site. Some believe that, even though the placement of mined and processed materials offshore can have a substantial impact on marine ecosystems, it might prove to be the best of a damaging set of options.[16]

Mined Rock Management

Mined rock consists of nonmineralized and mineralized materials that overlie the orebody or are interspersed with the ore. The geological and geochemical characteristics of these materials must be evaluated to identify their potential to generate acid or neutral drainage because such effluent can have serious short- and long-term impacts on the project and the environment.

Considerations/Characteristics

There is a significant body of research and practical applications with respect to the management of acid and neutral drainage; project management must make sure that personnel apply good practice in all aspects of this problem.

Acid drainage can present a number of potential problems during operations and closure:

- Degradation of mine water quality, limiting its reuse;
- Degradation of receiving surface waters;

- Impact on the aquatic ecosystem from acidity and dissolved metals;
- Impact on riverside communities;
- Possible impact on groundwater quality;
- Difficulties in stabilizing and revegetating mine waste; and
- Long-term water treatment costs.

One of the most serious aspects of acid drainage is its persistence in the environment. An acid-generating mine has the potential for long-term, severe impacts on rivers, streams, and aquatic life. Mined rock and processed materials that have not been properly deposited or closed can produce acid drainage for hundreds of years or more after mining has ceased. After the process of acid generation has started, it is extremely difficult to stop and can effectively kill most living organisms in an entire water system for years, creating a biological challenge and a significant economic burden.

The foundation conditions of the mined rock facilities must be characterized with respect to geotechnical and hydrological conditions. Specific attention must be given to the presence of springs and seeps, as well as the quality of the water from these. Understanding these baseline conditions is a prerequisite to the development of a management strategy for these materials.

The unit operations associated with the production and handling of mined rock are drill, blast, load, haul, and dump. There are a few mines where mined rock is crushed and transported by conveyor belt. Regardless of the transportation method, it is important to note that these materials are placed close to the open pit where they are produced. The surrounding natural topography should be used as guidance to the height and shapes of the final landforms resulting from placing these materials.

All mined rock facilities at a mine should be placed in one watershed as far as possible to simplify surface water controls and to limit the environmental footprint of the operation.

Sustainable Management Issues

As mentioned in the previous sections of this chapter, it is the author's recommendation that the seven themes to sustainability be reviewed on a regular basis for all projects. The seven themes can be used to establish the project goals and criteria, as well as auditing and reviewing the project with respect to its contributions to sustainability (see Chapter 5).

All economic, environmental, and social considerations must be evaluated on a project life-cycle basis. Short-term economic benefits must be weighed against longer-term impacts. Economic analyses should include life-cycle costing, considering all the aspects of project development, operations, and closure.

Ongoing mined rock material characterization, reviewing the results, and incorporating them into the modeling and other evaluations for rock disposal at a site are essential steps in the sustainability of mining projects. Continuous improvement must be an outcome at all mined rock management projects.

Concurrent reclamation must be evaluated and implemented at all mined rock management facilities. Concurrent reclamation activities provide a number of opportunities, such as ongoing evaluation and improvement of recontouring and revegetation practices, as well as establishment of final contours, which allows stakeholders to visualize the final site conditions.

Tailings Management

The prime function of a tailings management facility is the safe, long-term storage of process waste with minimal environmental or social impact.

Considerations/Characteristics

The design of each facility is specific to the mining operation and site conditions. The design life of a tailings management facility is effectively perpetuity, which means it should be able to survive in a stable form without human intervention. Tailings management facilities are constructed over a long period, and this must be reflected in the geotechnical stability evaluations. One of the most important issues is that stability analyses are done with the correct geotechnical assumptions.

Tailings dam facilities can be constructed using the coarser fraction of the tailings (tailings sand) or using "borrowed" materials for the embankment. The amount of coarse tailings material or suitable rock, as well as the regional seismicity, will govern the type of embankment constructed. The construction of the embankment is done in a series of lifts during the operational life of the mine. Geomembrane liners, compacted clay, or other components can be used to minimize seepage.

Historically, tailings management received little attention because it was considered a non-essential cost.[17] This has changed significantly as a number of significant tailings impoundment failures occurred during the last 50 years, which has sensitized the industry, designers, and regulators to the importance of carefully managing tailings impoundment design, construction, and operations. Ongoing research and review of failures also resulted in the compilation of a number of publications and manuals by various organizations on the design, operation, and failures of tailings management facilities. Table 9.1 presents a partial list of these publications, including manuals, summaries of failures, and guidance documents.

The amount of water in the tailings at the time of deposition determines the type of facility that must be constructed, the behavior of the facility during operations and closure, and the amount of water required to operate the mine. The three approaches are slurry tailings (typically having slurry densities from 30% to 55%), paste tailings, and dry tailings. The latter approaches result in dryer tailings; however, specific water contents (or slurry densities) cannot be stated because it depends on the material characteristics and behavior (e.g., paste tailings can flow in pipes but do not release free water upon deposition). In recent years, paste tailings have gained acceptance, as shown by the increasing number of paste tailings facilities used by the mining industry.[18]

TABLE 9.1 List of tailings management publications

Publications
1. *A Guide to the Management of Tailings Facilities*, Mining Association of Canada, 1998
2. *Developing an Operation, Maintenance, and Surveillance Manual for Tailings and Water Management Facilities*, Mining Association of Canada, 2002
3. *Leading Practice Sustainable Development Program for the Mining Industry: Tailings Management*, Commonwealth of Australia, Minister of Industry, Tourism and Resources, 2007
4. *Best Practice Environmental Management in Mining: Tailings Containment*, Commonwealth of Australia, Environmental Protection Agency, 1995
5. *Reference Document on Best Available Techniques for Management of Tailings and Waste-Rock in Mining Activities*, European Commission, 2004
6. *A Guide to Tailings Dams and Impoundments: Design, Construction, Use and Rehabilitation*, International Commission on Large Dams, 1996
7. *Guidelines on the Development of an Operationg Manual for Tailings Storage*, Western Australia Government, 1998
8. *Guidelines on the Safe Design and Operating Standards for Tailings Storage*, Western Australia Government, 1999
9. *Code of Practice for Mine Residue Deposits*, South African National Standards, 1998
10. *Tailings Dams: Risk of Dangerous Occurrences—Lessons Learnt from Practical Experience*, International Commission on Large Dams, 2001

Regional practices are extremely important as they typically develop as a result of local climatic conditions, regulatory requirements, and operational experiences. However, as a result of the international expansion of the mining industry, many regional practices are now being exported and adapted to other regions because of the experience of individuals and companies.

Sustainable Management Issues

As part of the Mining, Minerals and Sustainable Development (MMSD) project, a study on large volume waste was completed that involved a large number of international participants. The results of this study,[19] as well as the rest of the MMSD project, added much to the ongoing conversations about the application of sustainable development concepts to tailings management. The documents recently published by the Mining Association of Canada (*Developing an Operating, Management and Surveillance Manual for Tailings and Water Management Facilities*) and the Commonwealth of Australia (*Leading Practice Sustainable Development Program for the Mining Industry: Tailings Management*) are important milestones toward increasing the contributions that mining makes to sustainable development. Developing and implementing an operating manual helps to involve all the stakeholders in the production and management of tailings at a mine.

Dust from the dry beaches of tailings management facilities is a specific concern for nearby communities. Beaches of tailings management storage contain fine sand or silt-size particles that can be easily removed by wind. Wetting of the beach or using special products to stabilize the surfaces has been implemented for temporary wind erosion and dust control. Long-term stabilization requires a gravel cover or vegetation to be established. This problem is one of the major ones related to tailings management facilities and adjacent communities, especially in dry and windy climates.

Blowing dust can have an impact on health by hampering breathing and creating other problems, as well as on agriculture, through metals uptake by plants.

Water management is an essential element of all tailings management plans. Many of the tailings failures occurred either as a result of poor water management or because poor water management was instrumental in the magnitude of the failure. Mine management must make sure that sufficient emphasis is placed on water management at tailings management facilities.

Liner systems to contain seepage are successful at a number of tailings management facilities. Such systems must be considered when the sensitive groundwater systems are located at shallow depths below the facility with contaminants such as cyanide in the tailings. Composite liner systems are very effective in containing contaminants.

RECLAMATION AND CLOSURE*

Mine closure consists of all the activities following the end of operations, including decommissioning of processing plants, removal and disposal of equipment (reselling, recycling the metal, landfilling), demolition of buildings, recontouring of the land to create positive site drainage, placement of topsoil/growth medium, and revegetating the land. Ongoing monitoring and maintenance following closure is necessary to make sure that physical, chemical, and biological stability is maintained in the long term.

---

* This section was written by D. Van Zyl.

Considerations/Characteristics

Mining operations are finite economic activities, which are usually relatively short term. For a mining project to contribute positively to an area's development in any lasting way, closure objectives and impacts must be considered from project inception. Mine closure policy and planning defines a vision of the end result and sets out concrete objectives to implement that vision. To achieve this, a mine closure plan should be an integral part of a project life cycle and be designed to ensure that[20]

- Future public health and safety are not compromised;
- Environmental resources are not subject to physical and chemical deterioration;
- The after-use of the site is beneficial and sustainable in the long term;
- Any adverse socioeconomic impacts are minimized; and
- All socioeconomic benefits are maximized.

The process of establishing a closure plan starts during the design of a mine and proceeds throughout the life of the mine. The process involves several stages:

1. A study of closure options—looking at the feasibility of all aspects of possible outcomes;
2. A consultative process—involving all interested parties, to determine the preferred post-mining use for the mine site and associated facilities;
3. A statement of closure objectives—the mining company's commitment to the outcome of the closure of its activities;
4. An estimate of closure costs—the cost of achieving the stated objectives; and
5. A program of studies and test work—to confirm any assumptions inherent in the closure objectives.

The closure criteria and goals include the following:

- Prescribed criteria—regulatory criteria and decisions;
- Corporate criteria; World Bank and International Finance Corporation guidelines;
- Nonregulatory stakeholder desires; and
- Other performance criteria (e.g., final landform, postmining land use).

The closure planning process for a mine site typically consists of the closure designs for the various facilities. Design details and technologies differ for mined rock management facilities, tailings management facilities, spent ore facilities, open pits, and so forth. After developing designs for the individual facilities, the total site closure plan is focused around the long-term site water management plan.

The overall cost estimates for mine closure should be based on the facility-specific designs and best estimated unit costs. Mine closure cost estimates form the basis for financial assurance required by many jurisdictions. Regular updates of the closure plan and cost estimates are necessary; in Nevada, United States, this is done every 3 years.

Financial assurance or surety is the amount of money available to a government entity for closure of the mine in cases when the mine owner is not available to perform the work (such as bankruptcy), during operations, or any time thereafter. The financial surety can be provided by a variety of financial instruments or cash deposited in a bank. Miller[21] presents a thorough review of the different types of financial surety instruments. However, it is important to realize that the governmental policy and local financial markets might determine the type of instrument available for a specific location.

Another important concept is that of financial accruals by mining companies for closure. It is common to base this accrual on a unit production basis (such as dollar per ounce of gold produced). The total amount of the accrual is estimated from the environmental closure cost plus other liabilities at a specific mine (e.g., land holdings, personnel cost associated with the end of operations).

The following should be noted regarding financial surety and closure cost accruals:

- Conceptually, financial surety is in place during the total life of the mine and will only be released (in part or in total) after the regulatory agencies have established that reclamation and closure have been completed to their satisfaction. The financial surety might not be a fixed amount throughout the life of the mine, but can vary as environmental issues develop at a mine, as regulatory changes occur, and community expectations change. For example, acid drainage might not be considered an issue at the time that the mine design is developed but might become a major issue during operations. Obviously, the amount of financial surety can increase considerably after evidence of acid generation is found at a site.

- Closure cost accrual takes place over the life of the mine based on a planned mine life; it is not necessarily a linear function as it can also vary over the mine's life.

- In the United States and some other countries, financial surety is not available to a mining operation for closure work at the end of the mine life. It can be released shortly after the work has been done, but the mining company must be a going concern in order to perform, or contract some entity to perform, the required activities.

A few mining companies have established sinking funds to pay for the closure of a mine. Money from a sinking fund will be available in cash to pay for closure; in contrast, an accrual is an accounting allowance that is not liquid. Such sinking funds can be attractive because they are liquid; however, in the case of a bankruptcy, it becomes part of the assets of the company and will not be available to pay for closure.

Sustainable Management Issues

The long-term land uses at a mine site as well as the postmining landforms are important mine closure issues. The final landforms of the facilities should ideally fit in with the natural terrain and be shaped to maintain long-term stability against erosive forces. Experience gained from observing the behavior of concurrently reclaimed facilities can provide very useful information in this regard.

Postmining land uses that provide further economic activities at the closed mine should be evaluated with input from the local communities. Many options could be available depending on the location of the mine and the site conditions (e.g., solar or wind power generation, use of truck facilities).

MAINTENANCE MANAGEMENT*

AFNOR (Association Française de NORmalisation) defines maintenance as the actions to maintain or restore an item to a specific state, or to a state where it is able to provide a determined service. Maintenance management is defined as the approach to maintenance that allows operating at minimum cost.

---

* This section was written by J. M. Quintana and J. A. Botin.

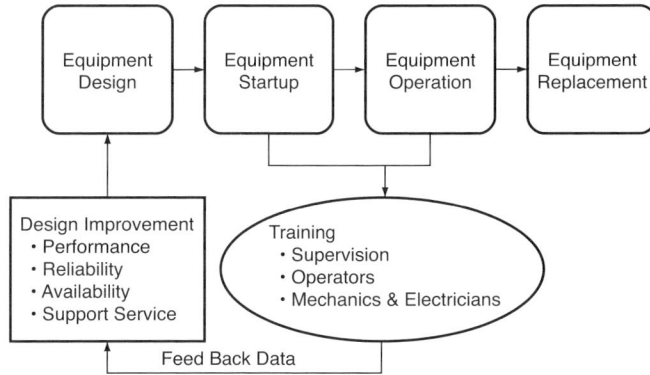

FIGURE 9.5   Maintenance life cycle

Therefore, maintenance and operating costs are linked through equipment price, performance, and operating cost. Maintenance can increase equipment availability and performance, but at the same time it implies a cost. The objective of maintenance management is to maximize operational efficiency through the optimum balance between equipment performance and operating cost.

Industrial Maintenance Concept

Maintenance management integrates the following concepts of equipment reliability, maintainability, and safety:[22]

- Reliability: the ability of a system or component to perform its required functions under stated conditions for a specified period of time;[23]

- Maintainability: the relative ease and economy of time and resources with which an item can be maintained or restored to a specified condition when maintenance is performed by personnel having specified skill levels, using prescribed procedures and resources, at each prescribed level of maintenance and repair;[24] and

- Equipment safety: the aptitude of a device to avoid the appearance of critical or catastrophic events.

Maintenance begins well before the first failure, right from the equipment design and startup stages (Figure 9.5), which are both concerned with the concepts of reliability and maintainability, and continues through the operation stage with maintenance strategies focusing on performance and cost. In the last stage of the cycle, the equipment is replaced when it becomes technically obsolete or its life-cycle cost is higher than that for a new unit.

When equipment is operating, failure can occur in many different ways (modes), each with a different probability of occurrence (failure mode reliability). For example, an ore haulage truck tire can fail by wear or by accident. The probability of wear failure increases with age (age failure), but the probability of failure by accident remains constant with time, because road conditions and other factors constantly present opportunities for damage (random failure).

When equipment has failed, each failure mode implies a specific repair task involving a specific repair time and cost. In the previous example, a wear failure can be predicted, so the repair can be performed in the workshop before breakdown, in a shorter time and at a lower cost than the repair of a tire failure by accident, which had occurred when the truck was in the pit.

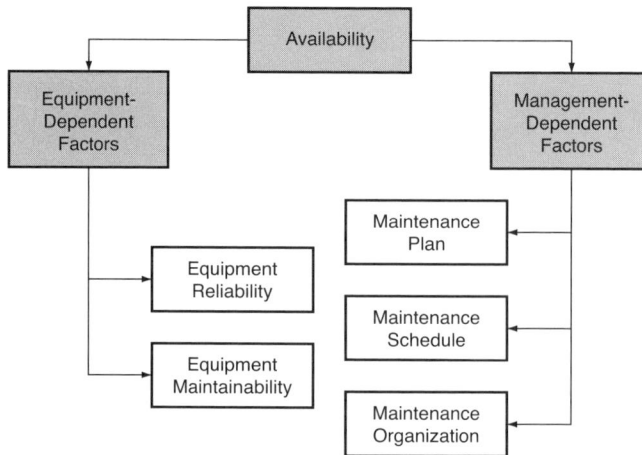

FIGURE 9.6   Availability versus maintenance management

In a simple two-state maintenance model, where a piece of equipment can only be either in operation or under repair, operating performance would only be dependent on the reliability and maintainability, the functions that characterize each of the equipment failure modes. For real-life multistate models, where the equipment might be at other states (i.e., idle, queuing for maintenance, held up by operating delays), maintenance management strategies become a decisive factor in equipment efficiency.

The overall maintenance performance is linked to the concept of functionability or the proportion of scheduled time during which the equipment or system satisfies the operating standards (i.e., can operate). Functionability is best characterized by the "availability" function.

Equipment availability (Figure 9.6) depends on reliability and maintainability, two equipment functions that are related to its design, and therefore are constant throughout its life cycle. In addition, availability depends on maintenance management factors such as maintenance strategy, plans, and schedules.

Equipment Maintenance Strategy

The concept of equipment maintenance strategy refers to the management objective for maintaining each piece of equipment or system. Maintenance strategies are specific for each piece of equipment or system and are a function of its process function.

Equipment strategy can be visualized using the three-dimensional (3-D) triangular model shown in Figure 9.7, where the strategy is characterized by its relative position with respect to three management objectives:

- Reliability
- Functionability
- Direct maintenance cost

As an example, in a mineral processing plant, the failure of the mill causes a shutdown of the entire process; therefore, the maintenance strategy for the mill must be to ensure functionability (i.e., maximum availability). In contrast, the failure of a sump pump does not cause any production losses; because of this, the maintenance strategy might focus on minimizing maintenance cost. Furthermore, certain failure modes can have safety consequences or can significantly

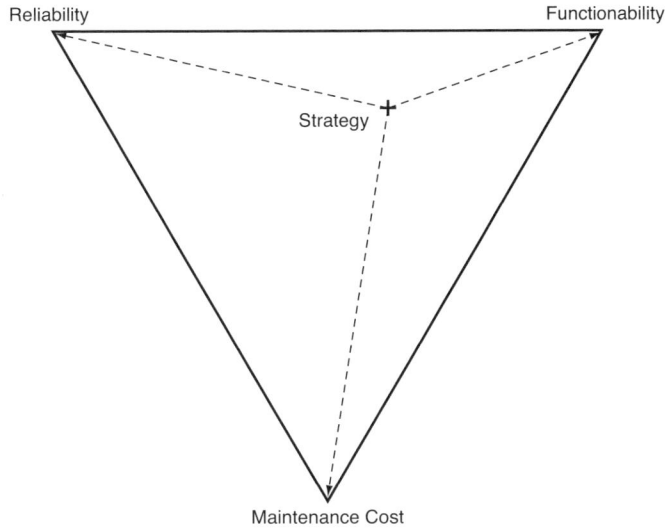

FIGURE 9.7   Maintenance strategy model

increase environmental risk (e.g., the failure of an automatic valve in the plant's fire suppression system); therefore, that strategy should be one of ensuring maximum reliability.

The total cost (C_t) associated with an equipment failure mode is the sum of the three cost elements:

- C_m = Costs associated with repair and maintenance work
- C_u = Costs associated with the unavailability of the equipment while it is shut down for maintenance
- C_r = Costs associated with increased risk (safety and environmental) as a result of the failure

Therefore, the optimum equipment maintenance strategy is one that minimizes the sum of these three cost elements over all equipment failure modes, as follows:

$$\text{Min}\, C_t = \sum_n (C_m^i + C_u^i + C_r^i) \qquad \text{for all failure modes } i \qquad \text{(EQ 9.2)}$$

An equipment maintenance strategy focusing fully on availability would make excessive use of condition-based maintenance (CBM) and opportunity-based maintenance (OBM), resulting in a very high equipment maintenance cost (C_m), which could only be justified for critical plant equipment (e.g., grinding mills, mine hoist), where the costs associated with unavailability (C_u) are very high.

An equipment strategy fully focused on reliability would require an excessive life-based maintenance policy with very conservative age criteria, leading to a very low availability and very high maintenance cost (C_m). Therefore, an excessive focus on reliability can only be justified when failure represents a high risk for the equipment itself, the safety of people, and/or the environment $(C_r$ very high). This is the case when maintaining the mine hoist ropes, the fire suppression systems, or the safety instrumentation of the rotary equipment and belt conveyors.

An equipment maintenance strategy fully focusing on direct maintenance cost would be based on failure-based maintenance (FBM) and would result in low availability, poor safety, and

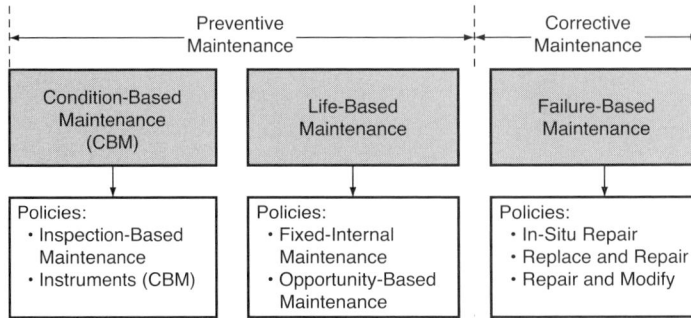

FIGURE 9.8 Maintenance policies

higher risk on assets and the environment. This strategy should only be applied to equipment in redundancy or where operational requirements are low and safety is not affected.

In summary, the detailed characterization of the equipment strategy requires (1) the identification and full characterization of all possible modes of failure for the equipment, and (2) the determination of a specific line of action for each failure mode, in alignment with the equipment strategy (reliability, functionability, or maintenance cost). These lines of action are referred to as a *maintenance policy* as described in the following paragraphs.

Maintenance Policy

A maintenance policy is the course of action or management decision criteria used to cope with a specific failure mode of a piece of equipment or system. For example,

- The failure mode caused by wear of the impeller of the mill pump is age related and predictable. Therefore, the course of action (policy) might be to replace the impeller before the pump fails (preventive maintenance).

- The failure mode of the pump electric motor caused by a power tension peak is unpredictable. Therefore, the logical course of action is to replace the motor after failure (corrective maintenance).

- The failure mode caused by wear of the drum bearings of a mine hoist is age related and predictable. But in this case, the consequences of failure and the bearing replacement cost are significant enough to justify the installation of temperature and vibration control instruments capable of anticipating failure so that the replacement action can be planned in advance when temperature or vibration levels reach a predetermined level (conditional maintenance).

In Figure 9.8, the different maintenance policies are classified into three groups:

- Failure-based maintenance (FBM) includes policies where the maintenance action in relation to a given failure mode is performed after the failure has occurred. Three different criteria can be used: (1) to repair the failed device while the equipment remains shut down; (2) to replace the device, restore operation, and then repair the device later; and (3) to change the design of the equipment by replacing the device with another that has higher reliability so that the mode of failure is eliminated or its frequency of occurrence reduced (modificative maintenance).

- Life-based maintenance (LBM) includes policies where the maintenance action takes place before failure and at a time determined by the age of the element related to the failure mode; this is the operating time elapsed since the previous failure mode maintenance action. A good example of this policy is the motor oil change at fixed hour-meter intervals (e.g., every 200 hours), made to prevent engine overheating failure. In some cases, the shutdown of a piece of equipment to apply LBM to a given failure mode is used as an opportunity to act on other failure modes that are close to requiring maintenance (OBM).

- Condition-based maintenance (CBM) includes polices where the maintenance action takes place before failure and timing is determined by a continuous discrete assessment of the operating condition of the equipment relative to a failure mode obtained from embedded instrumentation, portable measuring devices, or simple inspection, aiming to perform maintenance only upon evidence of need. Condition assessment is indirect; this is resulting from objective measurement of a variable or parameter (vibrations, temperature, pressure, oil analysis, etc.) or subjective assessment (look, smell, noise, etc.). For example, when a roller bearing is close to failure, the level of noise, vibration, and temperature increases.

Sustainable Management Strategies: The Reliability-Centered Maintenance Process

Analysis of maintenance policy in the airline industry in the late 1960s and early 1970s led to the development of reliability-centered maintenance (RCM) concepts. The principles and applications of RCM were first documented by Nowlan and Heap[25] in 1978; more recently, RCM has taken on a prominent role in the National Aeronautics and Space Administration's facility and equipment maintenance and operations program.

RCM is a standardized analytical process used to determine appropriate maintenance strategies that ensure safe and cost-effective operations of a physical asset in a specific operating environment. This method analyzes the equipment/system functions; identifies their safety, environmental, and economic priorities; and directs management focus toward those units that are critical from the reliability, safety, environment, and production point of view.

The key conceptual difference between conventional preventive maintenance (PM) and RCM is that although PM methodology focuses in avoiding failure, RCM focuses on avoiding failure consequences (not necessarily by avoiding failures).[26]

Equipment failure can have undesirable consequences in several aspects, such as

- Personal and equipment safety,
- Environmental health/compliance,
- Operations, and
- Economics.

RCM analysis is a powerful analytical process for optimizing maintenance strategies by the systematic assessment of management risks derived from failure consequences and determining the best maintenance policy to mitigate said risks. The process is based on answering seven questions about the equipment functions and the failure modes under review:

1. Equipment functions: What are the functions and associated performance standards of the equipment?

2. Functional failures: In what ways does it fail to fulfill its functions?

3. Failure modes: What causes each functional failure?

4. Failure effects: What happens when each failure occurs?

   5. Failure consequences: In what way does each failure matter?

   6. Possible preventive actions, if any: What can be done to predict or prevent each failure?

   7. Possible corrective actions: What should be done if a suitable proactive task cannot be found?

The maintenance policy selection process is outlined in Figure 9.9.

Analysis of Failure Modes of Complex Systems

From the maintenance perspective, industrial plants are a complex system (Figure 9.10) integrated at three functional levels (plant, equipment, and elements), subject to multiple failure modes. Industrial plants can have several processes, hundreds of pieces of equipment, and thousands of different failure modes. It is therefore impractical to attempt to apply RCM or any other failure analysis tool to each and every failure mode in a complex system.

Analyzing the grinding-flotation system of Figure 9.11, for example, and taking all 62 different failure modes of equipment, the first step in the analysis would be to identify those criteria that make a failure mode significant (e.g., the frequency of occurrence, equipment downtime, repair cost, safety, and environmental constraint).

In this example, it can be assumed that failures/year and failure-mode-related cost (US$/failure) are the two factors that really matter. The product of these two factors gives the annual requirement value (ARV)—in this case, the annual failure-related cost associated with a given failure mode, including direct repair and maintenance work, costs associated with the unavailability of the equipment, and costs associated with increased risk (plant assets, safety, and environment) associated with the failure.

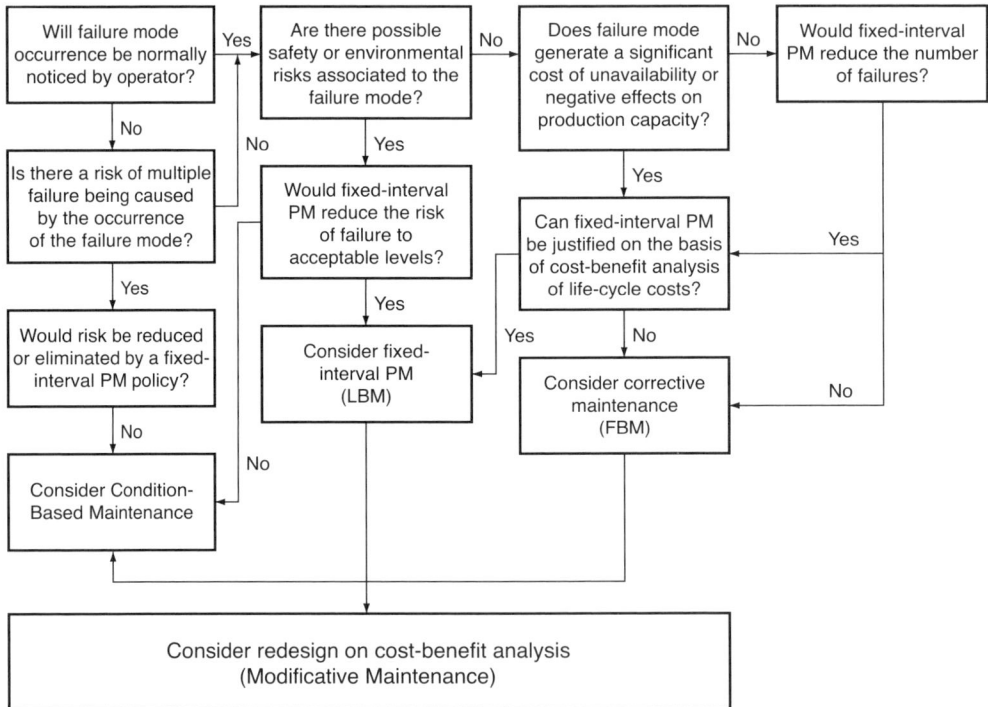

FIGURE 9.9   RCM process for the selection of maintenance policies

Using the ARV, the failure modes are ranked in descending order; the most important failure modes will appear at the top of the list. Then the cumulative ARV is plotted against failure mode name or rank number (1 through 62). The resulting graph, known as a Pareto diagram, will approximately follow the famous Pareto 80-20 rule (20%–25% of failure modes will account for approximately 75%–80% of the ARV). Alternatively, as shown in Figure 9.12, failure consequences are classified into three groups: group A (the critical failures) includes 14 failure modes that amount to 75% of the total ARV; group B (important failures) contains 13 failure modes that amount to an additional 15% of the total ARV; group C (minor failures) includes 35 failure modes that account for the remaining 10% of the ARV.

After all failure modes have been analyzed and classified by their consequences (ARV), a maintenance plan can be prepared along the following lines:

1. Apply RCM methodology to groups A and B (25 failure modes amounting to 90% of the ARV) and determine an optimum maintenance policy for each mode.

2. After a maintenance policy has been selected for groups A and B, develop a probabilistic failure-repair model for each failure mode in these groups.

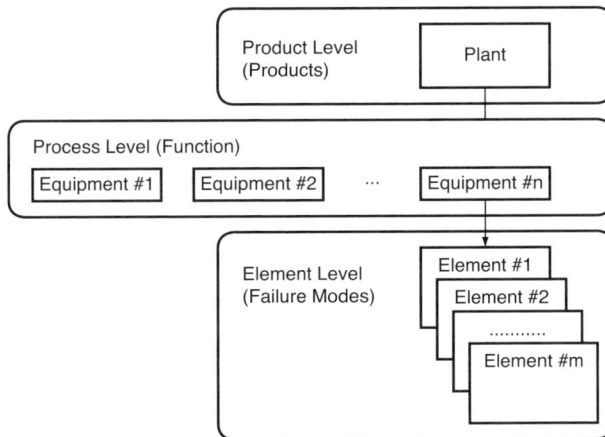

FIGURE 9.10   Plant maintenance model

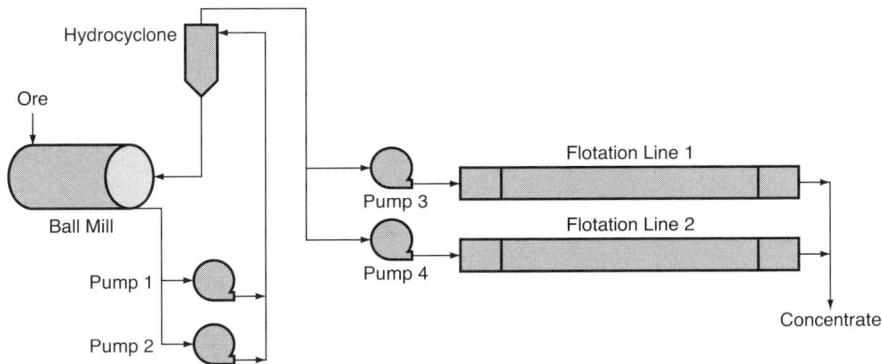

FIGURE 9.11   Grinding-flotation system flowchart

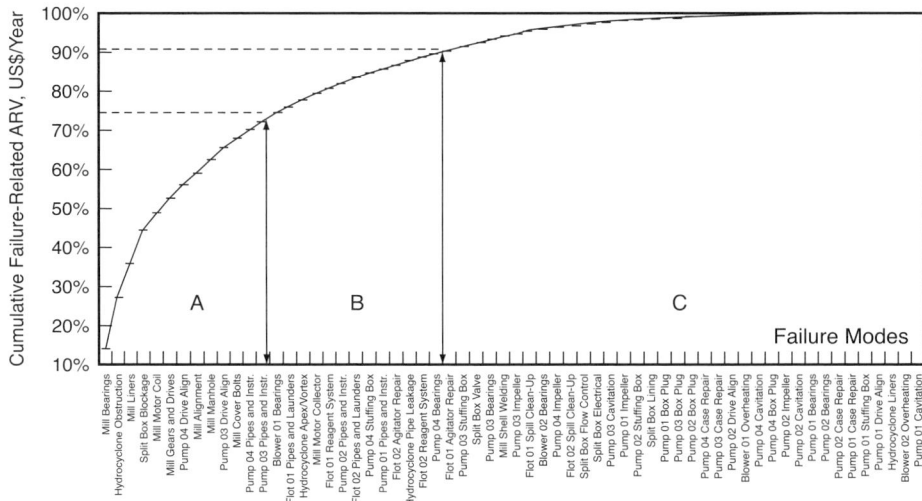

FIGURE 9.12   Pareto diagram of maintenance cost versus failure mode

3. Apply the FBM policy to all 25 minor failures without further analysis, and develop a global availability model to analyze the overall economic and operation consequences.

Failure Modeling

The behavior of a piece of equipment or a system can be predicted by probabilistic modeling of the functions of reliability and maintainability for each of the equipment failure modes. The reliability of the equipment with respect to a given failure mode can be modeled by a probability function $R(t)$, defined as the probability that the equipment that has been in continuous operation during a time t will not have a given failure mode during an infinitesimal time dt after t. Similarly, the maintainability of equipment with respect to a given failure mode can be modeled by a probability function $G(t)$, defined as the probability that the equipment that has been in under repair during a time t will be operational during an infinitesimal time dt after t. Note that reliability $R(t)$ represents the "probability of survival" and that the function $F(t) = 1 - R(t)$ represents the "probability of failure" relative to a given failure mode.

Two failure probability distributions are most commonly assumed: (1) the exponential function, mainly used to model accidental (random) failure with a constant failure rate with time and repair time (maintainability), and (2) the Weibull distributions mainly applied to model life-based (wear out) failure modes.

For a more detailed analysis of failure modeling, consult Blanchard,[27] Kelly,[28] and Moubray.[29]

Modeling of Major Failure Mode

When a failure mode has occurred enough times, a representative sample of times between failures and the repair times is available and may be used to fit a failure model for the probability of failure, $F(t)$, and the probability of repair, $G(t)$. Then, these models can be used to evaluate the costs and maintenance resources associated with a maintenance policy.

Probabilistic modeling is a powerful tool for maintenance planning but is time-consuming and requires failure data that is not always available. Therefore, probabilistic failure modeling

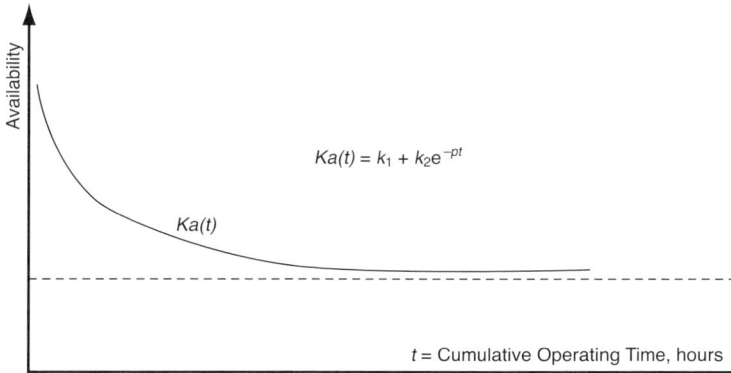

FIGURE 9.13   Modeling availability relative to minor failures

should only be used for major failure modes of high economic significance. Minor failure modes can be analyzed globally through their overall effect on availability, as described in the following paragraphs.

Modeling of Minor Failure Mode

The availability of a piece of equipment can be modeled by the probability function $A(t)$, defined as the probability that the equipment or system is operating after a time t after it was placed in service. Availability is adequate to characterize the overall performance of complex systems subject to repair and replacement. A practical approach to model availability is to compute the availability coefficient (Ka) or mean value of the availability $A(t)$ over finite periods (day, month, year, etc.).

$$Ka = \frac{T_o}{T_o + T_p} \tag{EQ 9.3}$$

where
  T_o = operating time for the period
  T_p = repair downtime

The Ka values for equal and consecutive period can then be fitted to a deterministic mathematical model allowing for the estimation of availability over time, relative to minor failure modes. Usually, minor failure availability fits well to a three-parameter exponential function (see Figure 9.13) of the type:

$$Ka(t) = k_1 + k_2 e^{-pt} \tag{EQ 9.4}$$

Although, in theory, this model is valid for estimating the overall availability of any complex system, its practical application is limited to minor failures (high frequency and low repair time). This is because major failures (low frequency and high repair time) would introduce a large error when including failure modes with frequencies of the same order of magnitude as the time period used in the model (normally 1 year).

Conclusions

Responsible mine maintenance management is critical to operations. Mine and mineral process-ing equipment can fail in many different modes, and, in some cases, failure has a negative impact on safety, equipment integrity, and the environment. In this regard, maintenance management should be based on the analyses of equipment functions, aiming to identify safety, environmen-tal, and economic priorities, and concentrate on the maintenance strategies for those pieces of equipment and systems that are critical from the reliability, safety, environment, and production point of view.

CASE STUDY: GRADE CONTROL SYSTEMS AT EL VALLE-BOINÁS MINE*

Grade control refers to the process of assessing economic areas during mining operations. The objective is to determine which geological resources become reserves and thus can be mined at a profit. Grade control also involves the supervision of the mining operation in order to make sure that economic ore is sent to the plant and waste to the waste dump.

The scope of the grade control function varies with the characteristics of the mine; there-fore, each mine will have its particular grade control system. Furthermore, for a specific mine, the grade control function cannot be defined before mining startup; it will be only after operation begins and several steps have been taken that the initial grade control system will reach its opti-mum state.

Depending on operational requirements, the grade control function can mean a simple visual delimitation of the ore–waste boundaries or a highly sophisticated process. In the present case study, in which visual delimitation of the ore–waste was not possible and grade variations were so vast, the ore–waste boundaries could only be defined through interpretation based in close pattern sampling and assaying.

This case study describes the scope and results of the grade control system that was imple-mented at the El Valle-Boinás open-pit gold mine in Spain, operated by Rio Narcea Gold Mines, S.A.

Project Goals and Objectives

The project had the following objectives:

- To establish a grade control system using state-of-the-art commercial software and hard-ware and also in-house software to optimize the process;
- To reduce the production cycle to a minimum to ensure that planned production was achieved;
- To implement the system gradually, thus avoiding taking the next step before the preced-ing one was verified and fully operational; and
- To assess system performance by means of reconciliation of mine and plant production data.

Work Methodology

The project methodology was planned along the following courses of action:

- Visiting similar mines in several countries
- Compiling existing public information on the subject, obtained from different publica-tions and the Internet
- Starting with a system based on RecMin software, creating a simple and functional soft-ware application capable of exporting and importing information generated by other commercial mining programs, such as Datamine

* This section was written by C. Castañon.

  - Defining the ideal work methodology for each unit of the production cycle:
    - Type of sampling
    - Drilling system (core drilling, direct or inverse circulation)
    - Sample sizes
    - Sample preparation process
    - Sample assay methodology
    - Calculation and interpolation method
    - Criteria for the definition of ore–waste boundaries
    - Staking and demarcation of ore–waste boundaries in the open pit
    - Loading/haulage systems
  - Implementing quality control systems for each of these units

Geographical and Geological Framework

The El Valle-Boinás mineral deposit is located in Asturias (northwest Spain), approximately 65 km west of the city of Oviedo, in a hilly region with heights ranging from 300 m to 750 m above sea level.

The deposit is located in the Cantabrian zone, the outermost area of the Variscan orogen of the peninsular northwest, and is located in the core of the Asturian arc. Its occidental boundary is the Narcea antiform, which separates it from the west Asturian-Leonese zone and marks the transition to the internal areas of the Orogenics. All the Paleozoic systems are represented in the Cantabrian zone, although there are important stratigraphical differences among units. The structure is epidermic and is essentially made up of connected thrusts and folds.[30] The internal deformation is small and cleavage is only present in a few areas. Evolution of most of the Cantabrian zone took place in diagenetic conditions, and only a few areas underwent a low-grade or very-low-grade metamorphism.

El Valle-Boinás Mine

The El Valle-Boinás gold mine started operations in 1997 by open-pit mining from three pits located relatively close together. Today, open-pit ore has been mined out and mining continues underground. The treatment plant was originally designed to treat 500,000 t/yr of ore. Plant capacity was increased to 750,000 t/yr after a series of improvements in the process and in the plant facility.

During the 7 years of open-pit mining, a total of 4,600,000 t of ore were mined with an average grade of 5.8 g/t of gold. The aerial photographs (Figure 9.14) show the evolution of the mining works.

Courtesy of Rio Narcea Gold Mines

FIGURE 9.14   Evolution of the El Valle-Boinás open pit

Mineral Deposit Model

The Au–Cu mineral deposit in El Valle-Boinás can be considered of the skarn type,[31] because approximately 90% of the gold and 100% of the copper contained in its ore reserves are of skarn genetic origin. On top of the skarn, there is an epithermal mineralization, which is associated with the existence of acid porfidic bodies. These bodies essentially produce the silicification and argilization of the pre-existing rock and also produce gold, which is estimated to represent 10% of the gold found in the deposit. The resulting complex mineralization results in areas with very high variations of gold grades, with or without the presence of oxidized and nonoxidized copper. In addition, the ore is often composed of an irregular mix (brechia) of very soft clay rocks in contact with very hard silicified materials. All of these aspects have a very relevant influence on the selection of the steps of the grade control process.

Block Model

Block models are an essential tool when studying a mineral deposit for the estimation of geological resources and reserves. The technique is based on modeling the area of interest by parallelepipeds (blocks) of the same dimensions, each of which will be assigned a record in a database. Information about grade and any other properties that are required for calculations and studies will be recorded (e.g., lithologies, densities, analysis data, geotechnical data, and hydrogeological data).

The main advantage of a block model is the simplicity it brings to the digitization of the orebody and the application of computerized methods for interpolations, simulations, and the planning and design of open-pit or underground mining operations.

The block model of the orebody was developed using Datamine software. For this purpose, a borehole core database was created with the following data:

- Borehole collar coordinates
- Hole deviation data
- Lithological description of core
- Assay data of mineralized zones

The lithological information was obtained by geologists from borehole logging and by grouping lithologies according to significant changes in density, alterations, and mineralizations. Lithology was used to generate the geological sections and for modeling the orebody. Finally, the assay data, the orebody topography, and the rock type's densities were incorporated.

Computer modeling work consisted of digitizing into the Datamine software the different geological plans and sections of the orebody as interpreted by the geologists. After the sections with the different rock types were created, they were three-dimensionally connected to form a wireframe model. This is a 3-D envelope containing all the blocks of a given type of rock. The geological model of the orebody integrates the wireframe models for all the different lithologies.

Ore-bearing and waste-rock lithologies were determined through the statistical correlation of grades and rock types present in the deposit. When mineralization is strongly controlled by lithology, then lithology wireframes can be considered as ore envelopes. Otherwise, ore wireframes must be designed crossing the boundaries of the different lithologies.

To standardize on a fixed unit core length, the samples of different lengths were used to compose 4-m core length intervals by weighted average methods (composites). The composite length of 4 m was selected to match the block height and also the planned mining bench height.

The block model (with 4 m × 4 m × 4 m blocks) was generated using two methods: (1) nearest neighbor, and (2) inverse distance.

In the nearest neighbor method, the grade assigned to a block is equal to the grade of the nearest composite interval falling within a "cylinder of search" that is parallel to the general trend and to the dip of the orebody, always within the limits of the ore envelope that has previously been defined. The geology of each block is assigned using the same method.

In the inverse distance method, the grade assigned to a block is calculated as a weighted average of the grades of the composite intervals from several boreholes. This method takes into account the internal dilution of the block and therefore is more precise. In this case, a lentil-shaped "ellipsoid of search" was used.

Each block in the block model was identified with its x, y, z coordinates; ore grades for Au, Ag, Cu, As, and Bi; and other types of information, like resource classification, lithology, and so forth.

Specific Issues in Gold Mining

Gold mining introduces additional complexities to the grade control process:

- Working with very low grades, measured in parts per billion (ppb), makes assaying very time-consuming and expensive.
- The density of gold is very high (19.3 kg/L), which means sample preparation methodology must follow a very thorough procedure to avoid segregation during sample grinding and splitting processes.
- The nugget effect—that is, the gold grade variance associated with large gold grains (nuggets)—can be significant, and this entails very fine sample grinding in order to limit sample preparation errors. In addition, splitting the finely ground samples requires the use of specific equipment and work procedures.

In view of these factors, the design of the sample preparation method requires the prior completion of very detailed statistical studies aiming to evaluate and control the error. These studies must be repeated for the different zones within the orebodies because they can differ in their characteristics (e.g., degree of oxidation, alterations, grain size).

Open-Pit Bench Height

Typically, open-pit mines operate with different bench heights for the ore—where grade control is necessary—and for the waste.

The higher the ore bench, the lower the operating cost, but also, the less controlled the grade and subsequently the higher the dilution of the ore blocks grade with the surrounding waste blocks. This dilution effect will be higher for more irregular orebodies, such as where orebodies are irregular both in shape and dip. To determine the optimum bench height for the mining of the ore, the following constraints have to be taken into account:

- High benches are more productive. For a given production plan, there will be a minimum bench height below which the required production rates would not be achievable.
- Size of the mining equipment. The smaller the equipment is, the higher the unit cost will be and the lower the performance rates achieved. Ideally, the best approach is to use the same equipment sizes for both ore and waste material.
- The factor limiting the production rate for a given work area is the "productive cycle," that is, the total time required for drilling, blasting, sample preparation and assaying, and for getting the grade control information ready to define the ore–waste boundaries for the next cycle.
- The number of work areas, that is, the different mining areas in which work can be done independently from each other.

Ore Types

Ore types are defined by the following characteristics:

- The shape of the ore–waste boundaries in the ore zones. The higher the regularity of these boundaries, the easier the grade control function will be. The regularity in the vertical direction is most important, because it significantly affects the choice of bench height.

- Ore hardness. This is a very important property because soft ores are easier and cheaper to blast and the amount of blast displacement—an important source of error when staking ore–waste boundaries after blasting—is low. When blasting produces large ore displacements, it might be necessary to blast the waste in a first stage, followed by separate ore blasting, considerably delaying the productive cycle.

- The variability of the grades. If there are important changes in the grade in a short distance, much more detailed grade control is needed. It is also important to determine if there are any waste areas isolated within the ore.

Mine Production Startup

When a mining operation begins, the initial grade control concepts are normally revised and optimized as discussed previously.

In the block model case, the initial bench height was 4 m, which was selected to be equal to the height of the model blocks. Also, the elevations of the benches were designed to coincide with the elevation coordinate (z) of the block base in the model to allow for an easier updating of the block and the borehole model databases as new information was generated.

As mentioned previously, it was not possible to delimit the ore–waste boundaries visually, which meant that ore sampling and assaying was required.

Taking into account that mining activities take place above the ore zones that will be mined with descent from one bench to another, there are several possible options for ore sampling:

- Collecting samples from the top surface of the bench where mining is taking place. Samples can be collected from panels, channels, or short inclined boreholes drilled in close pattern. The main problem with this system is the sample contamination caused by heavy equipment working on top of the bench. Furthermore, this method is not advisable for soft ores, when crushed rock fill might be required to repair working surfaces after heavy rains. With harder ores, it might be an acceptable system.

- Sampling from boreholes. In this case, three types of borehole drilling can be distinguished:

  - Core borehole drilling. Without doubt, this is the best sampling method because the entire core of rock drilled can be seen and it can be assayed in part or in full. However, the high price and the longer time required makes this method seldom used for grade control.

  - Reverse circulation drilling. This is a valid option when samples at different depths are sought in order to get assay data from several benches in advance of mining. In this drilling method, drill cuttings circulate inside the drilling pipes, thus eliminating contamination from the hole walls, which can be important in gold mining. However, the cost of this drilling method is around seven times higher than direct circulation drilling.

  - Direct circulation borehole drilling. This is the most used method, mainly because samples can be taken from blastholes with no additional drilling cost, but also because it allows for a closer drilling and sampling pattern wherever needed at a lower cost. As a drawback, the quality of the sample is less than when using inverse circulation methods, especially when attempts are made to take samples in the same borehole at different depths. However, this disadvantage is balanced by the possibility of reducing the sampling error by drilling and sampling with much closer patterns at a lower price.

Grade Distribution

One of the first steps to take at the beginning of the mining in one area, or whenever the ore type changes, is to perform enough testing and assaying to determine the distribution of gold values within mineralized zones. This can result in a significant change of grade control procedures that might require higher or lower levels of complication for each ore type.

The objective is to determine whether the gold contained by the orebody is distributed in a more or less homogeneous manner or, on the contrary, is concentrated in certain zones of the orebody. These can be small faults or alterations, contacts between different lithologies, specific rock types, and so forth. In the first case, grade control would be easy because it would consist of just delimiting the external boundaries of the orebody. In the second case, it would be more complicated because as ore is concentrated in some zones in the orebody, the amount of information and the sampling density must be increased to allow for the definition of ore–waste boundaries and to be certain to avoid sending ore to the waste dump or waste to the plant.

The procedure used at El Valle-Boinás was to define a volume of 27 m × 5 m × 5 m within the ore zone to be mined in the first bench of the mine and to obtain the following information from it:

- Lithological information and assay data from three lines of boreholes, each drilled with a different method (core holes, reverse circulation, and direct circulation);
- Detailed geological interpretation from several mining faces corresponding with longitudinal and cross sections of the volume under study, including type of alteration, joints, small faults, lithologies, and so forth; and
- Sampling from the said mining faces, using different sampling methods, such as panels of different sizes, linear channels, and spot samples from fault fill material, zones of alteration, lithological contacts, and so forth.

Figure 9.15 shows a section with the detailed geological information, the boreholes, and the sample assays taken from both the boreholes and the face samples. It was mainly concluded that there could have been small local zones with very high grades, as much as 950 g/t. Although this can be seen in borehole samples, it is clear that there is no homogeneity in the grades. This means that there must be a borehole sampling pattern as tight as possible and that benches of the minimum possible height must be mined.

Work Cycle

As mentioned previously, considering the great variability of gold grades and the small size of the open pits, which only allowed for a maximum of two independent work areas, the grade control system had to be based on systematic sampling and on the minimum bench height that would allow achievement of the planned production rates—which were around 14,000 t of ore per week using medium-sized mining equipment (the plant treated an average of 2,000 t/d, 7 days a week).

Mining equipment can operate at an average production rate of 300 m³/h when loading ore. However, it's important to remember that loading ore is not as productive as loading waste with the same type of equipment due to the need to mine ore and waste selectively at the same working faces.

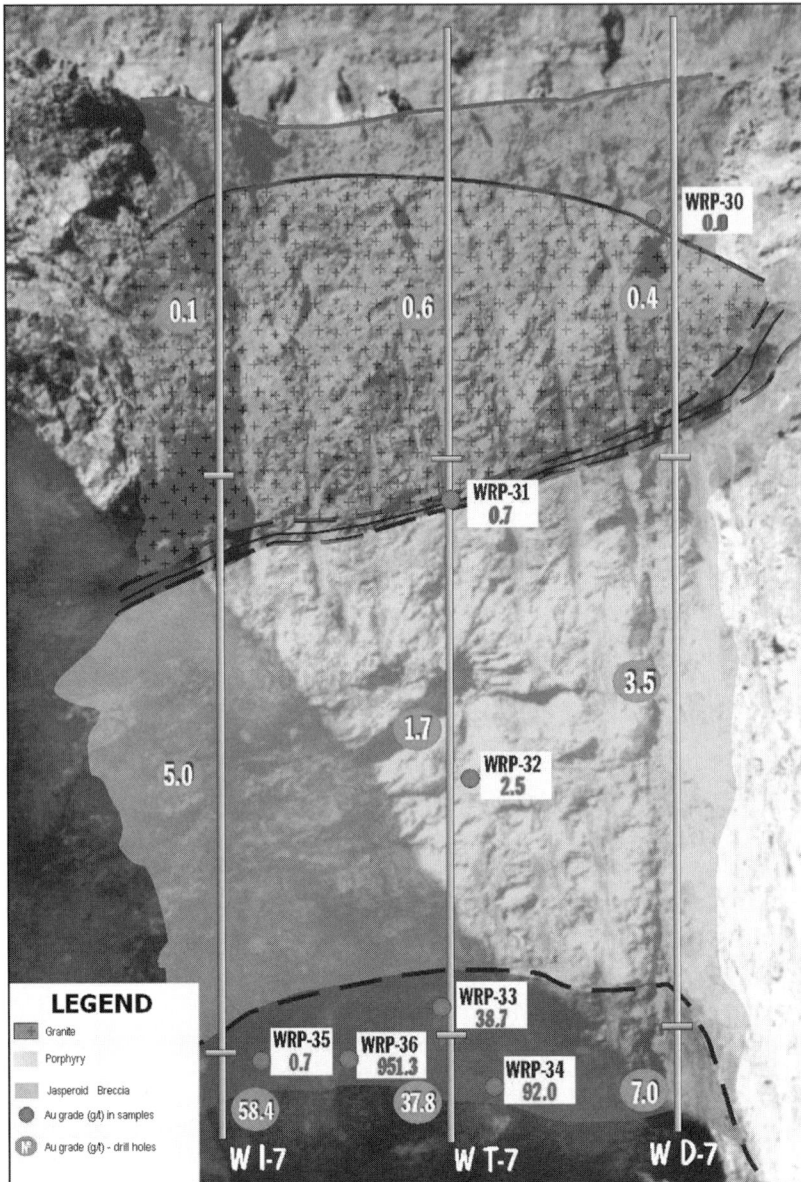

Courtesy of Rio Narcea Gold Mines

FIGURE 9.15   Detailed geological section

Bench height was fixed at 4 m, coinciding with the block model. Benches of less than 4 m would lead to loading problems and difficulties in achieving the desired production rate, among other problems.

Mining equipment works in 10-hour shifts, so a minimum of 3,000 m^3 must be staked for loading at the beginning of the shift. This means that for a bench height of 4 m, a minimum working area in ore of 750 m^2 for every shift is needed.

TABLE 9.2   Time chart of a complete mining cycle

Operation	Time, h
Drilling and sampling	8
Sample drying	20
Sample preparation	8
Sample assaying–ore grade	8
Sample assaying–plant samples	2
Staking ore zones by surveyors	2
Loading and haulage	10
Total	58

Source: Data from Rio Narcea Gold Mines

For grade control drilling and sampling, direct circulation borehole drilling was employed, using the same holes used for the blasting but increasing the sampling density by adding a hole in the center of the blasting pattern. As explained later, the resulting drill pattern was 3 m × 2.5 m, which means a total of 100 boreholes for the defined work area (750 m^2). Therefore, there were 100 samples to collect, prepare, and analyze.

Sample preparation was estimated to require 28 hours—20 hours for drying the ore, and 8 hours for sample preparation (grinding and splitting). Processing the samples at the laboratory was estimated to take 8 hours of work.

For data processing and the definition of ore–waste boundaries, a computerized system based in RecMin software (Mineral Resources) was arranged. The system processed all the information on hole survey data, lithology of drill cuttings, and sample assays data required for the interpretation and interpolation processes and the definition of the ore–waste boundaries in 2 hours.

The time for staking of ore–waste boundaries by the surveyors was also estimated at 2 hours. The total time to complete each production cycle is summarized in the timetable in Table 9.2.

Although the total duration of each cycle was estimated at 58 hours, not all the processes work on a 24-hour-per-day schedule, and some of them—like sample preparation and assaying—must be shared with other needs like exploration drilling, plant assaying, and so forth. Therefore, the drilling equipment was scheduled to work from 8:00 AM to 6:00 PM, Monday through Friday; sample preparation and laboratory worked 24 hours per day, 7 days a week; and geology and surveying worked from 9:00 AM to 6:00 PM, Monday through Friday.

The timetable shown in Table 9.2 outlines a complete example cycle, from the initial drilling in the work area to loading, transportation, and bench preparation for the initiation of a new cycle. Three days days were necessary to complete each cycle; therefore, three times the calculated area for each cycle was needed to be able to keep the loading and transportation teams working. Therefore, 3 × 750 = 2,250 m^2 was needed.

Given the use of two loading and transportation teams working in independent areas, a total of two areas was needed, each with 2,250 m^2 of available surface to achieve the planned production of 6,000 m^3/d that—at an average density of 2.2 t/m^3—would yield around 13,200 t/d. Therefore, it was necessary to drill and sample a total of 200 grade control boreholes per day.

Within the total surface area that is sampled and assayed for grade control, considering that only about 25% is defined as ore, the grade control system would have to allow for an ore production of around 3,200 t/d, 5 days a week; this is a total of 16,000 t/week.

The case described here is an ideal one. In reality, plans typically call for intensifying the work during the summer to increase the stock of ore, then slowing down in winter, when fewer natural-light work hours are available and more downtime can be expected because of weather conditions.

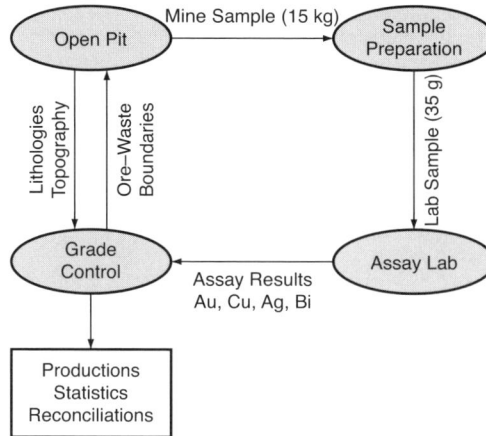

FIGURE 9.16 Grade control information flow

Information Management

Managing around 200 samples per day means that approximately 3,000 kg of samples must be transported, dried, prepared, and assayed every day and also that the grade control team must process the lithological and analytical information. This represents generating a great deal of information, such as borehole coordinates, borehole and sample coding, lithological information from each sample, analytical data from several elements in each sample, and so forth. In total, 3,200 pieces of data must be processed each day, not taking into account the information generated later when processing this data in terms of interpolation with the block model, marking ore–waste boundary lines for the economic areas, and so forth.

The diagram in Figure 9.16 outlines the type of information that is generated and where.

To allow managing all this information in a fast and safe manner, a system for coding, reading, and generating the information using barcodes was implemented and linked with the Rec-Min software. This system interconnects all departments and work places using a local intranet with wireless connection with the open pits, thereby allowing all the information to be centralized in a single database on the server. Bar coding was used to eliminate the need for each department to duplicate the process of keying in codes, thereby eliminating time losses and reducing the risk of making mistakes.

The coding was designed in such a way that the reports generated are automatically transferred to the corresponding place in the database. Sample codes were designed to be read from right to left, so that the last item to be read is the borehole code that represents where the sample was taken. It was designed that way so that it is the name of the borehole that varies in size; the code of the sample also begins with the name of the borehole, which makes identification easier.

In addition to the grade control sample codes, other barcodes were designed for other types of samples—like exploration boreholes, treatment plant samples, and so forth. All these allowed for easier information management and reduced operating errors.

Drilling and Sampling

The grade control boreholes were drilled with the same drilling equipment as the blastholes, that is, hydraulic drilling machines with a hammer at the head. These machines were modified in some areas so they could be used to collect samples.

The drilling pattern was defined as 2.5 m × 3 m, with 3 m in the longitudinal direction of the mineralization and 2.5 m in the transversal, with the objective of better limiting the transverse ore limits with more boreholes.

The holes were drilled vertically, 3.5 in. in diameter and 4 m long, which is equal to the bench height. In some areas, two benches were drilled in a single drilling pass, collecting two 4-m samples where each bench was mined separately. By doing this, the work cycle was shortened, but the quality of the sample from the lower bench might be reduced because it could become contaminated by the walls of the first 4 m.

The total quantity of drill cutting in each sample ranged between 52 and 68 kg, depending on the density of the material. In order to reduce the size of the samples, which was too large for handling at an acceptable cost, a sample splitter was prepared and placed under the cyclone of the machine. This method allowed separating a sample from 13 to 17 kg (one-fourth of the total cuttings) directly into a bag.

Part of the drill cuttings are lost and eliminated by the drill through another duct. It makes up around 5% of total cuttings and contains the finest part. To make sure that the error made discarding it is acceptable, tests were conducted comparing both the cyclone cuttings and the fine discarded cuttings. The results gave grades sufficiently close to consider that the error was not significant.

Geological Description and Sample Labeling

The drilling of grade control boreholes is subject to continuous monitoring by a geologist with the following functions:

- To check the 3 m × 2.5 m borehole pattern. This ensures it has been correctly staked and marked on the ground.
- To tag the stakes with the borehole codes according to the established order. These stakes will be next to the borehole permanently and will be used to identify the lithological and topographical information and the samples correctly.
- To tag the bags with samples that are produced by each drilled borehole with the corresponding codes. A copy of the code is placed inside the bag so it can be later used in the drying, preparation, and laboratory processes.
- To define the lithology for each sample. This information is entered in the database.
- To make sure that the boreholes are surveyed after they are finished.

Considering that samples are made up of drill cuttings, their description by the geologist is greatly assisted by the experience acquired from visual inspection of the ore drilled at the time of loading.

To enter the data, the geologist has a RecMin module in a personal digital assistant that allows for easy and quick input of the information, as well as for wireless transmission from the pit itself to the central database.

Sample Preparation

The preparation of samples for analysis in the laboratory is a delicate process for several reasons:

- The high density of gold (19.3 kg/L) facilitates the segregation of gold from ore. Because of this, sample handling and splitting processes must be done according to very specific rules.
- Sample grades are measured in parts per billion (ppb) of gold, and these are very small amounts, which require working with samples as large as possible in order to reduce sampling errors.

- The nugget effect is also important for sample size selection. For a given sample size, when gold is present in small grain sizes, sampling error is reduced; but conversely, if the gold grain sizes are large, sampling error will be larger and can only be reduced by taking larger samples, which in turn, would mean a more costly and labor-intensive sample preparation process.

If one imagines drilling a borehole every 2.5 m × 3 m × 4 m for a total of 30 m^3, or around 66 t, it becomes obvious that by sampling and assaying all 66 t, the sampling error would be nil. Given that this is impossible, then the sampling problem consists of taking a sample of a smaller size that is representative of the entire 66-t block of ore. Obviously, the smaller this sample is, the greater the chance of error.

The error is also dependent on the size of the grains of gold. The bigger these are, the larger the error. If all the gold is in just a few large gold grains, it will be very difficult to get a representative sample. Taking this idea to the extreme, if all the gold in the block was in the form of a single nugget, then the sample taken would either contain the nugget or would contain no gold. Conversely, if the grain size is small, the number of grains will be larger and any sample taken will be more representative of the whole.

To be able to determine the best sample preparation and assaying method, some comparison tests must first be performed to define the maximum margin of error that would be acceptable at a reasonable cost.

The tests were performed using the following procedure:

1. The first step involves looking at the size of the gold grains based on one or several representative samples. The gold grains are separated using gravimetric methods, then studied in the laboratory under a microscope in order to define the larger sizes that can be expected. With these tests, the size of the sample and the degree of grinding required for a sample as homogeneous as possible within a reasonable cost can be defined.

2. The 15-kg sample that is collected from the drill cuttings undergoes a process of drying and assay sample preparation, which yields a final sample of approximately 30 to 100 g, ground to a size of approximately 100 microns. This process entails three grinding processes, each followed by a sample splitting process. To optimize the sample preparation process, testing must be undertaken that involves comparing similar samples ground at different sizes and with different sample size reductions in each step. Using a statistical comparison of the results, a process can be designed that yields an acceptable error at a reasonable cost.

3. The drying process consists of drying the sample in an oven at 90°C for around 20 hours. This step aims to prevent contamination between samples caused by sample grains getting stuck to the grinding rollers and disks. Also, a sample of gold-free silica is ground between consecutive ore samples in order to clean the grinding mill.

4. The sample preparation laboratory has one roller mill, one disc LM5 mill, and one LM1 or LM2 disk or micronizing mill. In each of these mills, the size reduction tests necessary for the sample to be ground need to be conducted. The results will depend on the hardness and the size of the ore samples. If different types of ore are used, different processes must be defined.

5. The drill cuttings samples are smaller than 6 mm (>95%). For the roller mill, after it is adjusted, the size reduction required and the average amount of time needed for the preparation of the sample will be tested. With respect to the LM5 and the LM2 or LM1 disc mills, different grinding tests for each type of ore must be performed, varying the grinding

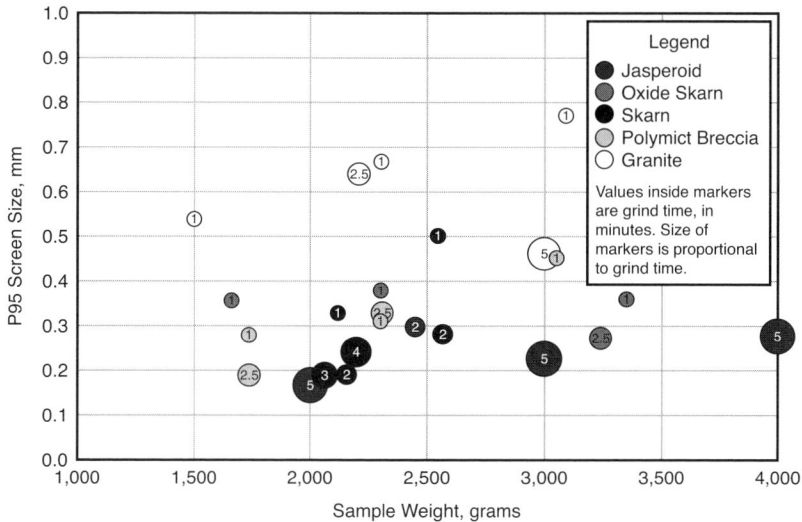

Courtesy of Rio Narcea Gold Mines

FIGURE 9.17 Sample preparation tests: size versus sample weight and grinding time

times and the attained sizes. In this way, graphical representations, as shown in Figure 9.17, can be obtained in which the size reductions in relation to the sample weight and the grinding time can be seen.

6. Finally, after the results of the tests have been studied, several sample preparation scenarios will be defined. The lower the cost of the preparation process, the larger the error that will be incurred. At the end, the scenario that yields the lower error at a reasonable cost will be chosen.

In the graph in Figure 9.18, the amount of error incurred in one of the cases studied can be seen. As shown, the process began with a large sample, approximately 90 kg, which was then split at the drill down to 17 kg. After drying, it was ground with the roller mill down to 2 mm maximum grain size (>95%). Then, the sample was split-reduced to 1,100 g, ground and split-reduced again to achieve a sample of 60 g. As shown in the graph, the total error incurred was around 25.1%, calculated as the sum of the errors obtained in the three grinding–splitting processes.

The sample preparation method finally selected is shown in Figure 9.19. The final sample size is 35 g and, as can be seen later, it will be divided in the lab into a 30-g sample for assaying of gold and a 2-g sample for assaying of copper and other elements. However, the sample bag that is sent to the laboratory with the prepared sample usually weighs around 100 g, of which the laboratory separates 30 g for gold and 2 g for copper and other elements; the rest is stored for several days in case it is needed for additional assays or if any assays need to be duplicated.

Sample Assaying

The size of the sample that will be assayed in the laboratory will be established after several tests with different sample sizes and a comparison study of the error incurred.

Usually, assay sample sizes will range between 30 and 100 g. In the case of gold, the assay procedure requires costly fusion and copellation processes; therefore, reducing sample size is important in reducing the final cost.

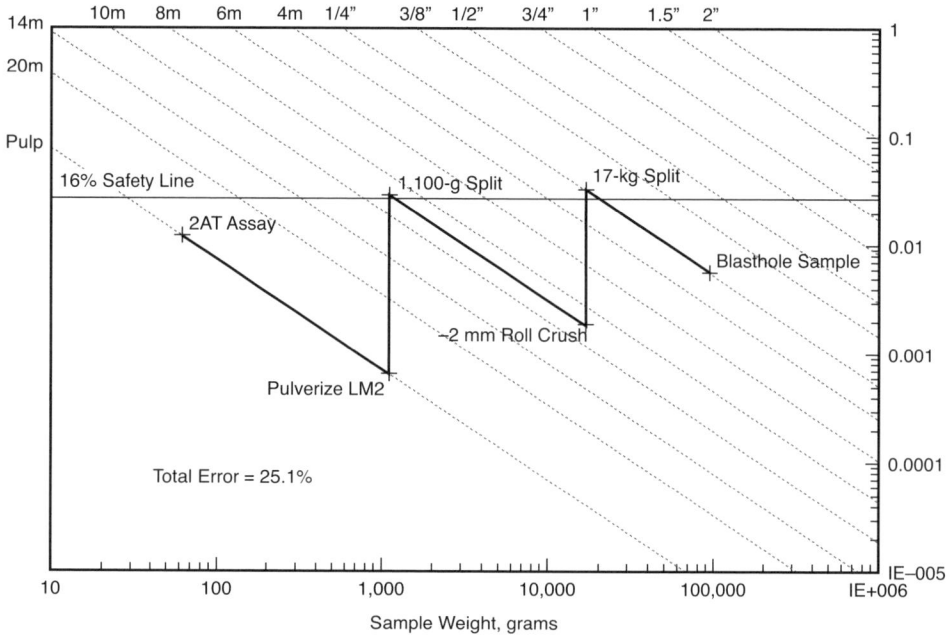

Courtesy of Rio Narcea Gold Mines

FIGURE 9.18   Sample preparation protocol required for acceptable accuracy on 4-m blasthole sample versus sample error

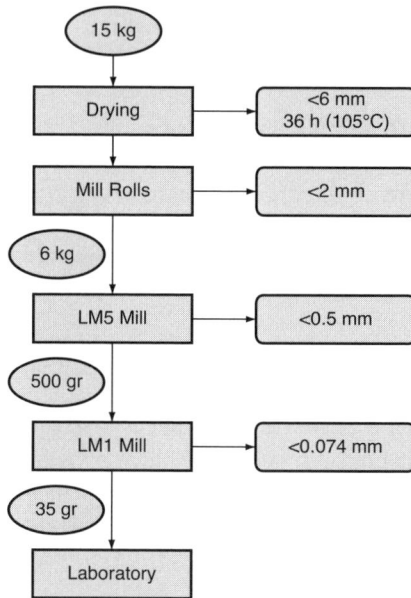

FIGURE 9.19   Sample preparation procedure

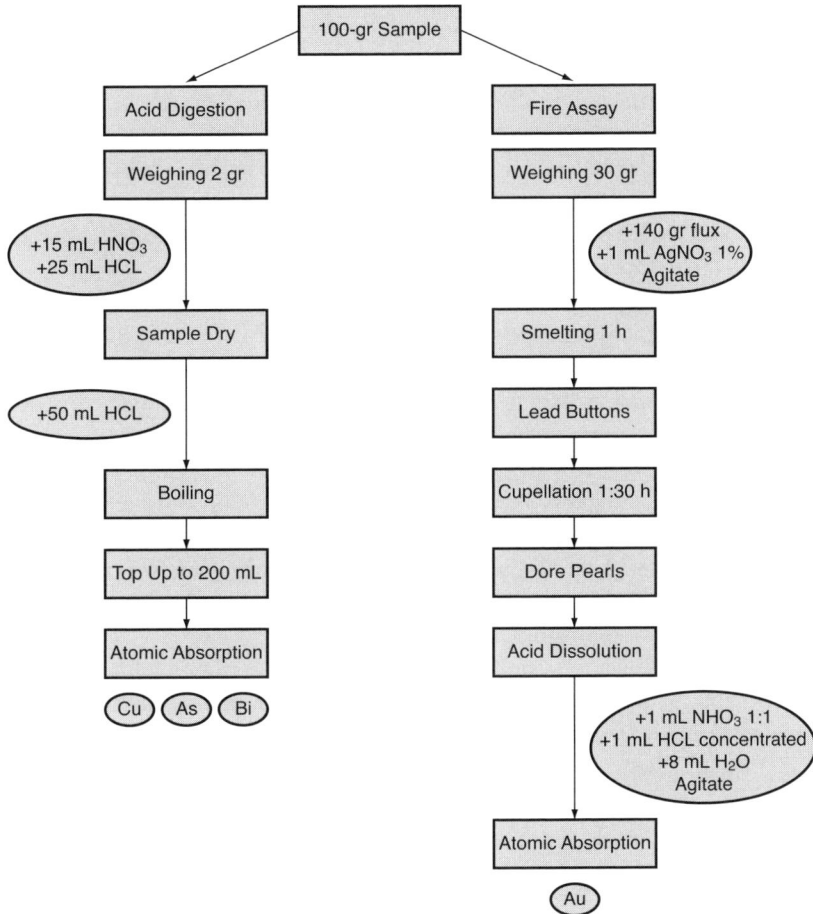

Courtesy of Rio Narcea Gold Mines

FIGURE 9.20   Grade control sample preparation procedure

The drawing in Figure 9.20 shows the process of analysis for the gold—which is the most important—and also for other elements like copper (Cu), arsenic (As), bismuth (Bi), and so forth.

From the results of the comparison tests, a standard process was defined for the gold assaying of 30-g samples. For high-grade gold samples and low continuity of the orebody, it was decided to perform two assays with two 30-g samples and calculate the average.

Data Interpretation

As mentioned before, all assay data is stored in a single database, which is managed using RecMin software. Every day, more than 3,500 data fields are input to the database and managed in a semi-automatic way by the grade control geologist, who also carries out the interpretation, the calculations, and the definition of the ore zones.

The definition of linking codes between the different data tables (topography, lithologies, assay data, borehole surveys, etc.) makes importing this information a simple and quick task. Graphic handling and interpolation are easily performed on the screen with color codes that simplify selecting the economic zones later and exporting the data to stake out positions in the pit.

The software program always works in three dimensions, which makes interpreting the information easier at all times.

Several methods were tested for defining ore zones. The simplest would be to draw lines to delimit the boreholes with grades above the cut-off grade and use those lines as ore–waste boundaries. In this case, ore grades would be calculated using the average grade value of the samples that fall within that drawn area. A more complicated method was attempted that involved doing a geostatistical grade estimation of block grades in a block model defined as $1 \times 1 \times 4$ m by applying a geostatistical interpolation method (Kriging) to the assay results, followed by the use of this block model to define the ore–waste boundaries.

Finally, interpolation by the inverse distance to a power was chosen. An ellipsoid of search was used to select which samples—from the bench being calculated and the bench below it—would be included in the interpolation.

The calculation parameters, like the power of the inverse distance and the size of the axes of the ellipsoid, are defined by means of geostatistical interpretation. These parameters are updated once or twice a year as more information becomes available. As an average, the power used at El Valle-Boinás was 3 and the distance from the main axes of the ellipsoid is 6 m for the principal axis, 3.5 m for the secondary axis, and 1.5 m for the third axis.

The spatial orientation of the ellipsoid of search is defined by the geologist on the basis of its position in previous benches and the dip of the orebody for each calculation area, given that it changes with the area that is being mined.

The definition of ore–waste boundaries is not as simple as drawing lines separating blocks according to a cut-off grade calculated from operating costs and metal prices. Other factors must be taken into account:

- There is an operational cut-off grade defining marginal ore that must be set aside for treatment at the plant when necessary. Given that a truck's load must be either taken to the waste dump or to the plant, the cut-off grade would be lower if just this cost of loading and transportation to the plant's stockpiles and the cost of treating it were considered. If the actual grade falls between both cut-off grades—the one that includes all the operation costs and the one just defined—this would be what is called marginal ore, and this must be separated and stored aside, because it contains enough gold to pay for the processing costs.

- In addition, the economic value of the ore is not solely from gold. The silver and the copper must be taken into account; for this reason, work is usually performed with a theoretical grade called *gold equivalent grade*. This grade is estimated as the grade of the principal element that alone would amount to the total net value of the block, taking into account the metal market value, the plant recovery, and the price paid by the smelter of the concentrate.

- The ore might also contain elements that, when present in the flotation concentrates above certain grades, are penalized by the smelter. This is the case for arsenic and bismuth. The grade control system makes it possible to prepare stockpiles with different concentrations of these elements so an "ideal blend" can be delivered to the plant to minimize penalizations. This is achieved, in some cases, by adding to the gold equivalent a negative gold value related to the grades of some of these elements. It could also happen that the hardness of the ore outlined by grade control is a problem for grinding in plant. In cases like this, it can be sent to a stockpile for future blending with softer ores to improve plant production.

- There are also some limits on the size of the ore zones. The minimum size is established by the grade control geologist, taking into consideration the size of the loader bucket, the direction of the load, and the geologist's experience in previous ore zones. Taking into account all this and the information available on the screen, the geologist adjusts the ore zone boundaries to a minimum in order to reduce dilution with the surrounding waste and, at the same time, achieve an acceptable loading performance.

- It is also important for the grade control geologist to have a precise idea of the value of each ore zone and to instruct the mining team accordingly. In some cases, there might be a small, very-high-grade ore zone (e.g., 400 g/t Au), which alone might be worth more that the rest of the ore in the bench. In cases like this, the grade control geologist must be able to convey to the mining team that the loading and haulage of the high-grade area must be done under the best possible conditions.

Summarizing all of these factors, the job of a grade control geologist is not limited to defining the ore–waste boundaries and where each ore zone is, but also to take into account other factors, like the ones described here. This makes experience a very important factor in the decision-making process.

Ore Loading

After the different ore types have been outlined and their destinations at certain stockpiles have been decided as described in the preceding paragraphs, the information is exported to the surveyors' team. These team members stake the ore zones on the ground and distribute the information to the operators and technical staff who are accountable for ore loading and haulage.

In addition to the usual operation tasks, special attention must be given to several other factors:

- Both the loader and the haulage truck units working with it must have a visible and easily identifiable flashing colored light, indicating whether a truck's load is intended as ore or as waste and the stockpile of destination. In the El Valle-Boinás mine, a red flashing light indicates that high-grade ore is being loaded and hauled to the plant's stockpiles, a green light indicates low-grade ore, and no light indicates waste. These security measures might not seem important, but they were decided upon after having experienced several cases where, perhaps because of fatigue, lack of information, or a desire to reduce truck cycle times, ore was dumped as waste, resulting in economic losses.

- Another way of controlling the movements of the ore would be to use a global positioning satellite dispatching system, linking trucks and loaders through a wireless network with the servers, so that information on the movements of the trucks, the type of ore, its destination, and so forth is available in real time. This system can also be complemented with the use of a warning signal for those cases in which the movement of the trucks and the truck dumping sites do not correspond to the type of mineral the truck carries.

- There must always be awareness of the need to handle zones of very-high-grade ore in a much more delicate manner, not only when loading and hauling the ore, but also when surveyors are staking high-grade ore zones, and when discharging the ore into the corresponding stockpiles.

- Careful monitoring and control of the elevation of the bench toe surfaces is critical. This can be done using topographic levels that are visible to loading operators, using automatic acoustic systems or manually. Toe elevation deviations of ±20 cm could be acceptable, which for a bench height of 4 m is equivalent to $\pm5\%$.

FIGURE 9.21   Typical ore zone definition and staking at El Valle-Boinás mine

The drawing in Figure 9.21 shows a typical example of the definition of ore areas for loading, marking in the pit, and the loading process.

Database and Report Generation

As previously mentioned, the block model is a database containing not only the information on the initial resource model but also all the data generated relating to interpolation, geology, type of mineral ore, calculations, destination, and so forth, which is entered as it becomes available.

All this information will allow for easy generation of all types of reports, including production, comparisons with expected outcomes, averages for certain periods, totals per mine, and so forth.

The statistical processing of all this information being generated allows for evaluation if the grade control system must be modified by comparing several scenarios or even conducting simulations.

Conclusions

This grade control system has evolved from the first year, adding improvements until the final state described in this case study was reached.

The main conclusion regarding grade control systems is that the concepts are far more important than the tools and the means. By this is meant that the mining concept of "what you want to achieve" is much more important than having sophisticated software systems without the concept. Attempts have been made to create very expensive custom-designed software applications for specific purposes that, in the end, were never used because they were too complex. Conversely, simple applications have been developed that were very good at accomplishing the functions for which they were created. In short, it is often difficult to explain to a software programmer what needs to be accomplished in mining with a given application because mining concepts are usually unfamiliar to them.

The development of this grade control system over these years has taught an important lesson: it is best to start with something simple that works before going on to more complex stages. Trying to implement a program with lots of objectives in a single step can easily turn into disaster.

A second important conclusion is that it is the actual mine operators and the users of software applications who must communicate what they need and determine whether an idea is feasible or not.

Perhaps the most important conclusion that can be drawn from all this testing is that money invested in grade control for this type of mineral deposit will make a profit through reduced dilution of the ore sent to the plant. Therefore, when increasing the amount of sampling and control, the limiting constraint is not cost but rather production cycle time, by increasing the time required to process grade control information and thus reducing production capacity.

Therefore, it is very important to have an automated grade control process, that is to say, to have all the information being generated processed by computer applications programmed with the logic of the process. For example, the atomic absorption machine, used for the last stage of gold assaying, initially operated independently from the rest of the lab equipment, so assay results were input manually and later were entered in spreadsheets and sent to the grade control staff. Today, when a package of samples arrives in the lab, these are entered in the database that links all the equipment using only the bar-code reader. Then the different machines, including the atomic absorption machine, link their results with that database through the interfaces in a way that eliminates the time needed for handwritten records and avoids the potential for input errors.

This example, just like others that have been mentioned, makes it possible to reduce the cycle drastically and to manage thousands of pieces of information efficiently and quickly every day.

Acknowledgments

The section author expresses his gratitude to Alberto Lavandeira for his contributions during the preparation of this section and gratefully acknowledges Rio Narcea Gold Mines for facilitating the publication of this case study.

CASE STUDY: RELIABILITY ASSESSMENT OF A CONVEYING SYSTEM AT ATLANTIC COPPER*

This section describes a methodology for improving maintenance practices based on the application of reliability-centered maintenance (RCM) and mathematical modeling for the conveyor belt system at Atlantic Copper.

RCM is a systematic consideration of system functions, the way functions can fail, and a priority-based consideration of safety and economics that identifies applicable and effective maintenance practices. So, RCM allows the focus of the maintenance efforts to be on those functions whose priority, in terms of production, safety, and protection of the environment, are higher, leaving aside other maintenance tasks that are not strictly necessary. This will lead to an increase in the effectiveness of the maintenance and better management of safety and environmental risk, thereby reducing costs.

The other main feature of this section is how mathematical modeling is used to assist in optimizing the maintenance tasks considered in the RCM work. A mathematical model provides the decision-maker with a powerful tool for knowing where the optimum is located and how far the current maintenance practice is from that optimum point. Frequently, RCM is used only as a systematic procedure to filter out unprofitable maintenance practices. However, without the assistance of modeling, the engineer tends to make decisions according to his/her experience or

---

* This section was written by E. Crespo and M. Palacios.

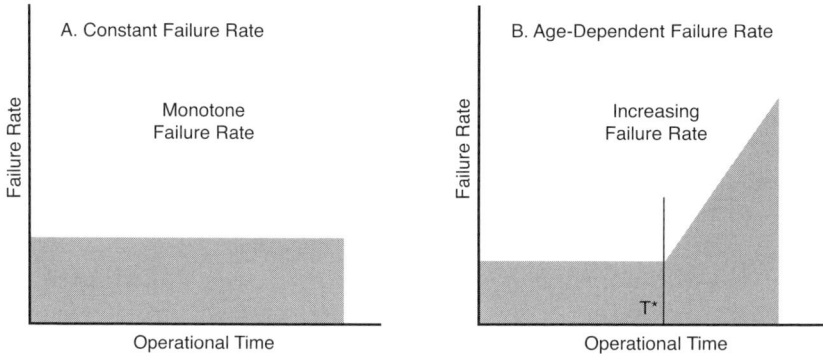

FIGURE 9.22   Constant failure and age-dependent rate modes

to use only that information reliably provided by manufacturers of the equipment, yet neither of these take into account where the optimum is; in other words, the state of the "best" is unknown, so that it is not possible to measure the efficiency of the expenditure on maintenance.

Here are typical questions for considering an RCM methodology:

- What are the functions of the system according to its operational context?
- What are the failures that keep the system from fulfilling the operational requirements?
- What happens when the system fails?
- What are the consequences of those failures?
- What maintenance could be done to prevent or predict the failure?
- What maintenance could be done if there was no effective preventive activity appropriate for that system?

Figure 9.22b shows a typical age-dependent failure rate curve. In contrast, industrial systems show different failure patterns, as in the constant failure rate (CFR) pattern shown in Figure 9.22a, so planned maintenance is far from efficient; thus, the failure rate can increase because of failures that emerge due to a bad or inappropriate repair. The conclusion could be not to perform preventive maintenance at all, which could be acceptable for noncritical failures; however, when the failure consequences are serious, it is mandatory to prevent them.

RCM can help to focus resources to avoid only those failures that can lead to severe consequences, which leads to an immediate improvement in efficiency of maintenance; in addition, safety and environmental protection are integrated in the maintenance analysis.

System Under Study

Several considerations have to be taken into account before starting an RCM analysis. First, RCM requires huge resources in terms of time from maintenance engineers, so that the RCM should be applied only to those systems that do not achieve satisfactory results from traditional maintenance planning. RCM is particularly recommended for those systems that are relevant for safety, environmental, and production losses.

Apart from that, an RCM study has to be limited in terms of the level of assembly; otherwise, the analysis will become tedious and useless. The level of assembly hierarchy could be summarized as follows:

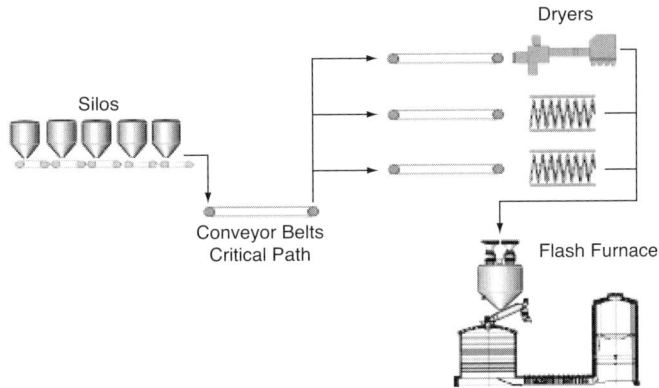

FIGURE 9.23   Ore feeding system under study

TABLE 9.3   Technical characteristics*

Subsystem	Length, m	Width, mm	Inclination	Engine Power, kW	Reducer	Coupling
1	68.5	650	0°	4	Type A	Rigid
2	43.5	650	0°	4	Type A	Rigid
3	68	650	30°	7.5	Type B	Rigid
4	160	650	35°	22	Type C	Fluid
5	17.6	650	0°	4	Type A	Rigid
6	128	650	0°	7.5	Type A	Rigid
7	78	650	0°	15	Type D	Rigid

*Performed by Atlantic Copper with a copper concentrate transport "critical path" system and conveyor belt subsystem.

- Plant: This is a set of systems that function together to provide some sort of output. The whole smelter would be an example of such a set of systems.
- System: This is a set of subsystems that perform a main function in the plant. For example, the "transport of copper concentrate system at the Atlantic Copper smelter" would be considered a system. This is a suitable level for starting the RCM work.
- Maintainable item: This is an item that is able to perform at least one significant function on its own, such as an electric engine.

Figure 9.23 shows the configuration of the feeding system. The conveyor belts that are studied in this case are those that form the critical path for the copper concentrate.

The system (Table 9.3) is designed to transport copper concentrate from the silos to the dryers. The first section of the transport chain is a series configuration of seven conveyors. After that section, the transport chain is divided into three parallel lines, one for each of the dryers. It is easy to understand that the second section is less important in terms of reliability for the whole transport system because only two dryers are needed to provide 100% of nominal production. Moreover, the dryers determine the maintenance of those parallel lines, so that the scope of this work does not cover those parallel lines.

The maximum speed is 180 t/h. In addition, the system should be able to regulate the material flow between 100 and 180 t/h. It should be operating 365 days a year, 24 hours a day, except during planned maintenance stops.

The feeding system ends in a 400-t bin that is located above the furnace. This system gives the maintenance crew about 2 hours to perform repairs when the bin is full.

Seven conveyors form the critical path; consequently, each conveyor band should be classified as a subsystem. The point to note here is that all the conveyors are similar; therefore, most of the components are the same.

For the sake of simplicity, the system hierarchy assembly will be a conveyor belt in general terms; however, for data analysis purposes, some specific components such as motors and reducers will be considered as different from one conveyor to another.

Functional Failure Analysis

The objective of functional failure analysis (FFA) is to identify the required functions of the system and, by that means, the ways in which the system can fail. As a result, the main functions and maintainable items should be identified for further analysis.

Steps to perform an FFA are as follows:

1. Define the operational modes of the system, for example, operating and standby.

2. Define all relevant system functions for each operational mode.

3. Define functional failures.

    – Include an estimation of consequences for each functional failure.

    – Include an estimation of consequences for the frequency of functional failures.

Consequences and frequency are taken into account to select those functional failures that are more relevant in terms of safety, environmental protection, or production losses.

Criticality can be categorized in this way:

*   S: critical for the Safety of workers

*   E: critical for the Environment

*   A: critical for production Availability

*   M: critical for Material loss

Additionally, each category can be ranked in terms of severity: high (H), medium (M), and low (L).

The frequency of the functional failure can also be classified according to the same categories; however, for the sake of simplicity, only one frequency measure is considered.

The frequency should be determined by the statistics on the data collected. However, this step only selects the more relevant functional failures and maintainable items so that a pseudo-qualitative measure can be admitted. Here's one possible way to classify frequency:[32]

*   Very unlikely: Once per 1,000 years

*   Remote: Once per 100 years

*   Occasional: Once per 10 years

*   Probable: Once per year

*   Frequent: Once per month or more

The FFA (Table 9.4) allows those functional failures that are not relevant in terms of safety, environment, availability, and production losses to be disregarded. As a rule of thumb, if all the criticality classifications are low and the frequency is also low, that particular functional failure can be ignored.

At this point items with a high failure rate, high repair cost, high maintainability, and a long wait-time for spare parts, as well as those items that require external maintenance personnel

TABLE 9.4 Functional failure analysis of a copper concentrate transport system in an, performed by Atlantic Cooper in an operating mode

System Function	Functional Failure	Consequences				Frequency MTTF
		S	E	A	M	
Concentrate Maximum rate 180 t/h	No transport: coupling broken	L	L	H	M	>1 year
	No transport: fluid-drive broken	L	L	H	M	>1 year
	No transport: reducer broken	L	L	H	M	>1 year
	No transport: electric engine broken	L	L	H	M	>2 years
	100% Blockage in hoppers	L	L	H	H	30 days
	No transport: belt broken	L	L	H	M	180 days
	No transport: drum bearings jammed	L	L	H	L	—
	No transport: drums worn	L	L	H	L	—
Regulate flow rate between 100–180 t/h	Frequency adapter malfunction	L	L	H	L	>2 years
Avoid spills of materials	Skirts worn	L	M	M	H	20 days
	Scalpers worn	L	M	M	H	45 days
Centered along conveyor	Head or tail drum, not horizontally aligned	L	L	M	L	40 days
To band	Rolls jammed	L	L	L	L	15 days
Stop conveyor if any blockage downstream	No stop, automatic locking malfunction	L	L	H	M	>1 year
	No stop, hopper level detector malfunction	L	L	H	M	>1 year
Stop if overcharge	No stop, electric protection malfunction	H	H	H	H	—
Provide light conveyor zone	Not enough light	H	L	L	L	—
Avoid contact with mobile parts	Safety protections broken	H	L	L	L	—
Stop in case of	Safety stop system malfunction	H	L	L	L	—

NOTES: S = critical for the Safety of workers   L = low
      E = critical for the Environment   M = medium
      A = critical for production Availability   H = high
      M = critical for Material loss

should be identified. As a conclusion, those items selected by the failure characteristics such as high frequency or criticality are termed functional significant items (FSIs). The items selected by their special maintainability characteristics are defined as maintenance cost significant items. Both become maintenance significant items (MSIs).[33]

Gathering Statistics

The objective of gathering statistics is to obtain reliability information on the MSIs to describe the failure process mathematically and thereby optimize the different maintenance tasks.

A complete lack of reliability data occurs in certain circumstances, such as when starting a new system. In those situations, information given by the equipment manufacturers and the experience of the engineers can be used to implement a tentative maintenance plan. In any case, an RCM procedure could provide very useful information and would be an excellent starting point to improve maintenance planning; thus, a former RCM job would provide the most significant items for collection and storage of reliability data.

The quality and quantity of data is always one of the key problems in every reliability analysis. In this case study, the failure data comes from a system that is still being maintained, so that most of the failure data is truncated (i.e., maintenance is performed before failure takes place, so

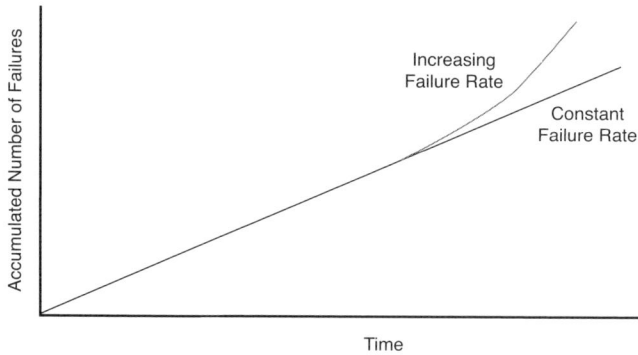

FIGURE 9.24   Nelson–Aelen plot

the actual time to failure cannot be measured). This limitation has to be taken into account in the statistical failure model.

Life Data Analysis

It is beyond the scope of this work to fully study all the techniques available for life data analysis, so only some important concepts for performing RCM will be discussed.

In particular, there is interest in knowing if there is an age-dependent failure.

It is important to know the failure model because those failures that have an increasing failure rate (IFR) trend are suitable for typical planned maintenance. On the other hand, the failures that show a linear trend, meaning a CFR, cannot be prevented with planned maintenance. The point here is to know what to do with those failures. This leads to the concept of "potential failure" that can be defined as the potential warning time in advance of a functional failure. The warning time is called the P-F interval or delayed failure time[34] and can be used to predict the failure.

Plotting the accumulated number of failures is a very useful tool for checking the trend in time for failure data. In reliability theory, this is called the Nelson–Aelen plot, as in Figure 9.24. Generally, the curve shows two different shapes: a convex curve that corresponds to an IFR model and a linear curve that represents a CFR distribution. However, when real data are used, the curve will take on ambiguous shapes. In particular, the analyst must be very careful when checking a linear trend, because there could be correlations between failures. In fact, when those correlations are present, an ordinary linear regression will show an apparently good fit, with high values of the regression coefficient, in spite of the fact that a linear model is not adequate.

The time to failure of an IFR model can be fitted using a probability distribution of the Weibull family. In contrast, the time to failure for a CFR model follows an exponential distribution. In order to fit the probability models, various commercial software packages can be used, for example, Weibull++ by ReliaSoft, which is widely used in life data analysis.

Another way to test if there is a relationship between operational age and failure rate is by comparing the mean and the standard deviation of time to failure. As mentioned previously, the time to failure of those processes that show a CFR follows an exponential probability frequency distribution, as a result of which an interesting property of exponential distributions can be used: the mean is equal to the standard deviation. Also, in linear models the mean time to failure (MTTF) is the slope of the Nelson–Aelen plot. In addition, on these models, MTTF is the only parameter needed to define the probability failure mode.

The model should be as simple as possible, especially when data are lacking. In each case, a good approximation can be made with a CFR model; in practice, it means knowing MTTF.

Failure Mode Effects and Criticality Analysis

At this point, the dominant failures of each MSI should be determined. In this particular case, most of the items have one dominant failure; consequently, the failure mode effects and criticality analysis (FMECA) study would be similar to the FFA performed previously, but it would also cover a criticality and a maintenance task for each functional failure.

Criticality can be defined as a relative measure of the consequences of a failure mode and its frequency. A variety of ways to estimate the criticality can be found in the main RCM and reliability literature. In this case study, the estimation of criticality is proposed by the following equation:

$$Cr_i = \%TF_i \cdot Csqindex_i \cdot UR(t)_{\text{mission time}} \tag{EQ 9.5}$$

where

Cr_i = criticality number for i

TF_i = the percentage of total failures accounting for i failure

$Csqindex$ = the consequence index that measures the relevance of the functional failure

$UR(t)$ = the unreliability of i functional failure for a specific mission time

The consequence index represents the conditional probability that the loss will materialize after the failure has happened. Here's one general means of classification:[35]

- Actual loss: 1
- Probable loss: 0.1–1
- Possible loss: 0–0.1
- Negligible loss: 0

The $UR(t)$ is estimated using the probability model fitted to the failure data. When data are lacking, in terms of quantity or quality, an exponential model for the time to failure data can be assumed. The exponential model only requires the parameter λ. The main mathematical properties of exponential distribution are shown as follows:

- Density function of the failure probability distribution, $f(t)$: $\lambda e^{-\lambda t}$
- Reliability function (i.e., cumulative probability of survival), $R(t)$: $e^{-\lambda t}$
- Unreliability function (i.e., cumulative probability of failure), $F(t)$: $1 - e^{-\lambda t}$
- MTTF: $1/\lambda$
- Failure rate: λ

Figure 9.25 shows the value of criticality for each functional failure. Notice that in the FFA analysis, there are four types of consequences, so that the criticality should be estimated for each of them; for the sake of simplicity, only one criticality figure for the highest value of consequence indicated in the FFA table has been included (Table 9.4). However, a conveyor belt is a very simple system, and that simplification can be adopted without losing accuracy. In any case, the objective of RCM is not to create paperwork but to analyze the main functional failures and their consequences.

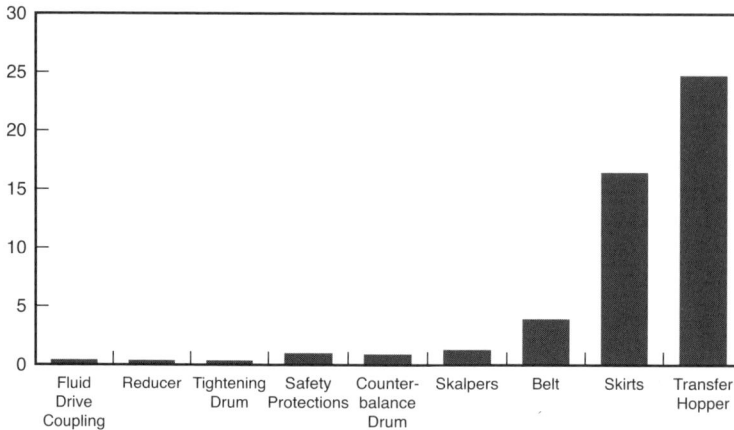

FIGURE 9.25   Criticality versus functional failure

Selection of Maintenance Actions

The point here is to determine if there exists effective preventive maintenance for avoiding the failure or its consequences. The effectiveness of a maintenance task depends on the characteristics of failure and the type of maintenance task.

There are generally three main reasons for doing a preventive maintenance task:

- To prevent a failure
- To detect a faulty condition (fault) that may lead to failure
- To discover a hidden failure

On the other hand, the following maintenance tasks should be taken into account.

Scheduled on-condition task.   The maintenance task is triggered based on the measurement of one or more variables correlated to a degradation, or loss of performance, of the system. As a result of these measurements, it is possible to predict the residual life of the system. The maintenance policy that covers this type of maintenance is called condition-based maintenance (CBM) or predictive maintenance. The following criteria must be met to set up an on-condition task:

- It must be possible to detect reduced failure resistance for a specific failure mode.
- It must be possible to define a potential failure condition that can be detected by an explicit task.
- There must be a reasonably consistent age interval between the time of potential failure (P) and the time of functional failure (F). The longer the P-F interval, the more time there is available to trigger a preventive maintenance task.

The concept of potential failure is a powerful tool in the design of preventive maintenance for those failures that are not clearly correlated to age. In fact, most of those failures show evidence that a future failure will happen. In practice, there are lots of ways to detect those potential failures (e.g., vibrations on a reducer, hot spots on outer surfaces of furnaces, visual inspections, and so forth).

The P-F interval can vary depending on the inspection technique used. For example, an analysis of vibrations will predict a failure months before it happens, whereas heat detected by touch can only provide a P-F interval of days. The inspection frequency is determined by the P-F interval and has to be at least two or three times shorter than the P-F. In cases where the P-F interval is technically very short, then continuous monitoring is needed.

Although CBM appears to be an excellent policy for maintenance, in some systems it is not economically efficient. As a result, both the technical and economical aspects must be weighed.

Scheduled overhaul.   A scheduled overhaul is performed at or before some specified age limit. There are several criteria for applying a scheduled overhaul:

- There must be an identifiable age at which there is a rapid increase in the failure rate function of the item.

- The age-dependent and CFR modes must be identified as outlined in Figure 9.22.

- A large proportion of the items must survive to that age.

- It must be possible to restore the original failure resistance of the item by reworking it.

Scheduled replacement.   Scheduled replacement is the replacement of an item (or one of its parts) at or before some specified age or time limit. There are several criteria for applying a scheduled replacement:

- The item must be subject to critical failure.

- The item must be subject to a failure that has major potential consequences.

- There must be an identifiable age at which the item shows a rapid increase in the failure rate function.

- A large proportion of items must survive to that age.

In those systems whose maintenance strategy is oriented to availability, as in this case study, the scheduled replacement of significant items is recommended instead of scheduled overhaul, because the time for performing a replacement is generally shorter than for performing an overhaul.

Scheduled function test.   This is a scheduled failure-finding task or inspection of a hidden function to identify failures. The preventive sense of the failure-finding task relies on preventing future surprises by revealing hidden functions. In these cases, two conditions must be fulfilled:

- The item must be subject to a functional failure that is not evident to the operating crew during the performance of normal duties.

- The item must be one for which no other type of task is applicable and effective.

Run to failure.   This is a deliberate maintenance decision because the other tasks are not possible or are economically less profitable.

Transfer Hoppers

Transfer hoppers drive the copper concentrate between conveyors. The hopper jams regularly and the system stops. The criticality is so high because of the high frequency of failure. The root cause is humidity in the copper concentrate and the impurities that come with it.

This functional failure is unpredictable, in which case, any scheduled maintenance is useless, so there is no effective maintenance task.

It was decided to redesign the vibrating system on all the critical feed-through hoppers.

Skirts, Rolls, and Scalpers

Looking at Table 9.5, it can be seen that skirts, rolls, and scalpers are the most frequent failures after hopper jamming. These failures are caused by wear.

Previously, maintenance practices were based on visual inspections. These failures could be inspected while the system was operating, so that the inspections did not cause unavailability. Every day, it was possible to halt the system to fix it: changing skirts, rolls, scalpers, and when necessary, lubricating the different bearings. These stops could be done only when the 400-t bin was full, so that about 2 hours were available for the maintenance crew.

The RCM team found that daily frequency was too high and, without the assistance of modeling, frequency was fixed at 15 days. Again, it is important to highlight that an excess of preventive maintenance is far from efficient, and is sometimes much worse.

To estimate the optimum frequency, a mathematical model was done to describe the failure behavior of skirts, rolls, and scalpers. The data were extracted from the maintenance database. As a result, the time between replacements was modeled as an exponential variable. In fact, a renewal process like this can be assumed to be a homogeneous Poisson process, and the time to the next event in these kinds of processes follows an exponential probability distribution.

The next point was to define a window of opportunity to perform the repair before the system failed. Because the system was visually inspected, a P-F interval for these inspections could be defined. A reasonable P-F interval was found to be 15 days for skirts and scalpers, while the P-F for rolls was estimated at 60 days. The reason for this longer interval is the low criticality of rolls because so many of them have to fail in order to stop the system. So, when some rolls are jammed, the window of opportunity to repair them is very large.

The model was simulated using the Matlab software environment and the results are shown in Figure 9.26. The main conclusion is that the optimum frequency is 2 weeks, as the RCM team formerly suggested, as shown in Figure 9.26a. Figure 9.26b shows the evolution of simulated failures when the frequency of stops is increased. No failure is predicted for 7- and 14-day frequency, but for 21-day frequency, the number of failures caused by skirts is five. The real data used for validating the model shows that for 7- and 14-day frequency, the failures caused by skirts were six and five, respectively. In the case of the scalper, the real failures were two for both intervals. There was a small difference between simulation and reality, but it was explained by the fact that while an inspection may be perfect in a simulation, it may be far from perfect on a real system.

However, Figure 9.26c shows the number of stops needed to perform repairs. At this point, the agreement between real data and simulation is quite high for 14-day frequency, but in the case of 7-day frequency, the number of stops made in reality is higher. Maintenance practice consisted of making stops with daily frequency; in contrast, the simulation only considered a minimum frequency of 7 days, so that the difference between simulation and real data is acceptable.

Notice also that in reality and in simulation, there is no need to stop if no fault has been detected. For example, in the case of 7-day frequency, the nominal number of stops for a year would be 53, and the maximum calculated is 31 and real data is 37.

Rigid Coupling and Fluid Drive

It has been decided not to perform any preventive task for rigid coupling. In the case of fluid drive, the level of oil will be measured opportunistically when the system stops.

Belt Breakage

The incidents related to this functional failure are caused by impurities carried by copper concentrate, for example, sticks, metal objects, and so forth. When that happens, the belt can be torn and the repair time ranges from 4 to 8 hours.

TABLE 9.5 Failure mode effects and criticality analysis*

Description of Item		Failure Mode	Effect of Failure, Consequence Class				MTTF	Consequence Index	Criticality Mission Time 1 Year	Failure Cause	Failure Mechanism	Maintenance Action
MSI	Function		S	E	A	M						
Rigid coupling	Drive gear engine/reducer	No transmission	L	L	H	L	>3 years	1	0.13	Mechanical breaking	Fatigue	Run to failure
Fluid drive coupling	Drive gear engine/reducer	No transmission	L	L	H	L	>2 years	1	0.27	Oil spills	Wear	Check level of oil (opportunistically)
Reducer	Increasing torque	No transmission	L	L	H	L	>2 years	1	0.27	Mechanical breaking	Fatigue	On-condition task, vibrations (monthly)
Feeding through hopper	Transfer material from one conveyor to another	Hopper jammed	L	L	H	M	21 days	1	24.7	Foreign object / Humidity	Foreign object / Humidity	Run to Failure
Skirts	Avoid spills of material	Material is spilled on transfer points	L	H	M	H	21 days	0.7	16.5	Friction skirt-belt	Wear	(Weekly) Visual inspections (15-days) Planned stops
Scalpers	Clean the belt avoiding spills	No clean belt and material is spilled on transfer points	L	M	M	H	42 days	0.7	1.25	Friction skalper-belt	Wear	(Weekly) Visual inspections (15-days) Planned stops
Head drum	Drive gear to the belt	Drum blocked	L	L	H	L	>4 years	1	0	Bearing jammed	Wear/lubricant	Run to failure
Tail drum	Roll back the belt	Drum blocked	L	L	H	L	>4 years	1	0	Bearing jammed	Wear/lubricant	Run to failure
Rolls	Support and slip the belt	Rolls jammed	L	L	L	L	17 days	0.1	0	Bearing jammed	Wear	Run to failure
Safety protections	Avoid contact with mobile parts	Safety protections broken	H	L	L	L	>1 year	1	0.87	Bad repair and use techniques	Human Factor	(Weekly) Visual inspections
Lights	Provide light to work zone	Lamps fused	H	L	L	L	No data	1	—	Unpredictable	—	(Weekly) Visual inspections. immediate repair
Emergency stop	Stop in case of emergency	System malfunction	H	L	L	L	No data	1	—	Hidden failure	—	Scheduled function test (6 months)
Belt	Transport material along conveyor	Belt broken	M	L	H	M	121 days	1	3.95	Foreign object	Bad screening	Run to failure
Tightening drum	Tight belt	Drum blocked	L	L	H	L	2 years	1	0.27	Bearing jammed	Wear/lubricant	Run to failure
Counterbalance drum	Tight belt	Drum blocked	L	L	H	L	1 year	1	0.87	Bearing jammed	—	Run to failure

NOTES:
S = critical for the Safety of workers
E = critical for the Environment
A = critical for production Availability
M = critical for Material Loss

L = low
M = medium
H = high

*Performed by Atlantic Copper with a copper concentrate transport system and conveyor belt "critical path" subsystem.

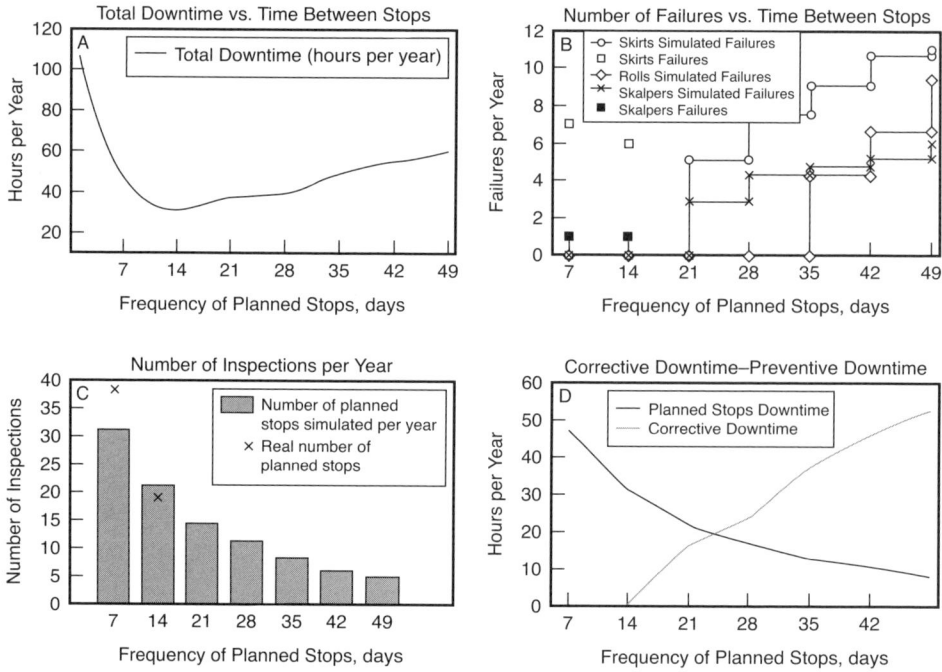

FIGURE 9.26   Simulation results

The failure is unpredictable, so that there is no preventive task effective in avoiding it. The possibility of duplicating two of the conveyors has been studied because there is enough free space on either side of them, but it is not economically feasible. As a result, the maintenance policy is run to failure.

Drum Failures

These failures can be detected by inspections, with enough time for planning a preventive maintenance task. Looking at the decision diagram (Figure 9.27) or the more detailed flowchart of Figure 9.9, it can be seen that an on-condition task is recommended.

The point is that the inspection has to be made with the system stopped because the internal gap of the bearings needs to be measured. So inspections increase the unavailability of the system, and eventually could cause the item to be damaged. Looking at the FMECA table (Table 9.5), it can be seen that the MTTF is quite high. As a result, the criticality for a mission time of 1 year is very low.

The RCM team has decided not to perform any preventive maintenance. Mathematically, it means that the run to failure cost is lower than any preventive maintenance cost.

Motors and Gear Reducers

Reducers have quite low criticality because of their low frequency. However, it has been decided to measure vibrations on all the conveyors belts on the critical path. Looking at the decision diagram, it can be seen that CBM can be employed. The cost of inspection is quite low because inspection is performed while the system is working, and so it is economicall feasible. The

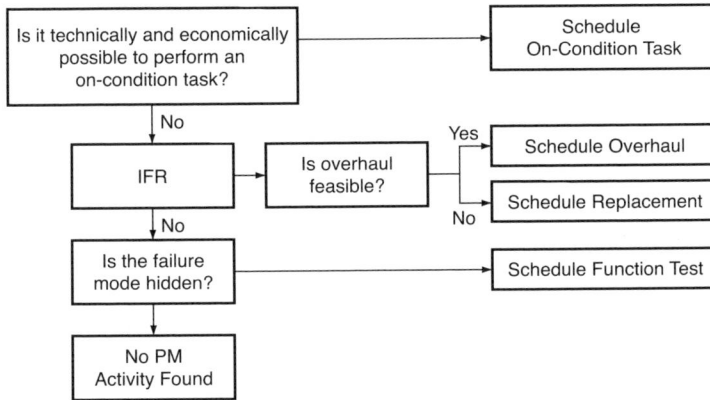

FIGURE 9.27 Selection of maintenance task

frequency of inspections is monthly. The P-F interval for a vibration measurement is about 4 months. Thus, there is a warning time of about 3 months to perform preventive repairs. So far, insufficient information to validate the frequency of inspection has been gathered.

Safety Protections and Lighting

Safety protections will be inspected weekly, and repaired as soon as the system stops, either opportunistically or at the 15-day frequency fixed in Table 9.5.

A functional failure in the emergency stop system can be considered a hidden failure. It has been decided to perform a scheduled function test every 6 months. The lights will be repaired as soon as a failure has been detected.

Conclusions

The main advantage of using the RCM methodology is that it integrates the environmental and safety functions with the maintenance objectives, thereby leading to sustainable maintenance management.

RCM gets the operators involved in the maintenance aspects of the system. That is particularly important because they are more familiar with the system. As a result of the analysis, they will become more concerned about the everyday safety risks, and the level of motivation at work will also increase.

In terms of cost efficiency, RCM reduces routine maintenance tasks considerably. For example, this case study has shown how a maintenance service performed daily can become one performed only once every 15 days. Mathematical modeling helps RCM to find the optimum maintenance plan. That is particularly important when deciding the scheduling of different maintenance tasks.

RCM provides the best system know-how in the operational context. Therefore, it puts the operation in a better position for acquiring new technologies. At the same time, continuous improvement programs can be developed based on the knowledge RCM provides. After the systems' structure has been broken down and their main functions analyzed, it is possible to generate an effective and efficient maintenance database.

In conclusion, RCM provides an integrated point of view for dealing with the key aspects related to industrial management.

FIGURE 9.28   Aerial view of Aznalcóllar tailings impoundment (after failure, looking north)

Acknowledgments

The section authors gratefully acknowledge the management of Atlantic Copper, S.A., for their assistance in the publication of this section.

CASE STUDY: OVERVIEW OF THE AZNALCÓLLAR TAILINGS DAM FAILURE*

This case study summarizes the findings and conclusions of the investigation carried out immediately after the failure of the Aznalcóllar tailings dam. The investigation was headed by the author of this case study[36] and drew upon the advice of a review panel of five internationally recognized experts.

The mine is located in the southwest of Spain and forms part of the Iberian pyrite belt, a mineral belt with a long mining history dating back to 3000 BC. The Aznalcóllar tailings dam (Figure 9.28) failed during the morning of April 25, 1998, causing an estimated 1.3 million m³ of tailings and 5.5 million m³ of tailings water to spill into the nearby River Agrio, and then flow downstream.

The failure event started shortly before 1:00 AM, and the maximum effluent outflow occurred sometime between 1:00 and 3:00 AM, as evidenced by the peak water level recorded at 3:30 AM by the staff gauge located 13 km downstream near Sanlúcar la Mayor. The outflow discharge ceased shortly after 8:00 AM, when the river flow at the staff gauge (still high) started to decline gradually until 6:00 PM, when the flow level returned to normal.

The dam failed as the result of a 60-m lateral displacement of a 700-m-long segment of the southeast section of the dam (Figure 9.29). The dam, together with the 4-m-thick alluvium terrace upon which it lies and the upper 10 m of the blue marl formation underlying the alluvium, moved along a bedding surface in the blue marl at a depth of 14 m below the original ground surface.

---

* This section was written by J. A. Botin.

FIGURE 9.29 Aerial view of east dam displacements (looking north)

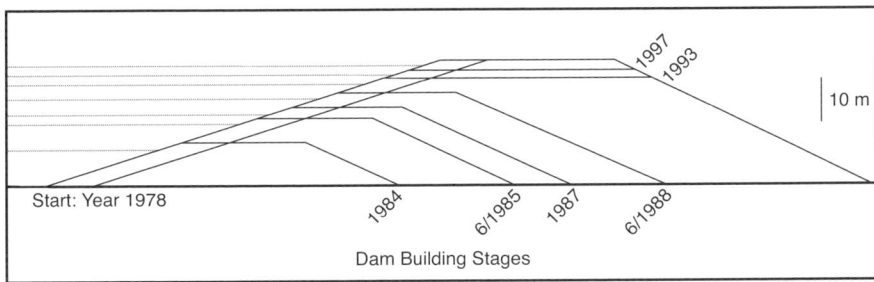

FIGURE 9.30 Typical section of tailings dam

Geometry of the Tailings Dam

Dam construction proceeded in stages over 20 years based on the operational needs of the mine, and impoundment filling closely followed crest rising. The building stages for a typical section in the failure zone are shown in Figure 9.30. At the time of failure, the dam had reached a maximum height of 27 m.

The main body of the dam was built with unclassified mine waste rock. The upstream slope was covered by a screen of compacted, low-permeability colluvium material named "red raña." The dam lies on a 3- to 5-m-thick permeable alluvium terrace underlain by 70 m of impermeable blue-gray clays. To control foundation seepage, a bentonite–cement cut-off wall was installed in a trench excavated 1.5 m into the marl formation.

Regional and Local Geology

The tailings impoundment is situated on the western side of the River Agrio, which lies in an erosive valley that formed as the regional drainage network become gradually embedded in the Miocene blue marls. The blue marls belong to the Tertiary marine depression of the Guadalquivir Valley, bounded by the Hercinian basement of the Iberian Massif (Sierra Morena) in the north, and the Bética Mountain Range in the south.

The blue marls (*margas azules del Guadalquivir*) are heavily overconsolidated, fissured, carbonate-rich clays of high plasticity. They contain random and discontinuous slicken sides of variable frequency. The marl mass also contains systematic subvertical jointing and displays a laminated texture associated with bedding structures. At a larger scale, the marl is generally homogeneous and no pre-existing failure planes were encountered in the investigation.

A study of the lithostatic load on the marls prior to the erosive cycle was attempted, considering the important effect of overconsolidation on the geotechnical behavior of the blue marls. It was concluded that, prior to the Pliocuaternary erosive cycle, the thickness of the layer of marls covering the present dam site was between 75 and 150 m (i.e., 75–150 m of overburden has been eroded from the dam site).

Morphology of Failure and Dam Displacement

Extensive forensic drilling was performed to define the shape and location of the sliding plane precisely. This work involved the drilling of 35 boreholes, totaling 1,240 m, and the excavation of five deep trenches.

An afterfailure survey of the east dam showed that a section of the dam more than 700 m long had been displaced to the east and east-southeast. Maximum displacement occurred in the central part of the 700-m section, where the maximum horizontal component of movement was 49 m at the dam crest and 67 m at the dam toe. This lateral displacement was accompanied by a reduction in elevation of the dam crest, which was a maximum of 2.4 m.

The failure involved translational sliding of the dam on a subhorizontal plane (2° from horizontal) about 8–10 m below the top of the marl formation. The translational nature of the movements is corroborated by the measured postfailure surface movements of the dam crest, which ranged up to 49 m in horizontal movement with only 2.4 m of movement in vertical subsidence. The translation involved a block formed by the dam, the 4-m-thick alluvium layer, and the upper part of the marl formation underneath.

The upstream end of the slide was nearly vertical and its morphology was most probably controlled by one of the dominant orientations of vertical jointing in the marl formation. This near-vertical detachment occurred close to the upstream toe of the dam, near the bentonite–cement cut-off wall.

The downstream exit of the slide was a shear-cut oblique to the marl strata as indicated by the observed ground displacements at the dam toe and by the 20°–40° inclined shear planes observed in boreholes drilled during the investigation.

At the northern end of the failure zone, near the southeast corner of the pyroclastic pond, the sliding block was sheared, probably by a tear displacement through the vertical jointing in the marls (Figure 9.29). In this area, the spill breach was initially formed and enlarged by erosion. The south end of the displacement was formed by gradual attenuation of movement and deformation.

Geotechnical Characterization

The blue marls are heavily overconsolidated, fissured, carbonate-rich clays of high plasticity. The marl mass contains systematic subvertical jointing and displays a laminated texture associated with bedding structures. Weathering and alteration is generally present in the upper 3–4 m in the marl formation. Alteration is identified by color change from bluish-gray to beige or yellowish-brown. Also, in some cases, alteration is present in the form of millimeter-sized halos along vertical jointing. The width of these halos decrease with depth, and they were never seen at depths greater than 15 m below the alluvium–marl contact.

The term *overconsolidated* that is applied to the blue marl formation refers to clays that, as the result of the sedimentary processes, were loaded with high stresses, which changed their normal structure and created diagenetic bonds between particles and then unloaded. A double

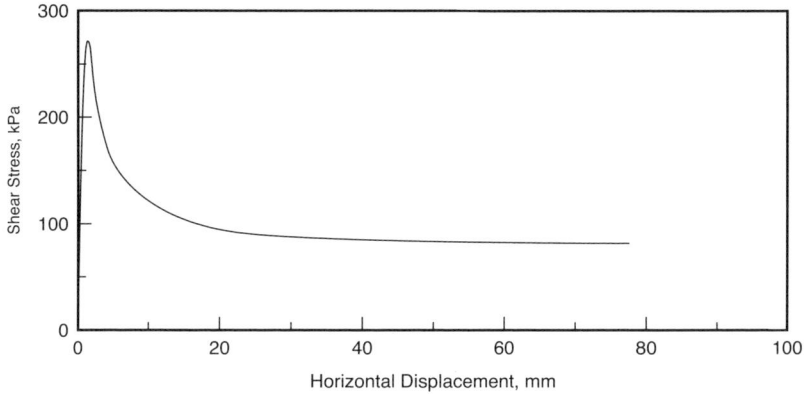

FIGURE 9.31   Results of ring shear test on marl clay

TABLE 9.6   Marl strength characteristics

Type of Marl	Peak Strength		Residual Strength, φ'_r
	c'(kPa)	φ'	
Intact bluish-gray marl	27	20°	11°–12°
Weathered, altered marl	28	18°	12°

process of recrystalization and cementation, which gives the clay high shear strength and makes it more brittle, generates these diagenetic bonds. These diagenetic bonds must be broken for these clays to fail, and, therefore, overconsolidated clays show higher peak shear strength than normally consolidated clays. However, an important feature of the marls is their brittle behavior (as exhibited in the ring shear test results shown in Figure 9.31), whereby a significant loss of frictional strength occurs when the marls are strained past their peak strength. This brittleness is associated with the progressive breaking of these bonds with straining and depends on time, weathering, and stress conditions.

The overconsolidation phenomena and the strain-weakening effects associated with it were first discussed by Bjerrum[37] and have been studied in relation to the failure of clay formations by Griffiths,[38] Chen and Morganstern,[39] Brooker and Peck,[40] Skempton and Vaughan,[41] Morgenstern,[42] Uriel and Fornes,[43] and other authors.

In general, the strength properties of the blue marls are quite homogeneous within the area of study, with no significant differences found among samples of the same marl type in relation to depth or location relative to the failure zone. From laboratory tests, the strength characteristics that are considered as representative for the different types of marl are summarized in Table 9.6.

It is important to remember, however, that laboratory tests carried out with small samples do not take into account the structure of the marl mass (i.e., its bedding and subvertical jointing); therefore, the strength parameters in the marl formation must be somewhat lower than those obtained from laboratory testing. In this case, the in-situ peak strength parameters of the marl formation are estimated to be cohesive in the order of 10 kPa to 20 kPa, and a friction angle of about 18°, which corresponds closely to the lower values obtained for the altered marls.

Two types of tailings were deposited: pyrite tailings and pyroclastic tailings. The pyrite tailings result from the milling and flotation of massive sulfide ores; the pyroclastic tailings result

TABLE 9.7   Material properties of pyrite tailings and pyroclastic tailings

	Pyrite Tailings	Pyroclastic Tailings
Size k80	45 μm	450 μm
Specific weight of particles	48 KN/m^3	30 KN/m^3
Saturated density in pond	29 KN/m^3–30 KN/m^3	20 KN/m^3–21 KN/m^3
Saturation moisture	20%	24%

from the treatment of a copper dissemination zone at the hanging wall of the orebody. The properties of these materials are outlined in Table 9.7.

Given the zero plasticity of these materials, an angle of internal friction of 25° and an undrained strength of 10 kPa were estimated. It is noted that the dam failure occurred in the pyrite pond, where the unusually high unit weight of the pyrite tailings was a contributing factor in the failure mechanism.

Piezometric Model

Instrumentation data prior to the failure showed that the water table within the rock fill section of the dam was located near the surface of the alluvial terrace overlying the blue marl.

As stated previously, the blue marl formation reaches a depth of approximately 70 m, below which there is a permeable limestone sand formation, acting as an enclosed aquifer (the Niebla-Posadas aquifer). The pressure measured in a piezometer installed in the formation was 860 kPa at a depth of 80 m, yielding a slight upward gradient. Thus, before the construction of the dam, initial pore pressures in the marl would have fit a line with a slope of 11.1 kPa/m of depth. This indicates that overpressures of about 1.1 kPa/m of depth already existed above the water table before the construction of the dam.

During the failure investigation, a total of 18 piezometers were installed in the marls to measure the pore pressures generated by the dam construction. Given the very low permeability of the marl formation, at the time of release of the final report of the investigation (November 1998), some of the readings taken from the piezometers installed in the marls had not fully stabilized. The final stabilized readings from 1999 were slightly higher than those released in 1998. The final stabilized pressures are described in the following paragraphs.

The piezometers located outside the failure zone gave pressure readings between 350 kPa and 550 kPa at a depth of 12 m in the marls, a depth similar to that of the failure surface. Lower values were obtained from piezometers located within the failure zone, but these pressures are attributed to pore pressure changes due to the shear dilation in the marls associated with large failure displacements.

Given that the initial pressure, prior to the construction of the dam, was approximately 150 kPa, an excess pressure of 200 kPa to 400 kPa exists because of the added weight of the dam. Considering that the weight of the dam is 570 kPa (27 m × 21 kN/m^3), the pore pressure increase is equivalent to 35%–70% of the weight of the dam.

From the analysis of the piezometric data, it was concluded that the blue marl formation was subject to excess pressures above the original pore pressure conditions as a consequence of the construction of the dam. At the moment of failure, it is reasonable to accept as correct the pore pressure readings obtained from the piezometers installed in the areas outside the influence of the failure. In other words, an excess pore pressure, above initial values, could be expected to exist before the failure occurred, the magnitude of which would match the values registered in those piezometers.

Dam Failure Mechanism

The limit equilibrium method of stability analysis was used to investigate the conditions for failure along a plane 10 m below the top of the marl formation. Using the measured pore pressures in the marl, it is estimated that, at the time of failure, the shear strength of the marls on the failure surface was approximately 15°, a value lower than the peak strength of the marls (18°–20°), but higher than the residual strength (11°–12°). Therefore, it can be concluded that, at the time of failure, the shear strength of the marls in the failure surface was reduced to less than the peak strength due to straining in the marls prior to failure.

The sample logging and testing performed on the 58 boreholes drilled through the upper 20–25 m in the marl formation under the dam and adjacent areas did not find evidence of sliding surfaces or any other type of continuous preconstruction planar weaknesses in the marls. This leads to the conclusion that the strain weakening of the marl was a result of the stresses induced by the tailings dam construction itself. As evidenced by the results of the ring shear tests performed, blue marls exhibit a significant loss of frictional strength when they are strained past the peak strength. This behavior, described in the literature as "strain weakening," is typical of over-consolidated clays like the blue marls (Figure 9.31).

Considering these findings, a study was undertaken to determine the evolution of the safety coefficient during the different stages of the dam construction process. The results confirm that critical stability situations, with safety coefficients close to unity, probably occurred between 1985 and 1989, as a result of a relatively fast dam construction process when the dam height was raised from 12.7 m to 21.2 m (see Figure 9.30). It is then possible that the strain-weakening process described previously might have started in this intermediate stage of dam construction, as a result of the excess pore pressures derived from the dam construction process and local yielding of the marls.

From the stability analysis, it can be concluded that failure initiation was caused by over-stressing and progressive failure on a bedding plane in the marl formation and that failure was mainly influenced by excess pore pressures induced in the marls by the dam construction process. Also, it is concluded that overstressing of the marl foundation, possibly occurring during the intermediate dam construction stages, allowed for the development of a strain-weakening process in the marl leading to progressive strength loss and ultimate dam failure.

Analysis of Postfailure Movements

Without a doubt, the most remarkable feature of the Aznalcóllar dam failure was the rapid and exceptionally large lateral displacements of the dam (up to 60 m in the center of the failure zone), which can only take place through a sharp reduction of the safety coefficient below unity.

Following this line of reasoning, the limit equilibrium method was used to calculate the safety coefficient at different times during the sequence of events. This study is based on the geometry of the displacement, as deduced from the boreholes, the evidence of the static liquefaction of the pyrite tailings behind the dam, and the sequential interpretation of the displacement described in the following paragraphs.

- In a first stage, at the onset of instability (safety coefficient = 1.0), the average mobilized shear strength of marl along the failure plane is back-calculated to be a friction angle of about 15°. At this moment, the tailings are in a drained state, with effective strength and a friction angle of 25°.

- In a second stage, after some 10 cm to 20 cm of movement, the tailings start to mobilize and generate active thrust, but working in an undrained state ($c' = 10$ KPa and $\varphi' = 0$). The friction angle along the shearing plane would also quickly decrease to the residual angle of $12°$, thus reducing the safety coefficient to 0.76.

- In a third stage, the displacing block of ground detaches from the upstream end and a near-vertical crack a meter wide opens up and is filled with the liquefied tailings. In this state, the safety coefficient decreases to 0.55, displacement accelerates, and the dam transversally breaks at the northern end of the displacement area along a subvertical plane, probably an upward extension of pre-existing vertical joints in the marls.

- In the fourth stage, the tailings start to flow out and to erode the breach at the north end, opening up the pyroclastic pond. Eventually, the reduction of thrust from the tailings, together with the increase of passive resistance at the toe of the dam, restores the dam equilibrium (safety coefficient > 1).

Conclusions

The investigation demonstrated that failure initiation was caused by overstressing and progressive failure on a bedding plane in the marl formation and that failure was mainly influenced by excess pore pressures induced in the marls by the dam construction process. Pore pressures were measured by installing piezometric sensors and have been estimated to reduce the operational strength of the foundation soils by about 50%.

From the parametric analysis carried out, it was concluded that the strength reduction induced by pore pressure in the marl allowed for the development of a strain-weakening process in the marl, whereby shear strength on the failure surface was progressively reduced to values lower than the peak or even the residual strength.

Furthermore, it has been demonstrated that the failure zone was located in the section of the dam where stability was closest to critical. In fact, the failure zone was in the pyrite pond, where tailings are heaviest, in a section of the dam where both dam height and tailings height are greatest and where marl bedding planes adversely dip $1°$ to $2°$ downstream. The combined effect of these three conditions is unique to the failure zone.

The failure mechanism developed, during an undetermined time period, from an initial stage of slow and progressive weakening along the failure plane, eventually reaching instability. With the onset of instability and increased movement velocity, the pyrite tailings liquefied, increasing loading on the dam as the resistance of the foundation decreased, thereby accounting for the exceptionally large movements.

The increased lateral thrust caused a large shearing displacement of the dam crest at the north end of the failure zone, probably along a subvertical joint in the marl foundation. A breach was opened and the spill of water and liquefied tailings began to flow out. Soon, the destabilization and erosion created by the rapid outflow of water and tailings caused the enlargement of the breach, as well as the internal communication between the two ponds.

Acknowledgments

The section author gratefully acknowledges the members of the review panel of experts, the project team, and the management and staff of Eptisa and Andaluza de Piritas, S.L., for their cooperation and assistance in the investigation.

NOTES

1. J. Eloranta, *Downstream Costs and Their Relationship to Blasting* (Tower, MN: Eloranta & Associates, 1999), www.elorantaassoc.com/downstream.pdf (Accessed November 15, 2008).

2. J. L. Floyd, "The Development and Implementation of Efficient Wall Control Blast Designs," in *Proceedings of the International Society of Explosives Engineers*, 1998, 77.

3. B. A. Kennedy, ed., "Haulage and Transportation," in *Surface Mining*, 2nd ed. (Littleton, CO: Society for Mining, Metallurgy, and Exploration, 1990).

4. N. Barton, R. Lien, and J. Lunde, "Engineering Classification of Rock Mass for the Design of Tunnel Support," *Rock Mechanics* 6, no. 4 (1974): 189–236.

5. E. Hoek and E. T. Brown, *Underground Excavations in Rock* (London: The Institution of Mining and Metallurgy, 1980), 7–13.

6. B. A. Kennedy, K. E. Niermayer, and B. A Fahm, "A Major Slope Failure at the Chuquicamata Mine, Chile," *Mining Engineering, AIME* 12, no. 12 (1969): 60.

7. Matamec Exploration, Inc., Glossary, www.matamec.com/contenu/investisseurs_glossaire_ang.cfm.

8. M. C. Fuerstenau and K. N. Han, eds., *Principles of Mineral Processing* (Littleton, CO: Society for Mining, Metallurgy, and Exploration, 2003), 1–2.

9. A. L. Mular, D. N. Halbe, and D. J. Barratt, *Mineral Processing Plant Design, Practice, and Control* (Littleton, CO: Society for Mining, Metallurgy, and Exploration, 2002), 566–584.

10. Ibid.

11. T. Napier-Munn and B. Wills, *Wills' Mineral Processing Technology: An Introduction to the Practical Aspects of Ore Treatment and Mineral Recovery* (Oxford: Butterworth-Heinemann, 2006), 203–223.

12. M. C. Fuerstenau, G. J. Jameson, and R. Yoon, *Froth Flotation: A Century of Innovation* (Littleton, CO: Society for Mining, Metallurgy, and Exploration, 2007), 93–337.

13. P. M. Sandman, *Responding to Community Outrage: Strategies for Effective Risk Communication* (Fairfax, VA: American Industrial Hygiene Association, 1993), ch. 1.

14. Ibid.

15. "Patagonia Under Siege," 2007, http://patagonia-under-siege.blogspot.com/2007/11/about-esquel-project-100-owned-by.html.

16. H. Thiel, M. V. Angel, E. J. Foell, A. L. Rice, and G. Schriever, *Environmental Risks from Large-scale Ecological Research in the Deep Sea: A Desk Study* (Brussels: Report for the Commission of the European Communities Directorate-General for Science, Research and Development, 1997), 180–200.

17. J. A. Botin, "The Problem of Mine Tailings Disposal," in *Technological Challenges Posed by Sustainable Development: The Mineral Extraction Industries*, Roberto C. Villas Bôas and Lélio Fellows Filho, eds. (Madrid: CYTED/IMAAC-UNIDO, 1999), 137–148.

18. R. J. Jewell and A. B. Fourie, eds., *Paste and Thickened Tailings—A Guide*, 2nd ed. (Perth: Australian Centre for Geomechanics, 2000), 1–3.

19. D. Van Zyl, M. Sassoon, C. Digby, A.-M. Fleury, and S. Kyeyune. *Mining for the Future—MMSD Large Volume Waste Report* (London: International Institute for Environment and Development, 2002), 11–18.

20. M. Sassoon, "Environmental Aspects of Mine Closure in Mine Closure and Sustainable Development," in *Mine Closure and Sustainable Development*, T. Khanna, ed., Proceedings of a Workshop organised by the World Bank and the Metal Mining Agency of Japan, Washington, March 2000 (London: Mining Journal Books, 2000).

21. G. Miller, *Financial Assurance for Mine Closure and Reclamation* (London: ICMM, 2005), www.icmm.com/library_pub_detail.php?rcd=176 (Accessed: April 27, 2008).

22. B. S. Blanchard, Dinesh C. Verma, Elmer L. Peterson, *Maintainability: A Key to Effective Service-ability and Maintenance Management*, 2nd rev. ed. (New York: Wiley-Interscience, 1995), 1–5.

23. IEEE 90: Institute of Electrical and Electronics Engineers, *Standard Glossary of Software Engineering Terminology*, s.v. "Reliability."

24. U.S. Department of Defense Handbook, *Designing and Developing Maintainable Products and Systems*, MIL-HDBK-470A, August 4, 1997, 2-1.

25. F. Stanley Nowlan and Howard F. Heap, *Reliability-Centered Maintenance* (United Airlines and Dolby Press, sponsored and published by the Office of Assistant Secretary of Defense, 1978).

26. J. Moubray, *Reliability-Centered Maintenance* (New York: Industrial Press, 2001).

27. B. S. Blanchard, Dinesh C. Verma, Elmer L. Peterson, *Maintainability: A Key to Effective Service-ability and Maintenance Management*, 2nd rev. ed. (New York: Wiley-Interscience, 1995).

28. A. Kelly, *Strategic Maintenance Planning* (Burlington, MA: Elsevier, 2006).

29. Moubray, *Reliability-Centered Maintenance*.

30. A. Pérez-Estaún and F. Bastida, "Cantabrian Zone: Structure," in *Pre-Mesozoic Geology of Iberia*, R. D. Dallmeyer and E. Martínez-García, eds. (Berlin-Heidelberg: Springer, 1990): 55–68.

31. L. D. Meinert, "Skarns and Skarn Deposits," in *Ore Deposit Models, Volume II*, P. A. Sheahan and M. E. Cherry, eds. (Toronto, ON: Geoscience Canada, 1993): 117–134.

32. MIL-STD-1629a, *Procedures for Performing a Failure Mode, Effects and Criticality Analysis*, 37–40.

33. Rausand M. Hoyland, *System Reliability Theory: Models, Statistical Methods and Applications* (New York: Wiley, 2004), 404-405.

34. A. H. Christer, "Developments in Delay Time Analysis for Modelling Plant Maintenance," *Journal of the Operational Research Society*, no. 50 (1999): 1120–1137.

35. MIL-STD-1629a, 37–40.

36. J. A. Botin, ed., *Investigation of the Failure of the Aznalcóllar Tailings Dam: Final Report* (Minas de Aznalcóllar [Sevilla], unpublished document, 1998).

37. L. Bjerrum, "Progressive Failure in Slopes of Overconsolidated Plastic Clay and Clay Shales," *Journal of the Soil Mechanics and Foundations Division*, ASCE 93, no. SM5 (1967): 1–49.

38. F. J. Griffiths and R. C. Joshi, "Identification of Cementation in Overconsolidated Clays," *Géotechnique* 38 (1989): 451–452.

39. Z. Chen and N.R. Morgenstern, "Progressive Failure of the Carsington Dam: A Numerical Study," *Canadian Geotechnical Journal* 29 (1992): 971–988.

40. E. W. Brooker and R. B. Peck, "Rational Design Treatment of Slides in Overconsolidated Clays and Clay Shales," *Canadian Geotechnical Journal* 30 (1992): 526–544.

41. A. W. Skempton and P. R. Vaughan, "The Failure of Carsington Dam," *Géotechnique 43*, no. 1 (1993): 151–173.

42. N. R. Morgenstern, "Managing Risk in Geotechnical Engineering," The 3rd Casagrande Lecture in *Proceedings, 10th Pan-American Conference on Soil Mechanics and Foundation Engineering*, Guadalajara, vol. 4 (1995): 102–126.

43. S. Uriel and J. Fornes "Back Analysis of a Landslide in Overconsolidated Tertiary Clays of the Guadalquivir River Valley (Spain)," in *13th International Conference on Soil Mechanics and Foundations Engineering*, New Delhi (1992).

About the Authors

M. N. ANDERSON
President
Norman Anderson & Associates
Vancouver, British Columbia, Canada

Since 1986, Norman Anderson has been president of Norman Anderson & Associates, a consulting firm to the mining industry. He is currently nonexecutive chairman of HudBay Minerals Inc. in Canada and a director of Cia de Minas Buenaventura in Peru. From 1978 to 1986, he was chief executive officer of Cominco Limited; three years prior to that, he was CEO of Fording Coal; and four years before that, he was vice president of AMAX Lead and Zinc Inc. Prior to 1970, he was with Cominco at many of their operations. Mr. Anderson has more than 50 years experience in the mining industry, holds a BS degree in geological engineering from the University of Manitoba, was a registered professional engineer in several provinces and states, and has held many offices and directorships over those 50 years. He has participated directly and indirectly in many new projects.

A. AUBYNN
Head of Corporate Affairs and Social Development
Gold Fields Ghana Ltd.
Accra, Ghana

Anthony (Toni) Aubynn is head of corporate affairs and social development at the Gold Fields Ghana division. Between 1998 and 2002, he worked as a human resources and local affairs manager of Abosso Goldfields Limited. He currently chairs the International Council on Mining and Metals' Working Group on Artisanal Small-scale Mining and is a member of the Strategic Management Advisory Group of the Communities and Artisanal Small-scale Mining. Prior to joining the mining industry, Toni worked as a lecturer/researcher at the universities of Helsinki and Tokyo, as well as the United Nations University. He is a product of the University of Ghana-Legon, where he earned his first degree; he pursued his tertiary education at the universities of Oslo (Norway), Tampere, and Helsinki (Finland). Toni was the first Ghanaian PhD Fellow at the United Nations University Institute of Advanced Studies in Tokyo, Japan.

J. A. BOTIN

Professor and Chair
Division of Management, Environmental Safety & Health
Universidad Politécnica de Madrid (Madrid School of Mines)
Madrid, Spain

Jose Botin is professor of mine management at Universidad Politécnica de Madrid. He was mine planning engineer for Placer Development in Canada (1970–1971), general mine foreman for FosBucraa in Western Sahara (1972–1973), project engineer for McKee Engineers (1974–1975), general manager of mining for Rio Tinto (1976-1982), chief operating officer for Cominco in Spain (1982–1987), and chief executive officer for Anglo American in Spain (1988–1991). He holds an EM (engineer of mines) degree (1971) and a PhD degree in mining (1987) from Universidad Politécnica de Madrid, an MSc (mining) degree from Colorado School of Mines (1976), and a PADE (diploma on high management) from IESE Business School–University of Navarra (1992).

T. BUCHANAN

Director, Energy and Extractives Practice
Business for Social Responsibility
San Francisco, California, USA

Tim Buchanan, director of the energy and extractives practice for Business for Social Responsibility, spent more than 20 years in the mining industry. He spent the first half of his career managing technical and production aspects of mining operations from development through closure, the second half managing environmental and social aspects. He has worked or consulted on the social and environmental aspects of projects located in Africa, Europe, North America, and South America. Tim is a registered professional engineer, holds a BS degree in mining engineering from Colorado School of Mines and an MS degree in resource management from the University of Nevada.

C. CASTAÑON

Independent Mining Consultant
Oviedo, Spain

Cesar Castañon is associate professor of mining engineering at the University of Oviedo's School of Mines and also an active consultant in the fields of mining engineering, ore resource estimation, and grade control. He has worked for Rio Narcea Gold Mines (1995–2004) as the manager of Asturias Operations (2004–2007), the mine manager of El Valle-Boinàs gold mine, and as an associate consultant. Cesar has also worked as a mine engineer for Anglo American Corporation at the Carles mine in Spain (1990–1991) and as a mine planning engineer for Cominco-Exminesa at the Rubiales mine in Spain (1987–1990). He holds BSc and PhD degrees in mining engineering from the University of Oviedo.

B. CEBRIAN
General Manager
Blast Consult, S.L.
Madrid, Spain

Benjamin Cebrian is a rock fragmentation consultant for operations worldwide. He worked as technical services manager, Europe, at UEE Explosives (now Maxam Corp.) until 2006. He holds an EM (engineer of mines) degree from Universidad Politécnica de Madrid and an MSc degree from Colorado School of Mines.

P. COSMEN
Manager, Environmental Management Systems
Cobre Las Cruces
Gerena, Sevilla, Spain

Paz Cosmen has been environmental manager at Cobre Las Cruces since 2000. Her previous experience includes work as environmental manager in Boliden Apirsa for one year; coordinator of the Environmental Quality Data Center at Consejería de Medio Ambiente in Andalusia (1997–2000); manager of the industrial wastewater treatment project at CIEMAT (Public Research Center in Energy and Environment; 1987–1996); manager of the technical department of Celestino Junquera, an electroplating company (1985–1987); and work in the field of hydrometallurgy, in the Research Center of Técnicas Reunidas (1979–1984). She holds a BSc degree in chemistry from Universidad Autonoma de Madrid (1989).

E. CRESPO
Research and Development Engineer
Atlantic Copper, S.A.
Huelva, Spain

Eloy Crespo is a mining engineer from Universidad Politécnica de Madrid (Madrid School of Mines) in Spain and has been a member of the Society for Mining, Metallurgy, and Exploration since 2005. He has been working since 2004 on a doctoral fellowship at the Freeport MacMoRan copper smelter in southwestern Spain (Atlantic Copper), where he is carrying out research and development work in mathematical modeling for maintenance management as part of the requirements for his PhD degree from Universidad Politécnica de Madrid.

G. A. DAVIS

Professor
Division of Economics and Business
Colorado School of Mines
Golden, Colorado, USA

Graham A. Davis is professor of economics and business at Colorado School of Mines. Prior to joining academia, he worked as a metallurgical engineer at metal mines in Canada and Namibia. He is a charter member of the Mineral Economics and Management Society and has been a member of the Society for Mining, Metallurgy, and Exploration for 27 years. His research focuses on the valuation and management of mineral and energy assets and on the impact of their extraction on developing nations. He is the author of one book and numerous academic papers related to the economics of the energy and minerals industries. Davis holds a BS degree in metallurgical engineering from Queen's University at Kingston (1982), an MBA degree from the University of Cape Town (1987), and a PhD degree in mineral economics from the Pennsylvania State University (1993).

M. G. DOYLE

Technical Director
Cobre Las Cruces, S.A.
Gerena, Sevilla, Spain

Mike Doyle is the technical director of Cobre Las Cruces. He began his career with Rio Tinto Zinc, working in various countries and projects, including the Neves Corvo mine in Portugal, mainly in exploration, geotechnical, and ore reserve areas. He led the team that discovered the Las Cruces project in 1994 and has been working on this project since then. He has a mining geology degree from the Royal School of Mines, an MEng degree from the Open University, and a postgraduate diploma in groundwater hydrology from Universidad Politécnica de Cataluña.

W. ECKLEY

Professor Emeritus
Department of Liberal Arts and International Studies
Colorado School of Mines
Golden, Colorado, USA

Wilton Eckley is professor emeritus of liberal arts and international studies at Colorado School of Mines, where he has taught since 1984 and where he held the position of department head until his retirement. Previously, he was professor of English at Drake University (1965–1984), where he served as department head for 15 years. He was a senior Fulbright professor at the University of Ljubljana, Slovenia (1972–1972), and at the Cyril and Methodius University, Bulgaria (1981–1982), and a visiting professor at Bilkent University in Turkey (1993–1994). He has lectured at universities in 12 countries and across the United States. He holds a PhD degree from Case Western Reserve University, an MA degree from the Pennsylvania State University, and an AB degree from Mount Union College.

R. G. EGGERT

Professor and Director
Division of Economics and Business
Colorado School of Mines
Golden, Colorado, USA

Roderick G. Eggert is professor and director of the Division of Economics and Business at Colorado School of Mines, where he has taught since 1986. Previously, he taught at the Pennsylvania State University and held research appointments at Resources for the Future (Washington, D.C.) and the International Institute for Applied Systems Analysis (Austria). Between 1989 and 2006, he was editor of *Resources Policy*, an international journal of mineral economics and policy. He has a BA degree in earth sciences from Dartmouth College, and an MS degree in geochemistry and mineralogy and a PhD degree in mineral economics from the Pennsylvania State University. His research and teaching have focused on various aspects of mineral economics and public policy, including the economics of mineral exploration, mineral demand, mining and the environment, microeconomics of mineral markets, and most recently mining and sustainable development. He served for two terms on the Committee on Earth Resources of the U.S. National Research Council.

J. A. ESPÍ

Professor
Department of Geological Engineering
Universidad Politécnica de Madrid (Madrid School of Mines)
Madrid, Spain

Jose-Antonio Espí is professor of geological engineering at Universidad Politécnica de Madrid where he has taught since 1989. Previously, he was director of mineral resources of IGME, the Geological Survey of Spain (1992–1997), and worked 15 years for several mining and exploration companies. He holds an EM (engineer of mines) degree (1971) and a PhD degree in mining (1977) from Universidad Politécnica de Madrid, and an MBA degree from Highlands University in New Mexico (1986).

L. W. FREEMAN

General Manager
Downing Teal, Inc.
Denver, Colorado, USA

Leigh Freeman is the general manager and a principal in Downing Teal, Inc. (2001–present), a global recruiting company. Previous positions include cofounder and president for Orvana Minerals (1986–1999), owner/consultant for Freeman & Associates (1985–2001), manager of project development for CoCa Mines Inc. (1981–1985), and chief geophysicist for the Placer-Dome Companies (1971–1981). Leigh is a director for Galway Resources, a trustee of the Society of Economic Geology, and serves on the industry advisory boards for the University of Arizona, Montana Tech, and South Dakota School of Mines. He holds a BS degree in geological engineering from Montana College of Mineral Science and Technology (1971).

M. G. HUDON
Senior Associate
Colby, Monet, Demers, Delage & Crevier, LLP
Montreal, Quebec, Canada

Michel G. Hudon has been a member of the Quebec Bar since 1968 and a graduate of Laval University (Quebec) Law School. As a practicing lawyer with the law firm of Colby, Monet, Demers, Delage & Crevier, LLP, based in Montreal (Quebec), his main areas of practice are corporate, transactional, securities, and equity financing law for mining companies and other clients. He has represented First Nations in the negotiation of impact benefits agreements in the context of hydroelectric and mining projects. He has also acted as legal advisor to certain African mining ministries in connection with mining capacity building and mining regulatory reforms on World Bank projects.

B. JOHNSON
Environmental Management Services
Council for Scientific and Industrial Research
Stellenbosch, South Africa

Brent Johnson heads up the environmental consulting group of the South African Council for Scientific and Industrial Research's Consulting & Analytical Services division (www.csir.co.za). He holds a postgraduate degree in environmental science and works principally in the African energy and mining sectors. He has a special interest in sustainable development, extractive industries, and the developing regional context of Africa. Acting principally in an advisory/consultancy capacity, the environmental consulting group works with environmental impact assessments, strategic environmental planning, and environmental reviews and audits, among other integrated environmental management tools. Recently, Brent was part of the team that assisted the South African government in developing a draft sustainable development strategic framework for minerals sector governance, making South Africa one of the first emerging economies to take this step. He lives in Cape Town, South Africa.

D. LIMPITLAW
Consulting Engineer
Johannesburg, South Africa

Daniel Limpitlaw is a mining engineer specializing in the assessment of both direct and indirect impacts of mining on the environment and surrounding communities. He has experience across the mining life cycle on projects in several southern African countries. He works on projects related to small-scale mining, management of mining impacts, and spatial assessment. Daniel consults extensively in the fields of mine closure, local government and local economic development, and impact assessment. Daniel was previously the director of the Centre for Sustainability in Mining and Industry at the University of the Witwatersrand.

H. B. MILLER
Associate Professor
Mining Engineering Department
Colorado School of Mines
Golden, Colorado, USA

Hugh Miller is an associate professor in the mining engineering department at Colorado School of Mines. Before joining Colorado School of Mines in 2005, he spent 6 years teaching at the University of Arizona and was the director of the San Xavier Mining Laboratory and co-director of the International Center for Mine Health, Safety, and Environment. He has also served on the boards of several companies and professional organizations, and regularly consults in the economic and technical evaluation of mining properties and mineral resources. Prior to entering academia, Hugh worked for 13 years for several mining and engineering companies, including 5 years as director of operations for International Engineering Technology, Inc. Hugh teaches courses and conducts research in a variety of specialized areas, including project feasibility/valuation, mine design and operations, and occupational health and safety. He received his undergraduate and graduate degrees from Colorado School of Mines.

N. MOJTABAI
Associate Professor and Chair
Mineral Engineering Department
New Mexico Institute of Mining and Technology
Socorro, New Mexico, USA

Navid Mojtabai is associate professor and department chair in the mineral engineering department at New Mexico Institute of Mining and Technology (New Mexico Tech). He teaches courses related to mining engineering, drilling and blasting, geomechanics, mine ventilation, and economic analysis. His research interest areas include rock fragmentation by blasting and geomechanics. He holds both BS (1982) and MS (1984) degrees in mining engineering from New Mexico Tech and a PhD degree (1990) in mining engineering from the University of Arizona.

L. E. ORTEGA
Manager, Environmental Planning and Development
Servicios Industriales Peñoles
Torreón, México

Enrique Ortega has been environmental planning and development manager for Grupo Peñoles since 1993 and a professor at Universidad Autónoma de la Laguna since 2005. Previously, he was an environmental engineer at Universidad Autónoma Metropolitana of México City (1989–1993) and was involved in environmental project evaluation for SEMARNAT (Secretaría de Medio Ambiente y Recursos Naturales) of Mexico (1991–1992). He holds an MSc degree in industrial engineering from Instituto Tecnológico de la Laguna (1998) and an MSc degree in education from Universidad Autónoma de la Laguna.

G. OVEJERO ZAPPINO
External Affairs Manager
Cobre Las Cruces, S.A.
Gerena, Sevilla, Spain

Gobain Ovejero Zappino is a mineral exploration and mine geologist with 38 years of professional experience in the private metallic mining sector, having worked for Peñarroya, Dupont de Nemours, Rocamat, Rio Tinto Group, MK Resources, Teck Cominco, and Cobre Las Cruces S.A. (Inmet Mining Group) in several countries. Over the last years, he has been external affairs manager for Cobre Las Cruces, dealing with permitting, land acquisition, and stakeholders for a new open-pit/hydrometallurgical copper project in an urban setting (Spain),and recently for a copper project in a jungle environment (Panama). He holds a degree in geological sciences from the University of Madrid, Spain.

M. PALACIOS
General Manager, Technology
Atlantic Copper, S.A.
Huelva, Spain

Miguel Palacios has been general manager of technology for Atlantic Copper, S.A., since 2005. He joined Atlantic Copper (then called Río Tinto Minera) in 1984, working as a metallurgist for the Huelva smelter. In 1987, he was appointed superintendent of the smelting section, and in 1996 he was promoted to manager of smelting. Previously, he worked as a process engineer for Unión Explosivos Río Tinto at the Ammonia-Urea factory in Huelva (1982–1984). He holds BSc and MSc degrees in chemical engineering from the University of Seville and a diploma in environmental engineering management from the Escuela de Organización Industrial.

J. M. QUINTANA
Senior Research and Development Engineer
Atlantic Copper, S.A.
Huelva, Spain

Jose Quintana is a senior research and development engineer at Freeport McMoRan–Atlantic Copper in southwestern Spain, where he has been leading operations efficiency improvement projects since 2000. Previously, he worked for General Electric Plastics in America and Spain as a process engineer and as a reliability engineer in the polycarbonate industry. His field of expertise is total quality management and continuous process improvement. He has been president of the Spanish National Association of Quality, Six Sigma Committee since 2003, and is a visiting professor of maintenance management at Universidad Politécnica de Madrid (Madrid School of Mines). He has an EM (engineer of mines) degree from Universidad Politécnica de Madrid.

J. L. REBOLLO

Visiting Professor
Division of Economics and Business
Colorado School of Mines
Golden, Colorado, USA

Jose L. Rebollo is a visiting professor and distinguished lecturer of the Division of Economics and Business at Colorado School of Mines, where he has lectured since 2004. He has spent 34 years in mining and metals. He started as an operations engineer in Penarroya-Espana. He was a member of the executive board and then chairman of the board of Metaleurop S.A. (Paris, France) and executive president of the Trappes Research Center (Yvelines, France), member of the board and chairman of the environment committee of Eurometaux, and French industry representative at the United Nations International Lead and Zinc Study Group. He served on the boards of the International Council for Metals and Minerals, the International Lead and Zinc Research Organization, and the International Zinc Association Europe. He holds an EM (engineer of mines) degree from Universidad Politécnica de Madrid (Madrid School of Mines) and has completed advanced programs at the Faculty of Economics at the University of Madrid and the École de Mines de Paris, as well as the Senior Executive Program at the Massachusetts Institute of Technology's Sloan School of Management.

A. S. RODRÍGUEZ-AVELLO

Professor
Materials Engineering Department
Universidad Politécnica de Madrid (Madrid School of Mines)
Madrid, Spain

Angel Rodríguez-Avello is professor of mineral processing at Universidad Politécnica de Madrid. Previously, he worked for 29 years in the mineral processing sector for WEMCO and EIMCO within the Envirotech Corp. and Baker-Hughes Process Group, in positions ranging from sales engineer to general manager and chief executive officer of the Spanish and Chilean subsidiaries. He holds BSc and PhD degrees from Universidad Politécnica de Madrid.

D. VAN ZYL

Professor of Mine Life Cycle Systems
Norman B. Keevil Institute of Mining Engineering
University of British Columbia
Vancouver, British Columbia, Canada

Dirk Van Zyl has more than 30 years experience in research, teaching, and consulting in tailings and mine waste rock disposal and heap leach design. Lately, much of his attention has been focused on mining and sustainable development. Dirk received a BSc degree in civil engineering in 1972 and a BSc (Honors) degree in 1974, both from the University of Pretoria, South Africa. He also received MS and PhD degrees in geotechnical engineering from Purdue University in 1976 and 1979, respectively. In 1998, he completed an Executive MBA degree program at the University of Colorado. He is a registered professional engineer in three states in the United States. Dirk became a Distinguished Member of the Society of Mining, Metallurgy, and Exploration in 2003. He received the Bureau of Land Management Sustainable Development Award in 2005 and the Adrian Smith International Environmental Mining Award in 2006.

A. ZOMOSA-SIGNORET

Manager, Sustainable Development
Servicios Industriales Peñoles S.A. de C.V
Torreon, Mexico

Andrea Zomosa-Signoret is manager of sustainable development at Servicios Industriales Peñoles S.A. de C.V. As part of her multidisciplinary and international background, she has a BA degree in international relations from El Colegio de México (Mexico City, Mexico) and an MA degree in law and diplomacy with a concentration in sustainable development at the Fletcher School of Law and Diplomacy (Boston, Massachusetts, USA). She has also undertaken some specialization studies at Sciences Po (Paris, France). Andrea has professional experience in the Organisation for Economic Co-operation and Development, the United Nations Economic Commission for Latin America and the Caribbean, and the Mexican public sector. As a Fulbright scholar from 2005 to 2007, Andrea did some academic research on the role of performance indicators in environmental assessments, especially in Latin America.

Index

NOTE: *f.* indicates figure; *t.* indicates table. Names or titles beginning with initial articles (The, A, El, La, etc.) are alphabetized under the first letter of the next word (e.g., La Araña Cement Quarry is alphabetized under A).